QUANTUM ELECTRODYNAMICS

QUANTUM ELECTRODYNAMICS

By
G. KÄLLÉN
Translated from the German by
C. K. IDDINGS & M. MIZUSHIMA

Springer Science+Business Media, LLC
1972

Gunnar Källén
Late Professor in the University of Lund

Prof. Carl K. Iddings and Prof. Masataka Mizushima
University of Colorado
Department of Physics and Astrophysics

Originally published by Springer-Verlag New York Inc. in 1972

Library of Congress Catalog Card Number: 76-172529

ISBN 978-3-642-88021-6 ISBN 978-3-642-88019-3 (eBook)
DOI 10.1007/978-3-642-88019-3

TRANSLATOR'S PREFACE

Källén's Quantenelektrodynamik provides a concise treatment of the subject. Its strong points are the careful attention to explanatory detail, the methodical coverage of all the major results and the straightforward, lucid style. Certainly it will be a valuable reference for one learning the subject or for one who requires the details of the practical results. Of course modern quantum field theory has now grown far beyond its dramatic beginnings in electrodynamics and we have therefore included some references to introduce the reader to the more recent and more specialized literature.

We have corrected some minor errors: we would appreciate it if readers would inform us of any others which they find.

We thank Professors Paul Urban and C. Møller for permission to use the biographical material on Källén. We also wish to thank Springer-Verlag for undertaking publication of this edition by an unorthodox method, but one which will reduce the cost to the reader. In particular, we are grateful to Dr. H. Mayer-Kaupp and Mr. Herb Stillman for their kind cooperation. Finally, we thank Mr. Michael Teague for reading and commenting on the first dozen sections and we thank Mrs. Joanne Downs for editing and typing the final manuscript.

May 1972

C. K. Iddings

M. Mizushima

Department of Physics
and Astrophysics
University of Colorado
Boulder, Colorado 80302

IN MEMORIAM PROFESSOR GUNNAR KÄLLÉN

On October 13, 1968, Professor Gunnar Källén of the University of Lund, Sweden, died in an airplane accident near Hanover. Undoubtedly Europe lost one of its most prominent theoretical physicists.

Let me first briefly state the significant dates in his remarkable scientific career:

Born on February 13, 1926 at Kristianstad, Källén studied physics in Sweden and completed his doctorate at the University of Lund in 1950. Starting as an assistant professor at Lund from 1950 to 1952, he got to the Theoretical Study Division of CERN at Copenhagen (1952-1958), worked at NORDITA (1957-1958) until he was offered a professorship for theoretical physics at the University of Lund. Moreover, Källén made several journeys for the purpose of research and took part in numerous scientific conferences in about 15 countries, including the U.S.A. and the U.S.S.R.

Let me now mention some details about his scientific work: Initially he studied electrical engineering but soon he changed over to physics, especially to the problems of quantum electrodynamics; in this field he achieved most important results in the years 1949 to 1955. His principal aim was the treatment of the theory of renormalization using, unlike other authors, the Heisenberg picture instead of the interaction picture and the relations now known as Yang-Feldman equations. Considering spectral representations for two- and three-point functions, he succeeded in separating the renormalization constants of quantum electrodynamics and in expressing them as integrals over certain weight functions; thus he could precisely formulate and try to solve the problem of the value of renormalization constants. Indeed, other authors are in doubt about his famous proof that at least one of the renormalization constants has to be infinite, but so far no definite answer to this question has been found.

Källén's authority at that time in the field of quantum electrodynamics is well illustrated by the fact that it was he who was requested to write the article on this topic in the Handbuch der Physik.

In connection with his work on quantum electrodynamics, he began to study closely the analyticity properties of three- and four-point functions and obtained a number of important results, partially cooperating with Wightman and Toll.

We must not forget a treatment of the Lee model, which Källén did together with Pauli and where they discovered and discussed the possibility of "ghost-states".

Källén was not only interested in the development of the general theory; he also treated many difficult concrete problems, such as in his works on vacuum polarization of higher order.

During the last years Källén performed fundamental work in the field of radiative corrections to weak decays. He took up the idea proposed by Berman and Sirlin in 1962, namely to take into account strong interactions by the introduction of suitable form factors. But there is one crucial difference from Berman and Sirlin:

Källén uses on-mass-shell form factors; that means quantities which are experimentally measurable in principle and can therefore be used as phenomenological parameters of the theory. By suggesting for the unknown form factor a behavior similar to the usual ones, one obtains higher powers of the photon momentum in the denominator; infinite integrals do not occur any more.

There are two important features in this method: first we get finite radiative corrections (although no exact numerical results can be expected because of the approximative character of the formalism), and secondly, an estimate of the cutoff is possible. This estimate shows that the assumption $\Lambda \approx m_p$ in the bare particle calculation was a very good approximation.

Källén's result -- finite radiative corrections by means of strong interactions -- is in striking disagreement with works of other authors, who included the modern concept of current algebra in the calculation of radiative corrections in β-decays.

Though Källén did not solve the problem of the influence of strong interactions on the convergence of radiative corrections in weak decays (by means of his form factor method), it turned out to be an important controversial question in this way, still lacking a final satisfactory solution.

Källén was one of the first who used reduction formalism, dispersion calculations and spectral representations in all his works, methods which became standard tools in modern physics. Surely Källén's works have contributed much to the fact that field theory is applied in elementary particle physics more than ever.

Furthermore, Källén has earned considerable merit in the field of elementary particle physics as the author of an excellent book in which many problems of strong and weak interactions are treated. Here, just as in his conference lectures, Källén proved his outstanding pedagogical talent. In his book he has shown excellently how much about mathematical methods and detailed calculations should be presented, enough to clear up the connection between theory and experiment, but not so extensively that the presentation could be spoiled.

I hope that to a certain extent I have been successful in doing justice to the personality of Gunnar Källén and his position in science. The reader will certainly agree if I emphasize again that his early death undeniably has left a gap among the most outstanding theoretical physicists of Europe.

Paul Urban

These remarks are a condensation of those appearing in "Particle Physics", Acta Physica Austriaca, Supplementum 6, Vienna, New York; Springer-Verlag (1969).

GUNNAR KÄLLÉN IN MEMORIAM

Gunnar Källén was born February 13, 1926, so he was only 42 years old when he died in the fatal airplane crash on October 13, 1968. In spite of the short span of time in which he was active in physics he left behind him a large number (about 60) of original papers, conference reports, lecture notes, and monographs on many different subjects of modern physics, in particular in the domains of quantum electrodynamics, quantum field theory in general, and elementary particle physics. It will not be possible, and in this circle also not necessary, to mention all these papers here, but I shall try to give an outline of his main contributions to our science in the different stages of his nineteen years of activity in physics.

As so many other physicists he started his career as an engineer. Twenty-two years old, he came to Lund to pursue his studies of theoretical physics at the University. With amazing speed he caught up with the problems and soon he was working at the front line of our knowledge at that time. The main subject of interest among the theoretical physicists in Lund and elsewhere at that time was the new method in quantum electrodynamics which was initiated by Kramers in 1947, and which seemed to make it possible to evade the disturbing divergence difficulties, inherent in the formalism of quantum electrodynamics, by a renormalization procedure. In the following years this program was successfully carried through by Tomonaga, Schwinger and Feynman by making use of the so-called interaction picture. Källén was fascinated by this difficult subject and by the challenge it represented. His first paper appeared in the Helv. Phys. Acta in 1949. It contained a treatment of the higher approximations in the vacuum polarization. This problem was suggested to him, during a short visit in Zürich, by Wolfgang Pauli who was much impressed by the young student's quick and independent mind. Källén, on the other side, admired Pauli immensely and took him as a model for his future work. The relations between the older and the young physicist developed into a life-long warm friendship, which also led to a fruitful collaboration between them in the later years.

After his return to Lund, Källén set himself the task to carry through the renormalization program without the use of the interaction representation which he regarded as an unnecessary mathematical complication. In a series of papers leading up to his

Inaugural Dissertation in 1950, he was able to show that the ideas of renormalization can easily be formulated in the original Heisenberg picture, and that many of the calculations are simpler and their physical interpretation more transparent in this picture. In these papers the notions of free "in"- and "out"-fields were defined clearly for the first time, and a method was developed which in the literature often has been called the Yang-Feldman method. The reason for this is probably that Yang's and Feldman's paper appeared in the Phys. Rev., while Källén's first paper on the subject was published in Ark. f. Fysik. Since these papers appeared nearly simultaneously and were produced independently, there is no room for any priority claims (and Gunnar would have been the last to make such claims), but one thing is certain: Källén made much more extensive use of his method for practical calculations, and soon he was also recognized by his colleagues everywhere as a master in his field. His brilliant appearance at international conferences, starting with the Paris Conference in the spring of 1950, contributed much to this. His elegant way of presenting his points of view and his sharp and witty dialogue in the discussions made him an excellent advocate for his ideas, which evoked the admiration of his older and younger colleagues. One of the latter was A.S. Wightman who later wrote about the early work of Källén: "At that time I was trying to puzzle out the grammar of the language of quantum field theory, and here was Källén already writing poetry in the language."

Gunnar Källén's connection with CERN dates back to the very first years of this organization. Already in 1952, when the site in Meyrin still consisted of a collection of deep holes in the ground and a few shacks, Källén became a Fellow of CERN's Theoretical Study Group, which at that time was placed at the Niels Bohr Institute in Copenhagen. I remember vividly his appearance there, which brought exciting new life to our group. He gave a series of admirable lectures on quantum electrodynamics, which clearly showed his superior mastery of the field and his exceptional gifts as a lecturer. Simultaneously, he pursued with characteristic energy a plan which he had conceived after the completion of his Doctor's thesis.

The current renormalization theory was based on a series expansion in powers of the fine structure constant and, although each term in this expansion was finite and showed a surprisingly good agreement with the experimental results, the convergence of the series had not been proved. Thus, it was still an open question whether renormalized quantum electrodynamics could be regarded as a consistent physical theory or whether it only represented a handy cookery-book prescription for getting useful results. The answer to this question was of great principle importance, but also so difficult to obtain that it required all the courage and tenacity of a Källén to attack and finally solve the problem.

By means of the formulation of the renormalization theory he had given in his thesis, he was able to define the renormalization constants without making use of perturbation theory. In a series of papers in Helv. Phys. Acta and in Physica he showed how this can be done and, in a final paper in the Proceedings of the Danish Academy (which later was reprinted in special collections both in Japan and U. S. A.) he proved that at least one of the renormalization constants had to be infinite. Thus, he had come to the conclusion that renormalized quantum electrodynamics could not be regarded as a completely satisfactory physical theory, in spite of the success of the perturbation theory version of the theory in accounting for the experimental results. On the other hand, the latter circumstance gives good reason for believing that the present formalism may be regarded as a limiting case of a future more complete theory.

Källén was a Fellow at CERN's Theoretical Study Group from October 1, 1952 to June 15, 1953. During this period, his professional ability and his personality had impressed us so much that we naturally tried to get him on the permanent staff of the Study Group. After he had finished a second longer stay in Zürich, he joined our staff in October 1954, where he remained until CERN's Theoretical Study Group finally moved to Geneva in September 1957. Thereafter, he accepted a chair as professor at the simultaneously established NORDITA in Copenhagen, where he stayed until a personal professorship was created for him at the University of Lund at the end of 1958.

Thus we had the privilege of having Gunnar with us as collaborator in Copenhagen during more than five of his perhaps most productive years. It is impossible in a few words to describe how much we owe him as a constant source of inspiration, as a teacher, and last but not least as an always alert critic. The ruthless honesty and objectivity of his criticism, which soon became legendary, recalled that of the young Pauli. It has even been said that Gunnar modelled his style on Pauli, but this was only partly true. I rather think that the similarity in their reactions was due to an inherent kinship of these two original personalities.

In Zürich Källén and Pauli had started a fruitful collaboration which was continued after Gunnar's arrival in Copenhagen. It resulted in a paper "On the mathematical structure of T. D. Lee's model of a renormalizable field theory" which was published in the Proceedings of the Danish Academy in 1955. Although the Lee model is non-relativistic, it is of great interest as an illustration of what might be hidden in the more complicated formalism of quantum electrodynamics. The advantage of the model is that it contains a renormalization of both the coupling constant and the mass, and still is so simple that it allows exact solutions. The main result in the just mentioned paper was the surprising discovery that the renormalized Lee model contains an unphysical state--the "ghost" state--which has a negative probability. It is quite possible that the formalism of renormalized quantum electrodynamics also contains such unphysical states, which further supports the view that quantum electrodynamics, when taken literally, does not represent a consistent physical theory. A fortiori, this holds for the current meson theories which, for obvious reasons, do not lend themselves to a perturbation treatment.

Therefore, in the following years and in particular after his return to Lund, Källén joined in the trend of research, which has been called the axiomatic way, and which was being pursued at several places in Europe and America. Instead of investigating the properties of a definite formalism, the idea was to see how far one can go by starting from a few general physically necessary requirements, such as relativistic invariance, causality and positive energy. With his usual energy, Källén threw himself on this seemingly infinite problem, which consisted in investigating the mathematical properties of the vacuum expectation values of the product of an arbitrary number n of field operators. In collaboration with Wightman, Toll and several of his students in Lund, he obtained many important results, in particular for $n = 3$ and $n = 4$. However, the general "n-point function" turned out to be so complicated that the problem could only be partially solved. This is probably the only problem, attacked by Gunnar, which he had to leave without having obtained a complete solution.

In parallel with these more mathematical investigations, he also worked on problems which had more immediate physical applications. I am thinking of his calculations of the radiative corrections to weak interactions, in particular to the nucleon-decay and the electromagnetic and nucleon form factors, a work which inspired his pupils in Lund to interesting investigations.

After Källén's appointment as a professor at the University of Lund, much of his time was occupied by teaching and guiding students, and his natural gifts as an educationalist were brought into full play. His review articles and monographs, in particular his article on Quantum Electrodynamics in the Handbuch der Physik (1958) and his book on Elementary Particle Physics (1964), are lasting witnesses of his pedagogical faculties. He gathered around him a large number of Swedish and foreign pupils, who benefited immensely by his profound knowledge, his lucid lectures and his objective criticism which fortunately was linked with a deep sense of humor. The latter most essential characteristic allowed him in the course of the years to develop a totally harmonious and well-balanced personality. A contributing factor in this respect was undoubtedly his happy family life with his charming children and his lovable wife Gunnel. She was an intelligent woman with a strong personality. The bravery and courage which she showed when her husband suddenly was taken away aroused the admiration and compassion of all her friends. On the day of the funeral she said to my wife: "After this I am not afraid of anything in the world, my only ardent wish is that I may keep my health". This wish was not fulfilled; only half a year later she followed her husband into the grave.

In the light of history, Gunnar Källén's appearance in the world of physics was like a shooting star. It was short, but so ardent, so shining that his name will be remembered, not only by those who knew him personally or even had the good fortune to become his friends, but also by the coming generations of physicists. He will be sorely missed.

C. Møller

This biography is taken from the Proceedings of the Lund Conference on Elementary Particles, 1969; G. von Dardel, editor.

TABLE OF CONTENTS

QUANTUM ELECTRODYNAMICS

CHAPTER I

GENERAL PRINCIPLES

1. Field Operators, State Vectors, Periodic Boundary Conditions

The general mathematical principles of quantum electrodynamics or any other quantum field theory are copied from those of the usual nonrelativistic quantum mechanics of (point) particles. Hence quantum electrodynamics also deals with a state vector in a "Hilbert space" and with a set of linear operators in this space, the "field operators". The latter are the dynamical variables of the theory and correspond to the classical fields, in the sense of the correspondence principle. As an example, certain operators correspond to the classical electromagnetic potentials. The four-dimensional space-time coordinates $x_1=x$, $x_2=y, x_3=z, x_4=ix_0=ict$, which we shall often designate simply by x, are not operators in the sense of the nonrelativistic quantum mechanics. Rather, they must be understood as "indices" or "labels" for the field operators. Thus with each point x there are associated a finite number of field operators (eight in quantum electrodynamics), which we shall designate as $\varphi_\alpha(x)$ in the first chapter. The index α distinguishes the various fields; for example, it distinguishes the four components of the electromagnetic potential and the four Dirac field operators. The quantities $\varphi_\alpha(x)$ and $\varphi_\alpha(x')$ must be regarded as two independent operators when they are referred to two different points x and x'. Since each finite volume V contains an infinite number of points x, our system has infinitely many operators or an infinite number of degrees of freedom. This is an important difference between a field theory and the usual mechanics of a point particle, where the number of degrees of freedom is always finite. The mathematical difficulties which we shall later encounter are always closely connected with summations over an infinite number of degrees of freedom.

For practical calculations, it is usually convenient not to work directly with the field $\varphi_\alpha(x)$ in x-space, but instead to employ the Fourier components of the field as the independent variables. For this purpose, we imagine the fields as being enclosed in a large cube of edge L and volume $V=L^3$, and we require the fields to be periodic in x-space with period L. Under these conditions, the field operators can be written as sums:

$$\varphi_\alpha(x) = \frac{1}{\sqrt{V}} \sum_p e^{ipx} \varphi_\alpha(p). \tag{1.1}$$

Here p stands for a four-vector with components $p_1=p_x$, etc., and the product px is to be understood as the scalar product of the vectors: $px = p_1 x_1 + \cdots + p_4 x_4$. In a similar way, we shall designate the square of a four-vector p by p^2. This latter quantity is therefore positive for space-like, zero for light-like, and negative for time-like vectors. When we use the spatial (three-vector) part of a four-vector, we shall designate it by boldface: $\mathbf{p}$. The summation in (1.1) is to be taken over all possible values of the spatial components of the vector p. Because of the periodic boundary conditions, only those values of $\mathbf{p}$ are allowed which have components satisfying the condition

$$p_i = n_i \frac{2\pi}{L} \qquad (i = 1, 2, 3), \tag{1.2}$$

where the n_i are positive or negative integers. In general, as a consequence of the equations of motion, the fourth component p_4 or $p_0 = -ip_4$, is a many-valued function of $\mathbf{p}$. Equation (1.1) must therefore be summed over all these possible values of p_0. The number of different values of p_0 is determined not only by the physical theory but also depends on the representation employed in Hilbert space. In Sec. 2 we shall give a more detailed discussion of the various possibilities for this representation.

In Eqs. (1.1) and (1.2) we have put $\hbar = 1$. This is only a matter of the units employed and does not have any physical significance. Likewise, we shall often take the velocity of light c equal to 1 and thus obtain a common unit for length, time, and reciprocal mass. For example, if the unit of length, the centimeter, is taken as the basic unit, then the unit of time is the time required for light to travel a distance of 1 cm. and the unit of mass is that mass with a Compton wavelength of 2π cm. These units are often referred to in the literature as "natural units".

For very large volumes V, all the results of the theory which are of physical interest must tend to finite limits. It often happens that a sum of the form

$$S = \frac{1}{V} \sum_{\mathbf{p}} f(\mathbf{p}) \tag{1.3}$$

occurs in a result where either $f(\mathbf{p})$ is independent of V or tends to a finite limit as $V \to \infty$. We see immediately from Eq. (1.2) that the number of states with spatial vectors $\mathbf{p}$ lying between $\mathbf{p}$ and $\mathbf{p} + d\mathbf{p}$ is given by

$$\left(\frac{L}{2\pi}\right)^3 dp_x\, dp_y\, dp_z = \frac{V}{(2\pi)^3}\, d^3p\,. \tag{1.4}$$

Thus it follows that as $V \to \infty$, the sum S tends to the limit

$$S \to \frac{1}{(2\pi)^3} \int d^3p\, f(\mathbf{p}). \tag{1.5}$$

The symbol d^3p in (1.4) and (1.5) means the three-dimensional volume element in p-space. Later we shall also use four-dimensional volume elements and designate these simply as dp (not as d^4p).

2. Different Pictures in Hilbert Space

It is well known that in ordinary quantum mechanics the time evolution of the system can be treated in several ways. For example, one can employ a picture where the operators are time-independent quantities, and all motion of the system is described by means of the state vector. Alternatively, one can regard the state vectors as constants and describe the evolution of the system by a time variation of the operators. In the first case, we have the "Schroedinger picture" and in the second, the "Heisenberg picture". These two possibilities are also present in field theory. In a way, they are extreme examples of the more general possibility in which the operators as well as the state vectors are treated as time-dependent quantities, and the precise "distribution" of the time dependence between the two is a matter of convenience. In the Schroedinger picture, the time variation of the state vector is given by the Schroedinger equation

$$i\,\frac{\partial|\psi(t)\rangle}{\partial t} = H\,|\psi(t)\rangle\,. \tag{2.1}$$

Here $|\psi(t)\rangle$ is the state vector, and the Hamiltonian operator H is a Hermitian operator depending on the dynamical variables. In a field theory, H depends upon the field operators. Under certain conditions, H can be time-dependent, but then the time must occur explicitly and not implicitly in the field operators. In this case, the system is not closed but is interacting with external sources. Later we shall frequently consider such systems; in Chap. I, for simplicity, we shall limit ourselves to closed systems, i.e., to time-independent Hamiltonian operators. The generalization of the results of this chapter to systems which are not closed can usually be made without difficulty. We may therefore assume

$$\frac{\partial H(\varphi_\alpha(\boldsymbol{x}))}{\partial t} = 0. \tag{2.2}$$

In a field theory, H usually contains the field operators at all points of three-dimensional space and can be written as an integral of a Hamiltonian density $\mathscr{H}$:

$$H = \int d^3x\,\mathscr{H}(\varphi_\alpha(\boldsymbol{x}))\,. \tag{2.3}$$

Although the operators are time-independent in this picture, a "time derivative" of an operator $F(\boldsymbol{x})$ can be defined by

$$\dot{F}(\boldsymbol{x}) = i\,[H, F(\boldsymbol{x})] \equiv i\,(H\,F(\boldsymbol{x}) - F(\boldsymbol{x})\,H)\,. \tag{2.4}$$

This operator is obviously not the actual derivative, but does have the property that its expectation value is the time derivative of the expectation value of the original operator:

$$\langle\psi(t)\,|\,\dot{F}(x)|\,\psi(t)\rangle = \frac{\partial}{\partial t}\,\langle\psi(t)\,|\,F(x)|\,\psi(t)\rangle. \tag{2.5}$$

The proof of (2.5) follows immediately from the definition (2.4) and the Schroedinger equation (2.1).

As is clearly evident from this discussion, the spatial coordinates x_k $(k=1,2,3)$ and the time x_0 are treated in quite different ways. Because of this, the relativistic covariance of the theory has been lost. However, it is often quite useful to employ this covariance in order to prove general results from considerations of symmetry, and sometimes it is even necessary to use such invariance considerations in order to give meaning to mathematically undefined expressions. It is appropriate to introduce another picture in which this invariance is clearly exhibited. The Heisenberg picture is completely symmetric in all coordinates. Here the state vectors $|\psi\rangle$ are time-independent, and there is no Schroedinger equation like (2.1). Instead, there are the equations of motion of the field operators,

$$\frac{\partial F(x)}{\partial t} = i\,[H, F(x)]. \tag{2.6}$$

The Eqs. (2.6) appear formally the same as Eq. (2.4); however, here we are dealing with time-dependent quantities, so that Eqs. (2.6) are the actual differential equations which can, in principle, be used to describe the time evolution of the system. Since the Hamiltonian operator commutes with itself, it follows directly from (2.6) that its time derivative vanishes. Accordingly, in the Heisenberg picture also, H is time-independent.

We can establish the connection between the Schroedinger and Heisenberg pictures with the formal solution of (2.1):

$$|\psi(t)\rangle = e^{-iHt}\,|\psi(0)\rangle\,. \tag{2.7}$$

If we designate the operators in the Schroedinger picture by $F(0)$, we see that the canonical transformation

$$F(t) = e^{iHt}\,F(0)\,e^{-iHt} \tag{2.8}$$

gives the operators in the Heisenberg picture. The operators $F(t)$ in (2.8) obviously satisfy the differential equations (2.6) with the boundary conditions that both pictures coincide for $t=0$. In the Heisenberg picture, the state vector is the quantity $|\psi(0)\rangle$ of (2.7).

Besides these two pictures, a third one is often used in the literature. Strictly speaking, it has been used for a long time in quantum electrodynamics; however, in recent years it has been particularly emphasized and has received the name "interaction representation" in the works of Tomonaga[1] and Schwinger[2,3]. To construct this picture, we begin with the Schroedinger picture and transform the operators by a canonical transformation which is not generated by the operator for the total energy, but only by a part of it, H_0. We write

$$F_W(t) = e^{iH_0t} F(0)\, e^{-iH_0t}, \tag{2.9}$$

where

$$H = H_0 + H_1 . \tag{2.10}$$

In this picture the operators as well as the state vectors are time-dependent. The latter quantities then satisfy a "Schroedinger equation" of the form

$$i \frac{\partial}{\partial t} |\psi_W(t)\rangle = e^{iH_0t} H_1 e^{-iH_0t} |\psi_W(t)\rangle = H_{1W}(t) |\psi_W(t)\rangle, \tag{2.11}$$

and the operators F_W obey the differential equation

$$\frac{\partial F_W(t)}{\partial t} = i\,[H_0, F_W(t)]. \tag{2.12}$$

The advantage of this picture is that with a suitable choice for H_0, Eq. (2.12) can be solved explicitly, and so part of the problem can be solved formally. The differential equation (2.11), which obviously contains the remainder of the problem, is usually so complicated that the solution of (2.12) has not really accomplished very much. Yet, in the historical development of modern quantum electrodynamics, this decomposition of the problem into two parts has played an important role. This is chiefly because the decomposition (2.10) often can be done so that H_1, the so-called interaction energy, is a three-dimensional integral over an invariant density function. The corresponding statement is valid neither for H nor H_0. In a sense, the differential equation (2.11) is invariant, and considerations of invariance and symmetry can be applied to it and its solutions. Tomonaga and Schwinger have particularly stressed this invariance; they do not write Eq. (2.11) as a time derivative, but introduce a system of spatial surfaces and consider the changes in the state vector with infinitesimal variations of these surfaces. The resulting formalism is extremely elegant and has played an important role historically. We shall not consider it further because we are not concerned with the historical development, and because there is no real advantage to the use of the spatial surfaces. Moreover, it is clear that one has real

---

1. S. Tomonaga, Progr. Theor. Phys. 1, 27 (1946) and later articles.

2. J. Schwinger, Phys. Rev. 74, 1439 (1948).

3. (Translator's note) The original term "interaction representation" is now generally replaced by the term "interaction picture".

covariance only if all four coordinates are treated in the same way, and this is the case only in the Heisenberg picture, which is therefore best suited for considerations of invariance.

The transformations (2.9) and (2.8) have been chosen so that all the pictures related by them coincide for $t=0$. This prescription obviously involves some arbitrariness, and if one starts with the differential equations (2.11) and (2.1), this arbitrariness appears in the boundary conditions. It is often convenient to choose the boundary conditions so that the interaction picture and the Heisenberg picture coincide for $t=-\infty$ rather than for $t=0$. Of course this is actually done by introducing a suitable limiting process. Although this requires special mathematical tricks, the physical interpretation of the theory is simplified so much, at least for scattering problems, that we will tolerate these mathematical difficulties. Later we shall return to this point in detail.

3. Lagrangian, Equations of Motion, and Canonical Quantization

In this section we shall consider only those theories which can be derived from a Lagrangian function. We therefore assume that there is a density function

$$\mathscr{L}\left(\varphi_\alpha(x),\ \frac{\partial\varphi_\alpha(x)}{\partial x_\mu}\right)$$

and that it is an invariant function of the field variables $\varphi_\alpha(x)$ and their first derivatives with respect to the four coordinates x_μ. From this density we can form the four-dimensional integral

$$L=\int dx\,\mathscr{L}\left(\varphi_\alpha(x),\ \frac{\partial\varphi_\alpha(x)}{\partial x_\mu}\right), \tag{3.1}$$

which is the Lagrangian function.[1] In a classical theory the equations of motion are obtained from the requirement that the integral (3.1) be stationary with respect to variations in the fields $\varphi_\alpha(x)$:

$$\sum_{\mu=1}^{4}\frac{\partial}{\partial x_\mu}\frac{\partial\mathscr{L}}{\partial\dfrac{\partial\varphi_\alpha(x)}{\partial x_\mu}}=\frac{\partial\mathscr{L}}{\partial\varphi_\alpha(x)}. \tag{3.2}$$

In a classical theory it is also possible to construct a canonically conjugate momentum $\pi_{\varphi_\alpha}(x)$ [abbreviated $\pi_\alpha(x)$] for each variable,

$$\pi_\alpha(x)=\frac{\partial\mathscr{L}}{\partial\dfrac{\partial\varphi_\alpha(x)}{\partial x_0}}, \tag{3.3}$$

1. (Translator's note) In customary usage the "Lagrangian" is defined as $L=\int d^3x\,\mathscr{L}$, while the integral $A=\int dx\,\mathscr{L}$ is known as the "action".

and hence to calculate the classical Hamiltonian function

$$H = \int d^3x \left[\sum_\alpha \pi_\alpha(x) \frac{\partial \varphi_\alpha(x)}{\partial x_0} - \mathcal{L} \right] = H\left(\pi_\alpha, \varphi_\alpha, \frac{\partial \varphi_\alpha}{\partial x_k}\right). \qquad (3.4)$$

In (3.4) it is necessary to express $\frac{\partial \varphi_\alpha(x)}{\partial x_0}$ as a function of $\pi_\alpha(x)$ by the use of Eq. (3.3) and to remember that it is this expression which is used in H. Thus the Hamiltonian function is understood to be a function of the momenta, the field variables, and their spatial derivatives. We must assume that the time derivatives of the field variables can actually be eliminated in this way. In the quantized theory we shall now require that the field operators and momenta, which are defined by (3.3), obey the canonical commutation relations

$$[\pi_\alpha(x), \varphi_\beta(x')] = -i\,\delta_{\alpha\beta} \cdot \delta(\boldsymbol{x} - \boldsymbol{x}'), \qquad (3.5a)$$

$$[\pi_\alpha(x), \pi_\beta(x')] = [\varphi_\alpha(x), \varphi_\beta(x')] = 0. \qquad (3.5b)$$

In a picture where the operators are time-dependent, <u>both times x_0 and x_0' must be the same</u>. This is the canonical quantization prescription, which we regard as a postulate.[1] Just as in ordinary quantum mechanics, it follows from (2.8) [or (2.9) in the case of the interaction picture] that, if the commutation relations are obeyed for one time, e.g., $x_0 = x_0' = 0$, then they also hold for arbitrary times.

Equation (3.4) will now be regarded as an operator equation, and the operator H defined there will be used as the Hamiltonian operator for the quantized theory. In principle, this prescription is not always unique, since the result can depend upon the order in which the operators are written. In actual use, however, this does not give rise to any essential difficulties.

We shall now show that in general the quantum mechanical equations of motion in the Heisenberg picture agree formally with Eqs. (3.2) and (3.3) if the latter are taken as operator equations. To do this, we calculate the commutator of H and $\pi_\alpha(x)$. Since two momenta commute with each other (for equal times) it follows that

$$[H, \pi_\alpha(x)] = \int_{x_0' = x_0} d^3x' \left\{ \sum_\beta \pi_\beta(x') \left[\frac{\partial \varphi_\beta(x')}{\partial x_0'}, \pi_\alpha(x) \right] - [\mathcal{L}(x'), \pi_\alpha(x)] \right\}. \qquad (3.6)$$

Obviously the commutator of a momentum and a function of the coordinates is equal to the derivative of the function with respect

---

1. J. Schwinger has attempted to deduce the canonical quantization prescription from a variational principle. We shall not discuss this point further here, but refer the reader to the original work: Phys. Rev. 82, 914 (1951).

to the coordinates multiplied by the commutator of the momentum and the coordinates. Using this result we can compute the last term of (3.6):

$$[\mathscr{L}(x'), \pi_\alpha(x)] = \frac{\partial \mathscr{L}}{\partial \varphi_\alpha(x')} i\,\delta(\boldsymbol{x}-\boldsymbol{x}') + \sum_{k=1}^{3} \frac{\partial \mathscr{L}}{\partial \dfrac{\partial \varphi_\alpha(x')}{\partial x'_k}} i \frac{\partial}{\partial x'_k} \delta(\boldsymbol{x}-\boldsymbol{x}') + \sum_\beta \frac{\partial \mathscr{L}}{\partial \dfrac{\partial \varphi_\beta(x')}{\partial x'_0}} \left[\frac{\partial \varphi_\beta(x')}{\partial x'_0}, \pi_\alpha(x)\right]. \tag{3.7}$$

From Eq. (2.6) we have

$$\frac{\partial \pi_\alpha(x)}{\partial x_0} = \frac{\partial \mathscr{L}}{\partial \varphi_\alpha(x)} - \sum_{k=1}^{3} \frac{\partial}{\partial x_k} \frac{\partial \mathscr{L}}{\partial \dfrac{\partial \varphi_\alpha(x)}{\partial x_k}} + i \int\limits_{x'_0=x_0} d^3x' \sum_\beta \left(\pi_\beta(x') - \frac{\partial \mathscr{L}}{\partial \dfrac{\partial \varphi_\beta(x')}{\partial x'_0}}\right) \times \left[\frac{\partial \varphi_\beta(x')}{\partial x'_0}, \pi_\alpha(x)\right]. \tag{3.8}$$

In a similar way we obtain the commutator of H and $\varphi_\alpha(x)$:

$$[H, \varphi_\alpha(x)] = \int\limits_{x'_0=x_0} d^3x' \left\{-i \frac{\partial \varphi_\alpha(x')}{\partial x'_0} \delta(\boldsymbol{x}-\boldsymbol{x}') + \sum_\beta \pi_\beta(x') \times \left[\frac{\partial \varphi_\beta(x')}{\partial x'_0}, \varphi_\alpha(x)\right] - [\mathscr{L}(x'), \varphi_\alpha(x)]\right\}, \tag{3.9}$$

$$[\mathscr{L}(x'), \varphi_\alpha(x)] = \sum_\beta \frac{\partial \mathscr{L}}{\partial \dfrac{\partial \varphi_\beta(x')}{\partial x'_0}} \left[\frac{\partial \varphi_\beta(x')}{\partial x'_0}, \varphi_\alpha(x)\right], \tag{3.10}$$

$$\frac{\partial \varphi_\alpha(x)}{\partial x_0} = \frac{\partial \varphi_\alpha(x)}{\partial x_0} + i \sum_\beta \int\limits_{x'_0=x_0} d^3x' \left[\pi_\beta(x') - \frac{\partial \mathscr{L}}{\partial \dfrac{\partial \varphi_\beta(x')}{\partial x'_0}}\right] \left[\frac{\partial \varphi_\beta(x')}{\partial x'_0}, \varphi_\alpha(x)\right]. \tag{3.11}$$

In Eqs. (3.6) to (3.11), when $\frac{\partial \varphi_\alpha(x)}{\partial x_0}$ appears on the right-hand side, it is to be regarded as the same function of $\pi_\alpha(x)$, $\varphi_\alpha(x)$ and $\frac{\partial \varphi_\alpha(x)}{\partial x_k}$ as one has in the classical theory. Equation (3.11) must then determine the corresponding function in the quantized theory. It obviously has the solution that the classical and quantum mechanical functions are equal. That is, (3.11) is equivalent to Eq. (3.3) considered as an operator equation. It then follows that (3.8) is the operator equation corresponding to Eq. (3.2). This concludes the proof. The difficulties which arise from the order of the operators in the Hamiltonian have not been taken into account. Since it will lead to no serious difficulties in the applications, we shall not pursue this question further.

Just as in the classical theory, there are also conservation laws for energy and momentum in the quantized field theory. We have already shown that the conservation of energy in quantum theory is a trivial consequence of the fact that H always commutes with itself, and therefore that the time derivative of the total energy must vanish. As in the classical theory, this conservation law can also be proved by direct calculation, starting from the operator equations (3.2) and (3.3).

Also as in the classical theory, the three spatial components of the momentum are defined by

$$P_k = -\int d^3x \sum_\alpha \pi_\alpha(x) \frac{\partial \varphi_\alpha(x)}{\partial x_k} \tag{3.12}$$

and, in the usual way, we have

$$[P_k, \varphi_\alpha(x)] = i \int_{x_0 = x_0'} d^3x' \, \delta(\boldsymbol{x} - \boldsymbol{x}') \frac{\partial \varphi_\alpha(x')}{\partial x_k'} = i \frac{\partial \varphi_\alpha(x)}{\partial x_k} , \tag{3.13}$$

$$[P_k, \pi_\alpha(x)] = -\int_{x_0' = x_0} d^3x' \, \pi_\alpha(x') \frac{\partial}{\partial x_k'} i \, \delta(\boldsymbol{x} - \boldsymbol{x}') = i \frac{\partial \pi_\alpha(x)}{\partial x_k} . \tag{3.14}$$

From these last two equations, it follows that any operator $F(x)$ which is constructed from $\varphi_\alpha(x)$ and $\pi_\alpha(x)$ satisfies

$$[P_k, F(x)] = i \frac{\partial F(x)}{\partial x_k} . \tag{3.15}$$

In the Heisenberg picture Eq. (3.15) can be understood as a counterpart of Eq. (2.6). Introducing the fourth component of the momentum by the definition

$$P_4 = i P_0 = i H , \tag{3.16}$$

we then have

$$[P_\mu, F(x)] = i \frac{\partial F(x)}{\partial x_\mu} \tag{3.17}$$

for $\mu = 1$ to 4. The Eqs. (3.17) are relativistically covariant, and will be very important in what follows. In Sec. 4 we shall derive them in another way.

If we insert the energy density $\mathscr{H}$ in (3.15) and use the periodic boundary condition, we obtain

$$[P_k, H] = i \int d^3x \frac{\partial \mathscr{H}}{\partial x_k} = 0. \tag{3.18}$$

As in the classical theory, the components of momentum are conserved quantities under the quantum mechanical motion of the system. In a similar way, any two spatial components of the momentum commute with each other. Thus it follows in general that

$$[P_\mu, P_\nu] = 0. \tag{3.19}$$

In principle it is therefore possible to choose a representation in which all four quantities P_μ are diagonal. We shall designate the state vectors of this representation by $|a\rangle$, $|b\rangle$, etc., and the corresponding eigenvalues of P_μ by $p_\mu^{(a)}$, etc. One can readily show that for an arbitrary $F(x)$ in (3.17),

$$\langle a|\,[P_\mu, F(x)]\,|b\rangle = [p_\mu^{(a)} - p_\mu^{(b)}]\,\langle a|F(x)|b\rangle = i\frac{\partial}{\partial x_\mu}\langle a|F(x)|b\rangle. \quad (3.20)$$

In this representation the x-dependence of each operator is therefore given by

$$\langle a|F(x)|b\rangle = \langle a|F|b\rangle\, e^{i\sum_\mu (p_\mu^{(b)} - p_\mu^{(a)})\, x_\mu}, \quad (3.21)$$

where the quantities $\langle a|F|b\rangle$ are independent of x. In Chap. VII we shall refer to this remark.

4. Transformation Properties of the Theory

In the canonical formalism developed so far, the original relativistic invariance of the theory has been lost. Certainly the equations of motion (3.2) and (3.3) have the desired covariant form as long as the Lagrangian is an invariant. However, the canonical commutation relations (3.5), which allow the proper transition to the quantized theory, are not covariant because the two times x_0 and x_0' are assumed equal. In this section we shall further investigate the properties of the quantized theory under Lorentz transformations. In particular, it will appear that Eqs. (3.17) not only form a covariant system, but also that they have a fundamental significance for the transformation properties of the whole theory.

We assume that we have two Lorentz frames, x and x' and that if a point P has coordinates x in the first system, its coordinates in the other system are given by

$$x'_\mu = x_\mu + \varepsilon_{\mu\nu}\, x_\nu + \delta_\mu\,. \quad (4.1)$$

Here, and in the following, we sum over repeated indices. For simplicity we further assume that the quantities $\varepsilon_{\mu\nu}$ and δ_μ are infinitesimal, so that their higher powers may be neglected. Since the Lorentz transformations form a group, it is sufficient to study the transformation properties of the theory under infinitesimal transformations. In order that (4.1) describe an actual Lorentz transformation without stretching the coordinate axes, $\varepsilon_{\mu\nu}$ must be chosen antisymmetric. The transformation properties of the classical fields are known and have the form

$$\varphi'_\alpha(x') = \tfrac{1}{2}\,\varepsilon_{\mu\nu}\, S_{\mu\nu,\alpha\beta}\,\varphi_\beta(x(x')) + \varphi_\alpha(x(x')). \quad (4.2)$$

The left side of this equation gives the field functions in the new coordinate system as functions of the new coordinates. On the right side are the field functions in the old coordinate system, evaluated at the same point P which is present on the left side. By solving (4.1) for x_μ,

$$x_\mu = x'_\mu - \varepsilon_{\mu\nu} x'_\nu - \delta_\mu \, , \tag{4.3}$$

one can likewise regard the right side of Eq. (4.2) as a function of the new coordinates. We shall now consider two points P and P' which are chosen so that the point P in the first system has the same coordinates as the point P' in the second system. The field functions $\varphi_\alpha(x)$ at the two points therefore differ by amounts $\delta\varphi_\alpha(x)$ which are given by

$$\left.\begin{aligned} \delta\varphi_\alpha(x) &= \frac{1}{2}\,\varepsilon_{\mu\nu} S_{\mu\nu,\alpha\beta}\,\varphi_\beta(x) + \frac{\partial\varphi_\alpha(x)}{\partial x_\mu}(x_\mu - x'_\mu) = \\ &= \frac{1}{2}\,\varepsilon_{\mu\nu} S_{\mu\nu,\alpha\beta}\,\varphi_\beta(x) - \frac{\partial\varphi_\alpha(x)}{\partial x_\mu}(\varepsilon_{\mu\nu} x_\nu + \delta_\mu). \end{aligned}\right\} \tag{4.4}$$

In a similar way we find the change in the quantities

$$\pi_{\alpha\mu}(x) = \frac{\partial \mathcal{L}}{\partial \dfrac{\partial\varphi_\alpha(x)}{\partial x_\mu}} \tag{4.5}$$

in going from one system to another:

$$\delta\pi_{\alpha\mu}(x) = -\frac{1}{2}\,\varepsilon_{\lambda\nu} S_{\lambda\nu,\beta\alpha}\,\pi_{\beta\mu}(x) - \frac{\partial\pi_{\alpha\mu}(x)}{\partial x_\lambda}(\varepsilon_{\lambda\nu} x_\nu + \delta_\lambda) + \varepsilon_{\mu\nu}\pi_{\alpha\nu}(x). \tag{4.6}$$

[The momenta (3.3) are special cases of these $\pi_{\alpha\mu}$.] The last term in (4.6) appears because the vector index μ in (4.5) must also be transformed. In particular, for the canonical momenta, we obtain

$$\delta\pi_\alpha(x) = -\frac{1}{2}\,\varepsilon_{\lambda\nu} S_{\lambda\nu,\beta\alpha}\,\pi_\beta(x) - \frac{\partial\pi_\alpha(x)}{\partial x_\lambda}(\varepsilon_{\lambda\nu} x_\nu + \delta_\lambda) - i\,\varepsilon_{4\nu}\pi_{\alpha\nu}(x). \tag{4.7}$$

Now if we first carry out the canonical quantization in the original coordinate system and then subsequently quantize in the new coordinate system, it is necessary that $\varphi_\alpha(x)$ and $\pi_\alpha(x)$ as well as $\varphi_\alpha(x)+\delta\varphi_\alpha(x)$ and $\pi_\alpha(x)+\delta\pi_\alpha(x)$ satisfy the commutation relations (3.5). The two procedures of quantizing are only equivalent if there exists a Hermitian matrix T for which

$$\varphi_\alpha(x) + \delta\varphi_\alpha(x) = e^{iT}\varphi_\alpha(x)\,e^{-iT}, \tag{4.8}$$

$$\pi_\alpha(x) + \delta\pi_\alpha(x) = e^{iT}\pi_\alpha(x)\,e^{-iT}, \tag{4.9}$$

with the same matrix T in both Eqs. (4.8) and (4.9). Since we have assumed that the Lorentz transformation (4.1) is infinitesimal, we can also take T infinitesimal and obtain

$$\delta\varphi_\alpha(x) = i\,[T, \varphi_\alpha(x)]\,, \tag{4.10}$$

$$\delta \pi_\alpha(x) = i\,[T, \pi_\alpha(x)]. \tag{4.11}$$

The existence of this matrix can be shown most simply by giving an explicit expression for it. We wish to prove that Eqs. (4.10) and (4.11) are satisfied if we choose the following form for T:

$$\left.\begin{aligned} T = \int d^3x \Big[& \frac{1}{2}\,\varepsilon_{\mu\nu}\, S_{\mu\nu,\alpha\beta}\, \pi_\alpha(x)\, \varphi_\beta(x) + x_\nu\, \varepsilon_{\nu k}\, \pi_\alpha(x)\, \frac{\partial \varphi_\alpha(x)}{\partial x_k} + \\ & + i\,(\varepsilon_{4k}\, x_k + \delta_4)\left(\pi_\alpha(x)\, \frac{\partial \varphi_\alpha(x)}{\partial x_0} - \mathscr{L}\right) - \delta_k\, \pi_\alpha(x)\, \frac{\partial \varphi_\alpha(x)}{\partial x_k}\Big]. \end{aligned}\right\} \tag{4.12}$$

The summation over the index k in (4.12) runs only from 1 to 3. We shall introduce the general convention that Latin characters like k , l , etc. can take only the values 1 to 3, while Greek characters like μ, ν, etc. can also take the value 4.

The proof for (4.10) and (4.11) is simple in principle, but somewhat laborious. We first give the calculation for $\varphi_\alpha(x)$. With the help of (3.5), one has

$$\begin{aligned} i\,[T, \varphi_\alpha(x)] = {} & \frac{1}{2}\,\varepsilon_{\mu\nu}\, S_{\mu\nu,\alpha\beta}\, \varphi_\beta(x) + x_\nu\, \varepsilon_{\nu k}\, \frac{\partial \varphi_\alpha(x)}{\partial x_k} + i\,(\varepsilon_{4k}\, x_k + \delta_4) \times \\ & \times \left[\frac{\partial \varphi_\alpha(x)}{\partial x_0} + \pi_\beta(x)\, \frac{\partial \dfrac{\partial \varphi_\beta(x)}{\partial x_0}}{\partial \pi_\alpha(x)} - \frac{\partial \mathscr{L}}{\partial \dfrac{\partial \varphi_\beta(x)}{\partial x_0}}\, \frac{\partial \dfrac{\partial \varphi_\beta(x)}{\partial x_0}}{\partial \pi_\alpha(x)} \right] - \delta_k\, \frac{\partial \varphi_\alpha(x)}{\partial x_k}. \end{aligned} \tag{4.13}$$

From (3.3), we conclude that the last two terms in the square brackets cancel. The other terms can be grouped to give

$$i\,[T, \varphi_\alpha(x)] = \frac{1}{2}\,\varepsilon_{\mu\nu}\, S_{\mu\nu,\alpha\beta}\, \varphi_\beta(x) + x_\nu\, \varepsilon_{\nu\mu}\, \frac{\partial \varphi_\alpha(x)}{\partial x_\mu} - \delta_\mu\, \frac{\partial \varphi_\alpha(x)}{\partial x_\mu}. \tag{4.14}$$

This proves (4.10). The calculation for $\pi_\alpha(x)$ proceeds in the same way:

$$\left.\begin{aligned} i\,[T, \pi_\alpha(x)] = {} & \frac{1}{2}\,\varepsilon_{\mu\nu}\, S_{\mu\nu,\beta\alpha}\, \pi_\beta(x) + \frac{\partial}{\partial x_k}\left(x_\nu\, \varepsilon_{\nu k}\, \pi_\alpha(x)\right) + i\,(\varepsilon_{4k}\, x_k + \delta_4) \times \\ & \times \left(- \pi_\beta(x)\, \frac{\partial \dfrac{\partial \varphi_\beta(x)}{\partial x_0}}{\partial \varphi_\alpha(x)} + \frac{\partial \mathscr{L}}{\partial \varphi_\alpha(x)} + \frac{\partial \mathscr{L}}{\partial \dfrac{\partial \varphi_\beta(x)}{\partial x_0}}\, \frac{\partial \dfrac{\partial \varphi_\beta(x)}{\partial x_0}}{\partial \varphi_\alpha(x)} \right) - \\ & - \frac{\partial}{\partial x_k}\left[i\,(\varepsilon_{4l}\, x_l + \delta_4)\, \frac{\partial \mathscr{L}}{\partial \dfrac{\partial \varphi_\alpha(x)}{\partial x_k}} \right] - \delta_k\, \frac{\partial \pi_\alpha(x)}{\partial x_k} = \\ = {} & - \frac{1}{2}\,\varepsilon_{\mu\nu}\, S_{\mu\nu,\beta\alpha}\, \pi_\beta(x) + x_\nu\, \varepsilon_{\nu k}\, \frac{\partial \pi_\alpha(x)}{\partial x_k} + i\,(\varepsilon_{4k}\, x_k + \delta_4)\, \frac{\partial \pi_\alpha(x)}{\partial x_0} - \\ & - \delta_k\, \frac{\partial \pi_\alpha(x)}{\partial x_k} - i\,\varepsilon_{4l}\, \pi_{\alpha l}(x) = \\ = {} & - \frac{1}{2}\,\varepsilon_{\mu\nu}\, S_{\mu\nu,\beta\alpha}\, \pi_\beta(x) + x_\nu\, \varepsilon_{\nu\mu}\, \frac{\partial \pi_\alpha(x)}{\partial x_\mu} - \delta_\mu\, \frac{\partial \pi_\alpha(x)}{\partial x_\mu} - i\,\varepsilon_{4\nu}\, \pi_{\alpha\nu}(x). \end{aligned}\right\} \tag{4.15}$$

The last form for the right side of (4.15) is identical to the right side of (4.7), and this proves (4.11).

The matrix T can be written in the following form:

$$T = \int d^3x \Big\{ (\varepsilon_{\mu\nu} x_\nu + \delta_\mu) \Big[-\pi_\alpha(x) \frac{\partial \varphi_\alpha(x)}{\partial x_\mu} - i \mathcal{L} \delta_{4\mu} \Big] + \\ + \frac{1}{2} \varepsilon_{\mu\nu} S_{\mu\nu,\alpha\beta} \pi_\alpha(x) \varphi_\beta(x) \Big\} . \tag{4.16}$$

Because of the antisymmetry of $\varepsilon_{\mu\nu}$, the expression[1]

$$-\frac{i}{2} \varepsilon_{\mu\nu} (S_{\nu 4,\alpha\beta} \pi_{\alpha\mu}(x) \varphi_\beta(x) + S_{\mu 4,\alpha\beta} \pi_{\alpha\nu}(x) \varphi_\beta(x)) \tag{4.17}$$

is equal to zero. We can rewrite the last term in (4.16) in the following way:

$$\begin{aligned} -\frac{i}{2} \varepsilon_{\mu\nu} (S_{\mu\nu,\alpha\beta} \pi_{\alpha 4}(x) \varphi_\beta(x) + S_{\nu 4,\alpha\beta} \pi_{\alpha\mu}(x) \varphi_\beta(x) + \\ + S_{\mu 4,\alpha\beta} \pi_{\alpha\nu}(x) \varphi_\beta(x)) \equiv -i \varepsilon_{\mu\nu} f_{4\mu\nu} = \\ = i \frac{\partial}{\partial x_\lambda} [f_{4\lambda\nu} (\varepsilon_{\nu\varrho} x_\varrho + \delta_\nu)] - i (\varepsilon_{\nu\varrho} x_\varrho + \delta_\nu) \frac{\partial}{\partial x_\lambda} f_{4\lambda\nu} = \\ = i \frac{\partial}{\partial x_k} [f_{4k\nu} (\varepsilon_{\nu\varrho} x_\varrho + \delta_\nu)] - i (\varepsilon_{\nu\varrho} x_\varrho + \delta_\nu) \frac{\partial}{\partial x_\lambda} f_{4\lambda\nu} . \end{aligned} \tag{4.18}$$

(The symbol $f_{\lambda\nu\mu}$ is antisymmetric in the first two indices.) From (4.16) and (4.18) we obtain

$$T = -i \int d^3x \, T_{4\nu} \, \delta x_\nu \ , \tag{4.19}$$

with

$$T_{\mu\nu} = -\pi_{\alpha\mu}(x) \frac{\partial \varphi_\alpha(x)}{\partial x_\nu} + \delta_{\mu\nu} \mathcal{L} - \frac{\partial}{\partial x_\lambda} f_{\lambda\mu\nu} \ , \tag{4.20}$$

and

$$\delta x_\mu = \varepsilon_{\mu\nu} x_\nu + \delta_\mu . \tag{4.21}$$

From the equations of motion (3.2) it follows that the tensor $T_{\mu\nu}$ satisfies the continuity equation

$$\frac{\partial T_{\mu\nu}}{\partial x_\mu} = -\frac{\partial^2}{\partial x_\lambda \partial x_\mu} f_{\lambda\mu\nu} = 0 . \tag{4.22}$$

The three-dimensional integrals

$$P_\mu = -i \int d^3x \, T_{4\mu} \tag{4.23}$$

1. F. J. Belinfante, Physica, Haag 6, 887 (1939).

are therefore constants of the motion. Since the last term in (4.20) for $\mu = 4$ can be regarded as a three-dimensional divergence, the P_μ of (4.23) are identical with the momenta defined in (3.12) and (3.16). If we now put $\varepsilon_{\mu\nu} = 0$ in (4.21), then with the help of (4.4) and (4.7) we can simplify Eqs. (4.10) and (4.11) to

$$[P_\mu, \varphi_\alpha(x)] = i\,\frac{\partial \varphi_\alpha(x)}{\partial x_\mu}, \tag{4.24}$$

$$[P_\mu, \pi_\alpha(x)] = i\,\frac{\partial \pi_\alpha(x)}{\partial x_\mu}. \tag{4.25}$$

Here we have recovered Eqs. (3.17) and can now regard them as expressing the invariance of the theory under infinitesimal translations. For this reason we shall also refer to the operators P_μ as displacement operators.

The conservation laws (4.22) and (4.23) can also be deduced in other ways. Since the operator H is obviously invariant under every translation, it follows from (4.10) and (4.11) that

$$\delta H = i\,[T, H] = i\,[P_\mu \delta_\mu, H] = 0. \tag{4.26}$$

The quantities δ_μ are arbitrary, and we can conclude from (4.26) that all P_μ commute with H and so have vanishing time derivatives in every coordinate system. This is only possible if the differential conservation law (4.22) holds. This consideration is particularly important because it is useful not only for translations but also, in slightly modified form, for "rotations" in four-dimensional space. For such transformations we have

$$\delta H = \frac{-i}{2}\,[\varepsilon_{\mu\nu} J_{\mu\nu}, H] = -i\,\varepsilon_{4\nu} P_\nu, \tag{4.27}$$

where

$$J_{\mu\nu} = i \int d^3x\,(T_{4\mu} x_\nu - T_{4\nu} x_\mu), \tag{4.28}$$

or

$$[H, J_{\mu\nu}] = -\delta_{4\mu} P_\nu + \delta_{4\nu} P_\mu. \tag{4.29}$$

Since the operators $J_{\mu\nu}$ are formed not only from the operators φ_α and π_α but also contain the coordinates x explicitly, we have

$$\frac{dJ_{\mu\nu}}{dx_0} = i\,[H, J_{\mu\nu}] - \int d^3x\,(T_{4\mu}\delta_{\nu 4} - T_{4\nu}\delta_{\mu 4}) = 0. \tag{4.30}$$

If Eq. (4.30) is to hold in every coordinate system, there must also be a differential conservation law here:

$$\frac{\partial}{\partial x_\lambda}(T_{\lambda\mu} x_\nu - T_{\lambda\nu} x_\mu) = 0. \tag{4.31}$$

Through comparison of (4.22) and (4.31) we obtain

$$T_{\mu\nu} - T_{\nu\mu} = 0\,, \tag{4.32}$$

i.e., the tensor $T_{\mu\nu}$ is symmetric. It is often referred to as the symmetric energy-momentum density in order to distinguish it from the tensor

$$\Theta_{\mu\nu} = -\pi_{\alpha\mu}(x)\,\frac{\partial\varphi_\alpha(x)}{\partial x_\nu} + \delta_{\mu\nu}\mathscr{L}\,, \tag{4.33}$$

which is known as the "canonical" energy-momentum tensor. This latter quantity usually gives the same displacement operators as $T_{\mu\nu}$ and therefore can also be interpreted as an energy-momentum density. For the construction of the angular momentum $J_{\mu\nu}$, however, it is essential to employ the symmetrical tensor $T_{\mu\nu}$.

CHAPTER II

THE FREE ELECTROMAGNETIC FIELD

5. Lagrange Function and Canonical Formalism

The Lagrange function of classical electrodynamics is obtained from

$$\mathscr{L} = -\tfrac{1}{4} F_{\mu\nu} F_{\mu\nu} \, . \qquad (5.1)$$

Here $F_{\mu\nu}$ is the electromagnetic field tensor:

$$F_{4k} = -F_{k4} = i\,\mathsf{E}_k \, , \qquad (5.2a)$$

$$F_{12} = -F_{21} = \mathsf{H}_3 \text{ , and cyclic permutations.} \qquad (5.2b)$$

In this Lagrangian we introduce the potentials $A_\mu(x)$ as the dynamical variables, rather than the field strengths. The field strengths are given by

$$F_{\mu\nu} = \frac{\partial A_\nu(x)}{\partial x_\mu} - \frac{\partial A_\mu(x)}{\partial x_\nu} \, . \qquad (5.3)$$

Regarding the potentials as the variables in the Lagrange function, we obtain the equations of motion according to (3.2):

$$\frac{\partial}{\partial x_\mu}\left(\frac{\partial A_\nu(x)}{\partial x_\mu} - \frac{\partial A_\mu(x)}{\partial x_\nu}\right) \equiv \Box\, A_\mu(x) - \frac{\partial^2 A_\nu(x)}{\partial x_\mu \partial x_\nu} = 0. \qquad (5.4)$$

This formulation of the classical theory is obviously Lorentz invariant as well as gauge invariant. That is, it is invariant under the transformations

$$A_\mu(x) \to A_\mu(x) + \frac{\partial \Lambda(x)}{\partial x_\mu} . \qquad (5.5)$$

Neither the field strengths nor the equations of motion are changed by these transformations. In the classical theory this invariance under general gauge transformations is often reduced by requiring that the potentials satisfy

$$\frac{\partial A_\mu(x)}{\partial x_\mu} = 0 \, . \qquad (5.6)$$

With an appropriate choice of the gauge function $\Lambda(x)$ in (5.5), it is always possible to achieve this; however, the gauge function is still not uniquely determined by this condition. It is always possible to carry out further gauge transformations, but only with functions which satisfy the wave equation

$$\Box \Lambda(x) = 0. \tag{5.7}$$

In this gauge, the equations of motion (5.4) simplify to

$$\Box A_\mu(x) = 0. \tag{5.8}$$

If we attempt to go to the Hamiltonian form of the theory, starting from this Lagrangian, we find that the canonical momenta (3.3) are

$$\pi_\mu(x) = i\, F_{4\mu}(x)\,. \tag{5.9}$$

The momentum conjugate to $A_4(x)$ vanishes identically and Eqs. (5.9) cannot be solved for all $\frac{\partial A_\mu(x)}{\partial x_0}$. Thus the whole method developed in Chap. I is useless. Instead of (5.1) we can start with the following Lagrangian:

$$\mathscr{L} = -\frac{1}{4} F_{\mu\nu} F_{\mu\nu} - \frac{1}{2} \frac{\partial A_\mu(x)}{\partial x_\mu} \cdot \frac{\partial A_\nu(x)}{\partial x_\nu}\,. \tag{5.10}$$

The equations of motion (5.8) are obtained, and the canonical momenta are

$$\pi_k(x) = i\, F_{4k}(x)\,, \tag{5.11}$$

$$\pi_4(x) = i\, \frac{\partial A_\nu(x)}{\partial x_\nu}\,. \tag{5.12}$$

In the Hamiltonian formulation of this theory, we do not obtain Eq. (5.6) as an equation of motion, but only the weaker condition

$$\Box \frac{\partial A_\nu(x)}{\partial x_\nu} = 0\,. \tag{5.13}$$

If we do not consider the most general solution of these equations but only those which satisfy the initial conditions

$$\frac{\partial A_\nu(x)}{\partial x_\nu} = \frac{\partial^2 A_\nu(x)}{\partial x_0\, \partial x_\nu} = 0, \text{ for all } \boldsymbol{x}, \tag{5.14}$$

and for some fixed time (e.g., $x_0 = 0$), then from (5.13) it follows that the quantity $\frac{\partial A_\nu(x)}{\partial x_\nu}$ must vanish for all times. As desired, we then recover the usual theory of electromagnetism.

In the quantization of the electromagnetic field, we shall first ignore completely the initial conditions (5.14) and use only the Lagrangian (5.10). For $x_0 = x_0'$ the canonical commutation relations become

$$[A_\mu(x), A_\nu(x')] = 0, \tag{5.15}$$

$$\left.\begin{aligned}[A_\mu(x), \pi_k(x')] = \left[A_\mu(x), \frac{\partial A_k(x')}{\partial x'_0} - i\frac{\partial A_4(x')}{\partial x'_k}\right] = \left[A_\mu(x), \frac{\partial A_k(x')}{\partial x'_0}\right] = \\ = i\,\delta_{\mu k}\,\delta(\boldsymbol{x} - \boldsymbol{x}')\end{aligned}\right\} \tag{5.16}$$

$$[A_k(x), \pi_4(x')] = \left[A_k(x), i\frac{\partial A_\nu(x')}{\partial x'_\nu}\right] = \left[A_k(x), \frac{\partial A_4(x')}{\partial x'_0}\right] = 0 \tag{5.17}$$

$$[A_4(x), \pi_4(x')] = \left[A_4(x), \frac{\partial A_4(x')}{\partial x'_0}\right] = i\,\delta(\boldsymbol{x} - \boldsymbol{x}'), \tag{5.18}$$

$$[\pi_k(x), \pi_l(x')] = \left[\frac{\partial A_k(x)}{\partial x_0}, \frac{\partial A_l(x')}{\partial x'_0}\right] = 0, \tag{5.19}$$

$$[\pi_k(x), \pi_4(x')] = \left[\frac{\partial A_k(x)}{\partial x_0}, \frac{\partial A_4(x')}{\partial x'_0}\right] = 0. \tag{5.20}$$

The Eqs. (5.15) through (5.20) can be summarized:

$$[A_\mu(x), A_\nu(x')] = 0, \tag{5.21}$$

$$\left[\frac{\partial A_\mu(x)}{\partial x_0}, A_\nu(x')\right] = -i\,\delta_{\mu\nu}\,\delta(\boldsymbol{x} - \boldsymbol{x}'), \tag{5.22}$$

$$\left[\frac{\partial A_\mu(x)}{\partial x_0}, \frac{\partial A_\nu(x')}{\partial x'_0}\right] = 0. \tag{5.23}$$

It must be emphasized that the commutation relations (5.15) to (5.23) are valid in a Heisenberg picture <u>only for equal times</u> x_0 <u>and</u> x'_0.

For what follows, it will be of interest to carry out the quantization in momentum space. We therefore introduce the expansion (1.1) for $A_\mu(x)$:

$$A_\mu(x) = \frac{1}{\sqrt{V}} \sum_k e^{ikx} A_\mu(k). \tag{5.24}$$

From (5.8) it follows that in (5.24) only those k can be present for which

$$k^2 = \boldsymbol{k}^2 - k_0^2 = 0. \tag{5.25}$$

Equation (5.25) has two solutions:

$$k_0 = \pm\omega, \qquad \omega = +\sqrt{\boldsymbol{k}^2}; \tag{5.26}$$

and we can write (5.24) as

$$A_\mu(x) = \frac{1}{\sqrt{V}} \sum_{\boldsymbol{k}} [e^{ikx} A_\mu(\boldsymbol{k}) + e^{-ikx} A_\mu^*(\boldsymbol{k})], \tag{5.27}$$

where

$$k_0 = \omega. \tag{5.27a}$$

From the requirement that $A_k(x)$ is a Hermitian operator and $A_4(x)$ is an anti-Hermitian operator, it follows that $A_k(\boldsymbol{k})$, $iA_4(\boldsymbol{k})$ and $A_k^*(\boldsymbol{k})$, $iA_4^*(\boldsymbol{k})$, respectively, are Hermitian conjugate operators. Moreover, since $A_\mu(x)$ is a vector, for every $\boldsymbol{k}$ there are four independent "possible polarizations", which are conveniently described by using a set of orthonormal, but otherwise arbitrary, "polarization vectors" $e_\mu^{(\lambda)}$, $\lambda=1,\dots,4$. If (5.27) is to be the most general possible solution of the equations of motion (5.8), we must sum over the four possible directions of polarization:

$$A_\mu(x) = \frac{1}{\sqrt{V}} \sum_{\boldsymbol{k}} \sum_{\lambda=1}^{4} \frac{e_\mu^{(\lambda)}}{\sqrt{2\omega}} [e^{ikx} a^{(\lambda)}(\boldsymbol{k}) + e^{-ikx} a^{*(\lambda)}(\boldsymbol{k})], \quad (5.28)$$

$$e_\mu^{(\lambda)} e_\mu^{(\lambda')} = \delta_{\lambda\lambda'}. \quad (5.29)$$

In (5.28) we have introduced an additional factor $\sqrt{2\omega}$ in the denominator. This is only a matter of notation with no special significance. The choice (5.28) gives the commutation relations for $a^{(\lambda)}(\boldsymbol{k})$ a very simple form.

It is not necessary for the vectors $e_\mu^{(\lambda)}$ to be independent of $\boldsymbol{k}$, and it is even advantageous to make the following $\boldsymbol{k}$-dependent choice:

$$e_4^{(1)} = e_4^{(2)} = e_4^{(3)} = 0 , \quad (5.30a)$$

$$e_l^{(1)} k_l = e_l^{(2)} k_l = 0 , \quad (5.30b)$$

$$e_l^{(3)} = \frac{k_l}{\omega} , \quad (5.30c)$$

$$e_l^{(4)} = 0 , \quad (5.30d)$$

$$e_4^{(4)} = 1 . \quad (5.30e)$$

In this case we shall refer to transversely polarized light for $\lambda=1$ or 2, to longitudinal polarization for $\lambda=3$, and to scalar polarization for $\lambda=4$. We now see that if we were to have Eq. (5.6) as an operator equation, only transversely polarized light could be present. Since Eq. (5.6) has been ignored so far, we have to take into account all four polarizations. With this choice of unit vectors $e_\mu^{(\lambda)}$, in addition to (5.29), we have

$$\sum_\lambda e_\mu^{(\lambda)} e_\nu^{(\lambda)} = \delta_{\mu\nu}. \quad (5.31)$$

Because of Eq. (5.30e) where the right side is 1 (rather than i, for example), Eq. (5.31) has formal Lorentz invariance. The reality conditions for $a^{(\lambda)}(\boldsymbol{k})$ and $a^{*(\lambda)}(\boldsymbol{k})$ are the same as those for $A_\mu(\boldsymbol{k})$ and $A_\mu^*(\boldsymbol{k})$. From Eq. (5.21), with the help of Eq. (5.28), we get

$$\left.\begin{aligned}[A_\mu(x), A_\nu(x')]_{x_0=x_0'} = \frac{1}{V}\sum_{\boldsymbol{k},\boldsymbol{k}'}\sum_{\lambda,\lambda'}\frac{e_\mu^{(\lambda)}e_\nu^{(\lambda')}}{2\sqrt{\omega\omega'}}\{e^{i(\boldsymbol{k}\boldsymbol{x}+\boldsymbol{k}'\boldsymbol{x}')-i(\omega+\omega')x_0}\times\\ \times[a^{(\lambda)}(\boldsymbol{k}), a^{(\lambda')}(\boldsymbol{k}')] + e^{-i(\boldsymbol{k}\boldsymbol{x}+\boldsymbol{k}'\boldsymbol{x}')+i(\omega+\omega')x_0}[a^{*(\lambda)}(\boldsymbol{k}), a^{*(\lambda')}(\boldsymbol{k}')] +\\ + e^{i(\boldsymbol{k}\boldsymbol{x}-\boldsymbol{k}'\boldsymbol{x}')-i(\omega-\omega')x_0}[a^{(\lambda)}(\boldsymbol{k}), a^{*(\lambda')}(\boldsymbol{k}')] + e^{-i(\boldsymbol{k}\boldsymbol{x}-\boldsymbol{k}'\boldsymbol{x}')+i(\omega-\omega')x_0}\times\\ \times[a^{*(\lambda)}(\boldsymbol{k}), a^{(\lambda')}(\boldsymbol{k}')]\} = 0.\end{aligned}\right\} \quad (5.32)$$

Since all ω are positive and since (5.32) is to hold for arbitrary times, we conclude

$$[a^{(\lambda)}(\boldsymbol{k}), a^{(\lambda')}(\boldsymbol{k}')] = [a^{*(\lambda)}(\boldsymbol{k}), a^{*(\lambda')}(\boldsymbol{k}')] = 0, \quad (5.33)$$

$$[a^{(\lambda)}(\boldsymbol{k}), a^{*(\lambda')}(\boldsymbol{k}')] = c^{(\lambda\lambda')}(\boldsymbol{k})\cdot\delta_{\boldsymbol{k},\boldsymbol{k}'}. \quad (5.34)$$

The quantity $c^{(\lambda\lambda')}(\boldsymbol{k})$ can depend on λ, λ' and $\boldsymbol{k}$ but only in such a way that

$$c^{(\lambda\lambda')}(\boldsymbol{k}) = c^{(\lambda'\lambda)}(-\boldsymbol{k}). \quad (5.35)$$

It follows from Eqs. (5.33) through (5.35) that (5.23) is identically satisfied. To get the specific form of c we use Eq. (5.22):

$$-i\,\delta_{\mu\nu}\,\delta(\boldsymbol{x}-\boldsymbol{x}') = \frac{1}{V}\sum_{\boldsymbol{k}}\sum_{\lambda,\lambda'}\frac{e_\mu^{(\lambda)}e_\nu^{(\lambda')}}{2\omega}(-i\omega)\,c^{(\lambda\lambda')}(\boldsymbol{k})\,2e^{i\boldsymbol{k}(\boldsymbol{x}-\boldsymbol{x}')}; \quad (5.36)$$

thus

$$c^{(\lambda\lambda')}(\boldsymbol{k}) = \delta_{\lambda\lambda'}. \quad (5.37)$$

Equation (5.34) then takes the simple form

$$[a^{(\lambda)}(\boldsymbol{k}), a^{*(\lambda')}(\boldsymbol{k}')] = \delta_{\lambda\lambda'}\cdot\delta_{\boldsymbol{k}\boldsymbol{k}'}. \quad (5.38)$$

6. The Hilbert Space of Free Photons

We obtain the Hamiltonian operator from the Lagrange function (5.10) after some partial integrations:

$$\left.\begin{aligned}H = \frac{1}{2}\int d^3x\left[\frac{\partial A_\mu(x)}{\partial x_0}\frac{\partial A_\mu(x)}{\partial x_0} + \frac{\partial A_\mu(x)}{\partial x_k}\frac{\partial A_\mu(x)}{\partial x_k}\right] =\\ = -\frac{1}{4V}\sum_{\boldsymbol{k},\boldsymbol{k}'}\sum_{\lambda,\lambda'}\int\frac{d^3x}{\sqrt{\omega\omega'}}[e^{ikx}a^{(\lambda)}(\boldsymbol{k}) - e^{-ikx}a^{*(\lambda)}(\boldsymbol{k})]\times\\ \times[e^{ik'x}a^{(\lambda')}(\boldsymbol{k}') - e^{-ik'x}a^{*(\lambda')}(\boldsymbol{k}')]\,\delta_{\lambda\lambda'}\,(\omega\omega' + \boldsymbol{k}\boldsymbol{k}') =\\ = \frac{1}{2}\sum_{\boldsymbol{k},\lambda}\omega\,\{a^{(\lambda)}(\boldsymbol{k}), a^{*(\lambda)}(\boldsymbol{k})\}.\end{aligned}\right\} \quad (6.1)$$

The last term on the right contains the anticommutator

$$\{a, a^*\} = a\,a^* + a^*a\,. \quad (6.2)$$

We introduce the Hermitian operators $q_{\boldsymbol{k}}^{(\lambda)}$ and $p_{\boldsymbol{k}}^{(\lambda)}$ by the definitions

$$\left.\begin{aligned} q^{(\lambda)} &= \frac{1}{\sqrt{2\omega}}\,(a^{(\lambda)} + a^{*\,(\lambda)})\ , \\ p^{(\lambda)} &= -\,i\,\sqrt{\frac{\omega}{2}}\,(a^{(\lambda)} - a^{*\,(\lambda)})\ , \end{aligned}\right\}\ \text{for}\ \ \lambda = 1,2,3, \tag{6.3}$$

$$\left.\begin{aligned} q^{(4)} &= \frac{1}{\sqrt{2\omega}}\,(a^{(4)} - a^{*\,(4)})\ , \\ p^{(4)} &= i\,\sqrt{\frac{\omega}{2}}\,(a^{(4)} + a^{*(4)})\ . \end{aligned}\right\} \tag{6.4}$$

The operators p and q satisfy the usual commutation relations for all λ and λ' :

$$[p^{(\lambda)}, q^{(\lambda')}] = -\,i\,\delta_{\lambda\lambda'}. \tag{6.5}$$

In these variables, the Hamiltonian becomes

$$H = \tfrac{1}{2}\sum_{\boldsymbol{k}}\left\{\sum_{\lambda=1}^{3}(p^{(\lambda)2} + \omega^2 q^{(\lambda)2}) - (p^{(4)2} + \omega^2 q^{(4)2})\right\}. \tag{6.6}$$

Thus the Hamiltonian is a sum of independent harmonic oscillators. The eigenvalues of the energy of this system are

$$E = \sum_{\boldsymbol{k}}\left\{\sum_{\lambda=1}^{3} n^{(\lambda)}(\boldsymbol{k}) - n^{(4)}(\boldsymbol{k})\right\}\omega. \tag{6.7}$$

In (6.7) the zero-point energy of the oscillators has been omitted. This is allowed here because the observed quantities are only energy differences. Formally this is apparent because we can always add an arbitrary c-number to the energy. The commutation relations (3.17) are not changed by this, and if the c-number is time-independent, the shifted energy is still a constant of the motion. One can also change the order of operators in (6.6) or (6.1) so that the zero-point energy automatically vanishes. This ordering is not determined by the correspondence principle, and therefore we have no particular reason for either choice. We shall designate the state of n particles of given k and given polarization direction simply by $|n\rangle$, momentarily suppressing all particles of other polarizations and $\boldsymbol{k}$, for simplicity. From (5.38), (6.1), and (6.7) we obtain

$$H\,a^*\,|n\rangle = [H, a^*]\,|n\rangle + a^*\,H\,|n\rangle = \left\{\begin{aligned} &(1+n)\,\omega\,a^*\,|n\rangle\ \text{for}\ \lambda \neq 4, \\ &(1-n)\,\omega\,a^*\,|n\rangle\ \text{for}\ \lambda = 4. \end{aligned}\right\} \tag{6.8}$$

Hence the state vector $a^*|n\rangle$ is also an eigenvector of the Hamiltonian, but with eigenvalue $(n+1)\,\omega$ for $\lambda=1, 2, 3$ and with eigenvalue $-(n-1)\omega$ for $\lambda=4$. With the help of Eq. (6.8) we can express all eigenvectors in terms of the vector $|0\rangle$, the state representing "no particles":

$$|n^{(\lambda)}\rangle = c_n^{(\lambda)}\,[a^{*\,(\lambda)}]^n\,|0\rangle \quad \text{for} \quad \lambda \neq 4, \tag{6.9a}$$

$$|n^{(4)}\rangle = c_n^{(4)}\,[a^{(4)}]^n\,|0\rangle. \tag{6.9b}$$

The normalization constants $c_n^{(\lambda)}$ can be determined from the condition

$$1 = \langle n^{(\lambda)}|n^{(\lambda)}\rangle.$$

For $\lambda \neq 4$, we obtain

$$\left.\begin{aligned}
\langle n^{(\lambda)}|n^{(\lambda)}\rangle &= |c_n^{(\lambda)}|^2\langle 0|[a^{(\lambda)}]^n\,[a^{*\,(\lambda)}]^n|\,0\rangle = \\
&= |c_n^{(\lambda)}|^2\langle 0|[a^{(\lambda)}]^{n-1}\,(a^{*\,(\lambda)}\,a^{(\lambda)}+1)\,[a^{*\,(\lambda)}]^{n-1}|0\rangle = \\
&= |c_n^{(\lambda)}|^2\{\langle 0|[a^{(\lambda)}]^{n-1}\,[a^{*\,(\lambda)}]^{n-1}|0\rangle + \langle 0|[a^{(\lambda)}]^{n-1}\,a^{*\,(\lambda)}\times \\
&\times (a^{*(\lambda)}\,a^{(\lambda)}+1)\,[a^{*\,(\lambda)}]^{n-2}|0\rangle\} = \cdots = \\
&= n\,|c_n^{(\lambda)}|^2\langle 0|[a^{(\lambda)}]^{n-1}\,[a^{*\,(\lambda)}]^{n-1}|0\rangle = n\cdot\frac{|c_n^{(\lambda)}|^2}{|c_{n-1}^{(\lambda)}|^2}\,,
\end{aligned}\right\} \tag{6.10}$$

with the solution

$$|c_n^{(\lambda)}| = \frac{1}{\sqrt{n!}}. \tag{6.11}$$

For $\lambda=4$, the calculation is similar and the result is the same:

$$\langle n^{(4)}|n^{(4)}\rangle = |c_n^{(4)}|^2\,(-1)^n\langle 0|[a^{*\,(4)}]^n\,[a^{(4)}]^n|0\rangle = n\,\frac{|c_n^{(4)}|^2}{|c_{n-1}^{(4)}|^2}\,, \tag{6.10a}$$

$$|c_n^{(4)}| = \frac{1}{\sqrt{n!}}. \tag{6.11a}$$

The most general eigenvector of (6.1) can now be written as a direct product of vectors (6.9):

$$\prod_k \prod_{\lambda=1}^{4} |n^{(\lambda)}(k)\rangle = \prod_k \prod_{\lambda=1}^{3} \frac{[a^{*(\lambda)}(k)]^{n^{(\lambda)}(k)}}{\sqrt{n^{(\lambda)}(k)!}}\,\frac{[a^{(4)}(k)]^{n^{(4)}(k)}}{\sqrt{n^{(4)}(k)!}}\,|0\rangle. \tag{6.12}$$

In (6.12) and all subsequent discussion, $|0\rangle$ is to be the eigenvector of the complete Hamiltonian which represents the state with "no particles". The arbitrary phases which could have appeared in (6.12) have been taken as zero, since they have no significance.

To recapitulate, we have found that the quantized electromagnetic field can be described by means of a system of oscillators and that the energy of such an oscillator is quantized in the usual way. This is the old (light) quantum hypothesis which is in a sense the source of the quantum theory and which we have recovered here as a consequence of our formalism.

From (6.12), matrix elements of the operators $a^{(\lambda)}(\boldsymbol{k})$ can be simply determined. Obviously these operators are diagonal in all quantum numbers other than $n^{(\lambda)}(\boldsymbol{k})$. We again suppress all indices with other quantum numbers and so obtain

$$\langle n|a^{(\lambda)}|n+1\rangle = \langle n+1|a^{*(\lambda)}|n\rangle = \sqrt{n+1} \quad \text{for} \quad \lambda \neq 4\,, \tag{6.13}$$

$$\langle n+1|a^{(4)}|n\rangle = -\langle n|a^{*(4)}|n+1\rangle = \sqrt{n+1}\,. \tag{6.14}$$

The other matrix elements are zero. The operators

$$\left.\begin{aligned} N^{(\lambda)}(\boldsymbol{k}) &= a^{*(\lambda)}(\boldsymbol{k})\, a^{(\lambda)}(\boldsymbol{k}) \qquad \text{for } \lambda \neq 4\,, \\ N^{(4)}(\boldsymbol{k}) &= -\, a^{(4)}(\boldsymbol{k})\, a^{*(4)}(\boldsymbol{k}) \end{aligned}\right\} \tag{6.15}$$

are therefore diagonal in the representation (6.12) and have the eigenvalues $n^{(\lambda)}(\boldsymbol{k})$. They are the operators which give the number of light quanta of given polarization and given momentum and are usually designated as the "number operators". In a similar way, we call the operators $a^{(\lambda)}(\boldsymbol{k})$ for $\lambda \neq 4$ and $-a^{*(4)}(\boldsymbol{k})$ the "annihilation operators" and the quantities $a^{*(\lambda)}(\boldsymbol{k})$ for $\lambda \neq 4$ and $a^{(4)}(\boldsymbol{k})$ the "creation operators".

By (6.15) we can write the energy as

$$H = \sum_{\boldsymbol{k}} \left\{ \sum_{\lambda=1}^{3} N^{(\lambda)}(\boldsymbol{k}) - N^{(4)}(\boldsymbol{k}) \right\} \omega\,, \tag{6.16}$$

where the zero-point energy is again neglected. This distinguishes (6.16) from (6.1) and (6.6) but, as already noted, there is no physical significance to this difference.

As an illustration of the interpretation of these results in terms of light quanta, it is interesting to compute the total momentum and the angular momentum. For the first quantity, we obtain from (3.12), (5.11), (5.12), and (5.28),

$$\left.\begin{aligned} P_k &= -\int d^3x \left\{ i\left(\frac{\partial A_l}{\partial x_4} - \frac{\partial A_4}{\partial x_l}\right)\frac{\partial A_l}{\partial x_k} + i\,\frac{\partial A_\nu}{\partial x_\nu}\,\frac{\partial A_4}{\partial x_k}\right\} = \\ &= -\int d^3x\, \frac{\partial A_\mu}{\partial x_0}\,\frac{\partial A_\mu}{\partial x_k} = \frac{1}{2}\sum_{\boldsymbol{k},\lambda} k_k \left\{ a^{(\lambda)}(\boldsymbol{k})\, a^{*(\lambda)}(\boldsymbol{k}) + a^{*(\lambda)}(\boldsymbol{k})\, a^{(\lambda)}(\boldsymbol{k}) - \right. \\ &\left. - a^{(\lambda)}(\boldsymbol{k})\, a^{(\lambda)}(-\boldsymbol{k})\, e^{-2i\omega x_0} - a^{*(\lambda)}(\boldsymbol{k})\, a^{*(\lambda)}(-\boldsymbol{k})\, e^{2i\omega x_0} \right\}. \end{aligned}\right\} \tag{6.17}$$

Since the operators $a^{(\lambda)}(\boldsymbol{k})$ and $a^{(\lambda)}(-\boldsymbol{k})$ commute with each other, the last two terms in brackets are symmetric in $\boldsymbol{k}$. The sum over both terms vanishes because of the antisymmetry of k_k, and Eq.

(6.17) simplifies to

$$P_k = \tfrac{1}{2} \sum_{\boldsymbol{k},\lambda} \{a^{(\lambda)}(\boldsymbol{k}), a^{*(\lambda)}(\boldsymbol{k})\} k_k \, . \tag{6.18}$$

With the introduction of the number operators (6.15), we can also write this expression as

$$P_k = \sum_{\boldsymbol{k}} k_k \left\{ \sum_{\lambda=1}^{3} N^{(\lambda)}(\boldsymbol{k}) - N^{(4)}(\boldsymbol{k}) \right\} . \tag{6.19}$$

In (6.19) the term with the zero-point momentum has been dropped because of the antisymmetry in k_k . No special ordering of the operators is necessary to obtain this result. It follows from (6.19) that each photon has a momentum $\boldsymbol{k}$ (or $-\boldsymbol{k}$ for $\lambda = 4$) in complete accord with the corpuscular interpretation of the radiation field.

In order to calculate the angular momentum, we begin with the symmetric energy-momentum tensor (4.20) and decompose it into two parts:

$$T_{\mu\nu} = T_{\mu\nu}^{(0)} + T_{\mu\nu}^{(1)} \, , \tag{6.20}$$

$$T_{\mu\nu}^{(0)} = \frac{\partial A_\lambda}{\partial x_\lambda} \frac{\partial A_\mu}{\partial x_\nu} - F_{\lambda\mu} \frac{\partial A_\lambda}{\partial x_\nu} + \delta_{\mu\nu} \mathscr{L} \, , \tag{6.21}$$

$$T_{\mu\nu}^{(1)} = - \frac{\partial}{\partial x_\lambda} \left[A_\nu F_{\mu\lambda} + (A_\mu \delta_{\nu\lambda} - A_\lambda \delta_{\mu\nu}) \frac{\partial A_\varrho}{\partial x_\varrho} \right] . \tag{6.22}$$

In a similar manner, we write the angular momentum as

$$J_{ij} = J_{ij}^{(0)} + J_{ij}^{(1)} \, , \tag{6.23}$$

$$\left. \begin{aligned} J_{ij}^{(0)} &= i \int d^3x \, (T_{4i}^{(0)} x_j - T_{4j}^{(0)} x_i) = \\ &= -i \int d^3x \left\{ \frac{\partial A_\lambda}{\partial x_\lambda} \left(x_i \frac{\partial A_4}{\partial x_j} - x_j \frac{\partial A_4}{\partial x_i} \right) + x_j F_{\lambda 4} \frac{\partial A_\lambda}{\partial x_i} - x_i F_{\lambda 4} \frac{\partial A_\lambda}{\partial x_j} \right\} , \end{aligned} \right\} \tag{6.24}$$

$$J_{ij}^{(1)} = -i \int d^3x \left\{ x_j \frac{\partial}{\partial x_k} \left[A_i F_{4k} + A_4 \delta_{ik} \frac{\partial A_\nu}{\partial x_\nu} \right] - x_i \frac{\partial}{\partial x_k} \left[A_j F_{4k} + A_4 \delta_{jk} \frac{\partial A_\nu}{\partial x_\nu} \right] \right\} . \tag{6.25}$$

The two expressions (6.24) and (6.25) may be simplified considerably. From appropriate combinations of the terms in (6.24), we obtain after integration by parts

$$\begin{aligned} \int d^3x \left\{ x_i \frac{\partial A_\lambda}{\partial x_\lambda} \frac{\partial A_4}{\partial x_j} - x_i \frac{\partial A_4}{\partial x_\lambda} \frac{\partial A_\lambda}{\partial x_j} \right\} &= \int d^3x \left\{ x_i \frac{\partial A_k}{\partial x_k} \frac{\partial A_4}{\partial x_j} - x_i \frac{\partial A_4}{\partial x_k} \frac{\partial A_k}{\partial x_j} \right\} = \\ &= - \delta_{ij} \int d^3x \frac{\partial A_k}{\partial x_k} A_4 + \int d^3x \, A_4 \frac{\partial A_i}{\partial x_j} . \end{aligned} \tag{6.26}$$

For the expression (6.24) it then follows that

$$J_{ij}^{(0)} = -i \int d^3x \left\{ x_i \frac{\partial A_\lambda}{\partial x_4} \frac{\partial A_\lambda}{\partial x_j} - x_j \frac{\partial A_\lambda}{\partial x_4} \frac{\partial A_\lambda}{\partial x_i} + A_4 F_{ji} \right\} . \tag{6.27}$$

In a similar way, we can transform (6.25) to

$$J_{ij}^{(1)}=i\int d^3x\left\{A_iF_{4j}-A_jF_{4i}\right\}=-i\int d^3x\left\{A_j\frac{\partial A_i}{\partial x_4}-A_i\frac{\partial A_j}{\partial x_4}+A_4F_{ij}\right\}. \tag{6.28}$$

The last two terms in (6.27) and (6.28) cancel upon addition and we have

$$J_{ij}=-\int d^3x\left\{x_i\frac{\partial A_\nu}{\partial x_0}\frac{\partial A_\nu}{\partial x_j}-x_j\frac{\partial A_\nu}{\partial x_0}\frac{\partial A_\nu}{\partial x_i}+A_j\frac{\partial A_i}{\partial x_0}-A_i\frac{\partial A_j}{\partial x_0}\right\}. \tag{6.29}$$

We introduce the expansion (5.28) into (6.29). The first two terms contain the expression $e_\nu^{(\lambda)}e_\nu^{(\lambda')}=\delta_{\lambda\lambda'}$ as a factor. This operator is therefore independent of the possible polarizations of the photons. Consequently, it must be interpreted as the orbital angular momentum. From the last two terms we get

$$\begin{aligned}\int d^3x\left\{A_j\frac{\partial A_i}{\partial x_0}-A_i\frac{\partial A_j}{\partial x_0}\right\}&=\frac{i}{2}\sum_{\boldsymbol{k},\lambda,\lambda'}e_j^{(\lambda)}e_i^{(\lambda')}\left[\{a^{(\lambda)},a^{*(\lambda')}\}-\{a^{(\lambda')},a^{*(\lambda)}\}\right]=\\&=i\sum_{\boldsymbol{k},\lambda,\lambda'}e_j^{(\lambda)}e_i^{(\lambda')}\left[a^{*(\lambda')}a^{(\lambda)}-a^{*(\lambda)}a^{(\lambda')}\right].\end{aligned} \tag{6.30}$$

Equation (6 .30) is certainly not diagonal in the representation introduced earlier; however, a simple transformation enables us to diagonalize simultaneously one component of the angular momentum (6.30) and the energy. To do this, let us write

$$a_+=\frac{1}{\sqrt{2}}\left(a^{(1)}-i\,a^{(2)}\right), \tag{6. 31}$$

$$a_-=\frac{1}{\sqrt{2}}\left(a^{(1)}+i\,a^{(2)}\right). \tag{6.32}$$

The new operators a_+ and a_- satisfy the usual canonical commutation relations

$$[a_+,a_+^*]=[a_-,a_-^*]=1\ , \tag{6.33}$$

$$[a_+,a_-]=[a_+,a_-^*]=\cdots=0. \tag{6.34}$$

In these variables, the energy becomes

$$H=\sum_{\boldsymbol{k}}\omega\left(a_+^*a_++a_-^*a_-+a^{*(3)}a^{(3)}+a^{(4)}a^{*(4)}\right)\ , \tag{6.35}$$

and the component of the angular momentum in the direction of $\boldsymbol{k}$ is

$$J^{(3)}(\boldsymbol{k})=a_+^*a_+-a_-^*a_-. \tag{6.36}$$

Therefore, if we choose a representation in which the operators $N_+ = a_+^* a_+$, $N_- = a_-^* a_-$, $N^{(3)}$, and $N^{(4)}$ are diagonal, we see from (6.36) that this component of the intrinsic angular momentum or "spin" can have the three values zero (longitudinal or scalar photons) and ± 1 (circularly polarized photons). If the square of the spin is calculated from the above expressions, it can be shown that the scalar photons have spin zero and the others spin one.

Indeed, serious objections can be raised against the formalism developed here. The energy is not positive definite; the last term in (6.7) has a negative overall sign. Furthermore, the theory we have quantized here is evidently not the "right" Maxwell theory, since the classical Lorentz condition (5.6) does not hold in the quantized theory. We shall later see that these two points are connected and that if we take account of the problem of the Lorentz condition, the states with negative energy will not occur. Before further discussion of these points, we wish to study a different problem in Sec. 7, but one which will be of importance in introducing the Lorentz condition.

7. Commutators for Arbitrary Times, the Singular Functions

The canonical quantization method prescribes the commutation relations between the field operators and their time derivatives only if the time coordinates of both the operators are equal. For unequal times there is no immediate recipe. In this section we will show that the commutators can be calculated for two arbitrary points from the solution (5.28) and the commutation relations (5.38) and that the result has a relativistically covariant form. Thus if we compute the commutator of $A_\mu(x)$ and $A_\nu(x')$ at two completely arbitrary points x and x', we obtain

$$\left.\begin{aligned}
[A_\mu(x), A_\nu(x')] &= \frac{1}{V} \sum_{\boldsymbol{k}, \boldsymbol{k}', \lambda, \lambda'} \frac{e_\mu^{(\lambda)} e_\nu^{(\lambda')}}{2\sqrt{\omega\omega'}} [e^{ikx} a^{(\lambda)}(\boldsymbol{k}) + \\
&\quad + e^{-ikx} a^{*(\lambda)}(\boldsymbol{k}), e^{ik'x'} a^{(\lambda')}(\boldsymbol{k}') + e^{-ik'x'} a^{*(\lambda')}(\boldsymbol{k}')] = \\
&= \frac{1}{V} \sum_{\boldsymbol{k}, \lambda} \frac{e_\mu^{(\lambda)} e_\nu^{(\lambda)}}{2\omega} [e^{ik(x-x')} - e^{-ik(x-x')}] = \\
&= \frac{\delta_{\mu\nu}}{(2\pi)^3} \int \frac{d^3k}{2\omega} [e^{ik(x-x')} - e^{-ik(x-x')}].
\end{aligned}\right\} \tag{7.1}$$

With the introduction of the δ-function

$$\delta(k^2) = \delta(\boldsymbol{k}^2 - k_0^2) = \frac{1}{2|\boldsymbol{k}|} [\delta(|\boldsymbol{k}| - k_0) + \delta(|\boldsymbol{k}| + k_0)], \tag{7.2}$$

Eq. (7.1) can be written as

$$[A_\mu(x), A_\nu(x')] = -\frac{\delta_{\mu\nu}}{(2\pi)^3} \int dk\, e^{ik(x'-x)} \delta(k^2)\, \varepsilon(k), \tag{7.3}$$

where

$$\varepsilon(k) = \frac{k_0}{|k_0|}. \tag{7.4}$$

The relativistic covariance of the commutation relations is clearly evident from the form (7.3). Furthermore we see that the commutator of the two field operators is a c-number even if the times are not equal. This is not a consequence of the canonical quantization but a special property of a field theory without interactions. We shall later be concerned with the general problem where there are two fields interacting with each other and then the commutator is no longer a c-number but is a q-number if the two times are not equal.

The right side of (7.3) will appear very frequently in what follows, and we therefore introduce the special symbol

$$[A_\mu(x), A_\nu(x')] = -i\,\delta_{\mu\nu} D(x'-x), \tag{7.5}$$

where the real function $D(x)$ is defined[1] by

$$D(x) = \frac{-i}{(2\pi)^3} \int dk\, e^{ikx}\, \delta(k^2)\, \varepsilon(k). \tag{7.6}$$

The function $D(x)$ has the following properties:

$$\left.\begin{aligned} D(x) &= 0 && (7.7\text{a}) \\ \frac{\partial D(x)}{\partial x_0} &= -\delta(\boldsymbol{x}) && (7.7\text{b}) \end{aligned}\right\} \quad \text{for} \quad x_0 = 0.$$

Because of relativistic invariance, $D(x)$ vanishes not only for $x_0=0$, but, in general, for $x^2 > 0$. It satisfies the differential equation

$$\Box D(x) = 0 \tag{7.8}$$

and can be defined as the solution of (7.8) which has (7.7) as initial conditions for $x_0=0$. Equations (7.7) and (7.8) can be shown to hold by means of the integral representation (7.6). For $x_0=0$ the integration over k_0 can be carried out immediately and, on the basis of symmetry, the result is zero. From this we have (7.7a). Alternatively, if (7.6) is first differentiated with respect to x_0 and then x_0 is set equal to zero, the integration over k_0 gives 1; and the three-dimensional integral gives the spatial delta function on the right side of (7.7b). Finally, Eq. (7.8) follows from (7.6),

---

1. In the literature, other notations than that of (7.5) are also used. In recent years the notation of (7.5) has been used most frequently, and we shall commit ourselves to this. In the older literature one can find the symbol $\Delta(x)$ used instead of $D(x)$; however, we shall reserve this for a more general function. (See below.)

because $k^2 \cdot \delta(k^2) = 0$.

Subsequently we shall be interested in the anticommutator of two field operators. This quantity is not a c-number but a q-number. We calculate its expectation value for the state $|0\rangle$, obtaining

$$\left.\begin{aligned}\langle 0|\{A_\mu(x), A_\nu(x')\}|0\rangle = \frac{1}{V}\sum_{\boldsymbol{k},\boldsymbol{k}',\lambda,\lambda'} \frac{e_\mu^{(\lambda)} e_\nu^{(\lambda')}}{2\sqrt{\omega\omega'}} \{e^{ikx+ik'x'}\langle 0|\{a^{(\lambda)}, a^{(\lambda')}\}|0\rangle + \\ + e^{ikx-ik'x'}\langle 0|\{a^{(\lambda)}, a^{*(\lambda')}\}|0\rangle + e^{-ikx+ik'x'}\langle 0|\{a^{*(\lambda)}, a^{(\lambda')}\}|0\rangle + \\ + e^{-ikx-ik'x'}\langle 0|\{a^{*(\lambda)}, a^{*(\lambda')}\}|0\rangle\}.\end{aligned}\right\} \quad (7.9)$$

From the representations (6.13) and (6.14) we obtain

$$\langle 0|\{a^{(\lambda)}, a^{(\lambda')}\}|0\rangle = \langle 0|\{a^{*(\lambda)}, a^{*(\lambda')}\}|0\rangle = 0, \quad (7.10)$$

$$\langle 0|\{a^{*(\lambda)}(\boldsymbol{k}), a^{(\lambda')}(\boldsymbol{k}')\}|0\rangle = \delta_{\lambda\lambda'}\delta_{\boldsymbol{k}\boldsymbol{k}'} \quad \text{for } \lambda \text{ and } \lambda' \neq 4, \quad (7.11)$$

$$\langle 0|\{a^{*(4)}(\boldsymbol{k}), a^{(\lambda)}(\boldsymbol{k}')\}|0\rangle = -\delta_{\boldsymbol{k}\boldsymbol{k}'}\delta_{\lambda 4}, \quad (7.12)$$

and we can write (7.9) as

$$\left.\begin{aligned}\langle 0|\{A_\mu(x), A_\nu(x')\}|0\rangle = \frac{1}{V}\sum_{\boldsymbol{k}} \frac{1}{2\omega}\left[\sum_{\lambda=1}^{3} e_\mu^{(\lambda)} e_\nu^{(\lambda)} - e_\mu^{(4)} e_\nu^{(4)}\right] \times \\ \times \left[e^{ik(x-x')} + e^{-ik(x-x')}\right] = (\delta_{\mu\nu} - 2\delta_{\mu 4}\delta_{\nu 4})\, D^{(1)}(x'-x),\end{aligned}\right\} \quad (7.13)$$

where the function $D^{(1)}(x)$ is defined by

$$D^{(1)}(x) = \frac{1}{(2\pi)^3}\int dk\, e^{ikx}\,\delta(k^2). \quad (7.14)$$

The principle difference between (7.14) and (7.5) is that the sign-function $\varepsilon(k)$ is lacking in (7.14). Because of this, the function $D^{(1)}(x)$ has properties that are quite different from those of the function $D(x)$. For example, the expression (7.14) does not vanish for $x_0 = 0$, although its time derivative is exactly zero over this surface. We note that the right side of (7.13) depends upon the coordinate system used and does not have simple transformation properties. This is because of the minus sign in (7.12) or (6.14) and is therefore associated with the difficulty of indefinite energies. We shall return to this point later, but in this section we wish to study only the properties of the functions $D(x)$ and $D^{(1)}(x)$.

Because of Eq. (7.7), the function $D(x)$ can be used to find the solution of the inhomogeneous wave equation

$$\Box\,\psi(x) = f(x), \quad (7.15)$$

with the prescribed initial conditions

$$\left.\begin{aligned} \psi(x) &= u(\boldsymbol{x}) \\ \frac{\partial\psi(x)}{\partial x_0} &= v(\boldsymbol{x}) \end{aligned}\right\} \quad \text{for} \quad x_0 = T \tag{7.16}$$

for an arbitrary time T. One can readily show that the desired solution is

$$\psi(x) = \int_T^{x_0} dx' D(x-x') f(x') - \int_{x_0'=T} d^3x' \left[D(x-x') v(\boldsymbol{x}') + \frac{\partial D(x-x')}{\partial x_0} u(\boldsymbol{x}')\right]. \tag{7.17}$$

The first term on the right side of (7.17) contains a four-dimensional integral in which the limits of integration are shown only for x_0. The integration over the three spatial variables is over the whole three-dimensional space. In what follows we shall employ this notation quite frequently. This integral alone is a solution of (7.15). Both this integral and its time derivative vanish for $x_0 = T$. By itself, the last integral in (7.17) is a solution of the homogeneous wave equation [because of (7.8)] and has the correct initial values. Together these terms give the desired solution.

In many applications, we shall pass to the limit $T \to -\infty$ in (7.17). With the assumption that the three-dimensional integral tends to a finite limiting value $\psi^{(0)}(x)$, we obtain

$$\psi(x) = \psi^{(0)}(x) + \int_{-\infty}^{x_0} dx' D(x-x') f(x') = \psi^{(0)}(x) - \int D_R(x-x') f(x') dx'. \tag{7.18}$$

In (7.18) we have introduced the retarded D-function $D_R(x)$ by the definition

$$D_R(x) = \begin{Bmatrix} -D(x) & \text{for} & x_0 > 0\,, \\ 0 & \text{for} & x_0 < 0\,. \end{Bmatrix} \tag{7.19}$$

If no limits of integration are written, for example, as in the last term of (7.18), the integration is to be taken over the whole four-dimensional space.

By analogy to (7.19), we can also introduce an "advanced" D-function $D_A(x)$ by the definition

$$D_A(x) = \begin{Bmatrix} 0 & \text{for} & x_0 > 0\,, \\ D(x) & \text{for} & x_0 < 0\,. \end{Bmatrix} \tag{7.20}$$

Half the sum of $D_R(x)$ and $D_A(x)$ is often designated by $\bar{D}(x)$ in the literature:

$$\bar{D}(x) = \tfrac{1}{2}[D_R(x) + D_A(x)]. \tag{7.21}$$

The last function is therefore related to $D(x)$ by

$$\overline{D}(x) = -\frac{1}{2}\frac{x_0}{|x_0|}D(x) = -\frac{1}{2}\varepsilon(x)D(x). \tag{7.22}$$

It will often be useful to have the Fourier representations for $\overline{D}(x)$, $D_R(x)$, and $D_A(x)$, analogous to (7.6) and (7.14). Starting with $\overline{D}(x)$, we write

$$\varepsilon(x) = \frac{x_0}{|x_0|} = \frac{2}{\pi}\int_0^\infty \frac{d\tau}{\tau}\sin(\tau x_0) = \frac{1}{i\pi}P\int_{-\infty}^{+\infty}\frac{d\tau}{\tau}e^{i\tau x_0}. \tag{7.23}$$

From this we obtain

$$\left.\begin{aligned}\overline{D}(x) &= \frac{-1}{2\pi i}P\int\frac{d\tau}{\tau}e^{i\tau x_0}\frac{(-i)}{(2\pi)^3}\int dk\, e^{ikx}\,\delta(k^2)\,\varepsilon(k) = \\ &= \frac{1}{(2\pi)^4}\int dk\, e^{ikx}P\int\frac{d\tau}{\tau}\delta(\mathbf{k}^2-(k_0+\tau)^2)\frac{k_0+\tau}{|k_0+\tau|} = \frac{1}{(2\pi)^4}P\int\frac{dk}{k^2}e^{ikx}.\end{aligned}\right\} \tag{7.24}$$

In (7.23) and (7.24) the letter P in front of the integral sign indicates that the principal value is to be taken. With this, we have the integral representation for $\overline{D}(x)$. From the equations

$$D_R(x) = \overline{D}(x) - \tfrac{1}{2}D(x), \tag{7.25a}$$

$$D_A(x) = \overline{D}(x) + \tfrac{1}{2}D(x), \tag{7.25b}$$

we obtain representations for $D_R(x)$ and $D_A(x)$:

$$D_R(x) = \frac{1}{(2\pi)^4}\int dk\, e^{ikx}\left\{P\frac{1}{k^2} + i\pi\,\delta(k^2)\,\varepsilon(k)\right\}, \tag{7.26a}$$

$$D_A(x) = \frac{1}{(2\pi)^4}\int dk\, e^{ikx}\left\{P\frac{1}{k^2} - i\pi\,\delta(k^2)\,\varepsilon(k)\right\}. \tag{7.26b}$$

For completeness, we give a few definitions:

$$D^{(+)}(x) = \frac{1}{2}\left(D(x) - iD^{(1)}(x)\right) = \frac{-i}{(2\pi)^3}\int dk\, e^{ikx}\,\delta(k^2)\,\frac{1}{2}\left[\varepsilon(k)+1\right], \tag{7.27}$$

$$D^{(-)}(x) = \frac{1}{2}\left(D(x) + iD^{(1)}(x)\right) = \frac{-i}{(2\pi)^3}\int dk\, e^{ikx}\,\delta(k^2)\,\frac{1}{2}\left[\varepsilon(k)-1\right], \tag{7.28}$$

$$D_F(x) = \frac{2}{i}\overline{D}(x) + D^{(1)}(x) = \frac{2}{i}\frac{1}{(2\pi)^4}\int dk\, e^{ikx}\left\{P\frac{1}{k^2} + i\pi\,\delta(k^2)\right\}. \tag{7.29}$$

Of these, (7.29) has found wide application.[1] All the D-functions given here can also be written as complex integrals with the integrand $1/k^2$. The various functions are then distinguished from each

1. In the literature, $D_F(x)$ is often designated by $D_C(x)$ and is sometimes defined without the factor $2/i$ in (7.29).

other by different contours of integration in the complex k_0-plane. For practical purposes, only the complex representation for $D_F(x)$ is essential. From (7.29) one sees directly that this function is given by

$$D_F(x) = \frac{2}{i} \frac{1}{(2\pi)^4} \int_{C_F} dk \frac{e^{ikx}}{k^2}, \tag{7.30}$$

where the path C_F is shown in Fig. 1.

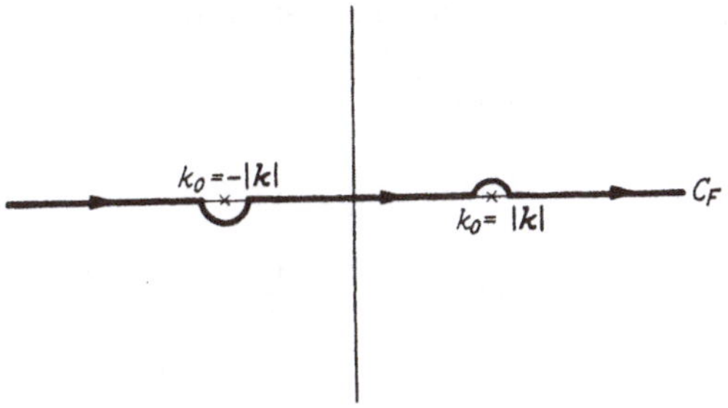

Fig. 1. The path C_F in the complex k_0-plane.

From the representations (7.24), (7.26), and (7.29), it follows that the four functions $\overline{D}(x)$, $D_R(x)$, $D_A(x)$, and $D_F(x)$ satisfy the inhomogeneous wave equations

$$\Box \overline{D}(x) = \Box D_R(x) = \Box D_A(x) = -\delta(x), \tag{7.31}$$

$$\Box D_F(x) = 2i\,\delta(x). \tag{7.32}$$

Instead of Eq. (7.15) we shall consider the more general wave equation

$$(\Box - m^2)\,\psi(x) = f(x) \tag{7.33}$$

with a mass m and introduce a system of singular functions for this equation. We shall denote them by $\Delta(x, m^2)$, etc. or by $\Delta(x)$ if we are discussing only a single mass. We have, for example,

$$(\Box - m^2)\,\Delta(x) = (\Box - m^2)\,\Delta^{(1)}(x) = 0, \tag{7.34}$$

$$\left.\begin{aligned} \Delta(x) &= \frac{\partial \Delta^{(1)}(x)}{\partial x_0} = 0 \\ \frac{\partial \Delta(x)}{\partial x_0} &= -\delta(\boldsymbol{x}) \end{aligned}\right\} \quad \text{for } x_0 = 0, \tag{7.35}$$

$$(\Box - m^2)\,\overline{\Delta}(x) = (\Box - m^2)\,\Delta_R(x) = (\Box - m^2)\,\Delta_A(x) = -\delta(x), \tag{7.36}$$

$$\overline{\Delta}(x) = -\tfrac{1}{2}\varepsilon(x)\,\Delta(x), \quad \text{etc.} \tag{7.37}$$

From the previous integral representations, we can easily find the representations for the various Δ-functions. To do this, we replace $\delta(k^2)$ by $\delta(k^2+m^2)$ and $P\frac{1}{k^2}$ by $P\frac{1}{k^2+m^2}$ everywhere.

In order to obtain the x-dependence of the various Δ-functions, the evaluation of the integral representations given here can be carried out in terms of Bessel functions. These expressions do not have any great significance for subsequent applications, and usually it is simplest to work directly with the integral representations. We shall not state explicit formulae for the singular functions, but instead refer the reader to the original work.[1] The result is simple only for the case $m=0$. We therefore give the result for the functions $D(x)$ and $D^{(1)}(x)$:

$$D(x) = -\frac{1}{2\pi}\varepsilon(x)\,\delta(x^2), \tag{7.38}$$

$$D^{(1)}(x) = \frac{1}{2\pi^2}P\frac{1}{x^2}. \tag{7.39}$$

The function $D(x)$ not only vanishes outside the light cone but also inside it. It is different from zero only for $x^2=0$, and there it is as singular as a delta function. The function $D^{(1)}(x)$ is different from zero everywhere and becomes singular on the light cone like x^{-2}. These results hold only for the case $m=0$. In a theory with a particle mass different from zero, the function $\Delta(x)$ is also different from zero inside the light cone. However, the strongest singularities on the light cone are the same for the functions $D(x)$ and $\Delta(x)$ and for $D^{(1)}(x)$ and $\Delta^{(1)}(x)$.

8. The Subsidiary Condition: First Method

As we have frequently noted, the theory which we have developed so far is incomplete because the Lorentz condition (5.6) has not been taken into account. This condition is essential in the classical theory if the equations of motion are to have the form (5.8). It is quite clear that we cannot simply interpret (5.6) as an operator equation because this would contradict the canonical commutation relation (5.18):

$$[A_4(x), \pi_4(x')]_{x_0=x_0'} = i\left[A_4(x), \frac{\partial A_\nu(x')}{\partial x_\nu'}\right]_{x_0=x_0'} = i\,\delta(\boldsymbol{x}-\boldsymbol{x}'). \tag{8.1}$$

In order to obtain the classical theory in the limit, it is not necessary that all the classical Maxwell equations correspond to operator identities in the quantized theory. It is sufficient to require only that the expectation values of the electromagnetic

---

1. Comprehensive discussions have been given by J. Schwinger, Phys. Rev. 75, 651 (1949), appendix; and by W. Heitler, Quantum Theory of Radiation, Third Ed., p. 71-76, Oxford, 1954. In Heitler the terms D and Δ are interchanged, and a different choice of sign is used in the definition.

field quantities satisfy these equations for every physically realizable state. It is sufficient to limit ourselves to the states $|\psi\rangle$ which satisfy the equation

$$\frac{\partial A_\nu}{\partial x_\nu}|\psi\rangle = 0 \qquad (8.2)$$

as a subsidiary condition.[1] Equation (8.2) is equivalent to

$$[a^{(3)}(\boldsymbol{k}) + i\,a^{(4)}(\boldsymbol{k})]\,|\psi\rangle = 0\,, \qquad (8.3)$$

$$[a^{*(3)}(\boldsymbol{k}) + i\,a^{*(4)}(\boldsymbol{k})]\,|\psi\rangle = 0\,. \qquad (8.4)$$

The subsidiary condition is thus only a requirement for longitudinal and scalar photons; it has no effect on the transverse photons. An allowable state vector $|\psi\rangle$ can be written as a product:

$$|\psi\rangle = |\psi_T\rangle \prod_{\boldsymbol{k}} |\Phi_{\boldsymbol{k}}\rangle\,, \qquad (8.5)$$

where $|\psi_T\rangle$ contains only transverse photons and where $|\Phi_{\boldsymbol{k}}\rangle$ can be expressed as

$$|\Phi_{\boldsymbol{k}}\rangle = \sum_{n^{(3)},\,n^{(4)}} \alpha_{n^{(3)}n^{(4)}}\,|n^{(3)}, n^{(4)}\rangle = \sum_{n^{(3)},\,n^{(4)}} \alpha_{n^{(3)}n^{(4)}}\,\frac{[a^{*(3)}]^{n^{(3)}}\,[a^{(4)}]^{n^{(4)}}}{\sqrt{n^{(3)}!\,n^{(4)}!}}|0\rangle. \qquad (8.6)$$

Substitution in (8.3) and (8.4) gives

$$\sum \alpha_{n^{(3)}n^{(4)}}\Big[\sqrt{n^{(3)}}\,|n^{(3)}-1, n^{(4)}\rangle + i\sqrt{n^{(4)}+1}\,|n^{(3)}, n^{(4)}+1\rangle\Big] = 0\,, \qquad (8.7)$$

$$\sum \alpha_{n^{(3)}n^{(4)}}\Big[\sqrt{n^{(3)}+1}\,|n^{(3)}+1, n^{(4)}\rangle - i\sqrt{n^{(4)}}\,|n^{(3)}, n^{(4)}-1\rangle\Big] = 0. \qquad (8.8)$$

The most general solution of (8.7) and (8.8) is

$$\alpha_{n^{(3)}n^{(4)}} = c\,\delta_{n^{(3)},\,n^{(4)}}\,(-i)^{n^{(4)}}\,, \qquad (8.9)$$

where the constant c is to be determined from the normalization requirement

$$\langle\Phi_{\boldsymbol{k}}|\Phi_{\boldsymbol{k}}\rangle = 1\,. \qquad (8.10)$$

The subsidiary condition (8.2) therefore furnishes an exact prescription for the mixing of longitudinal and scalar photons which may be present in a physically realizable state. In a sense, these degrees of freedom of the field are eliminated and only the transverse photons distinguish the physically realizable states from each other. For many applications it is therefore sufficient to regard $|\psi_T\rangle$ alone as the state vector and to consider the dynamical variables to be $a^{(\lambda)}(\boldsymbol{k})$ only for $\lambda = 1$ and 2. With sufficiently careful calculation, one can actually obtain correct results in this manner. For example, the total energy (6.7) is just equal to that

1. E. Fermi, Rev. Mod. Phys. 4, 87 (1932).

of the transverse photons since, because of (8.9), the energy of the longitudinal photons is exactly cancelled by the energy of the scalar photons. The total energy of a state (8.5) is therefore positive definite, and the state of lowest energy--the vacuum--has no transverse photons and only the mixture (8.6), (8.9) of the others. A similar result occurs for the spatial P_k and J_{ij}, whose eigenvalues can be expressed in terms of only the transverse photons. In particular, it turns out that for the component of the angular momentum in the direction $\boldsymbol{k}$ only the two values +1 or -1 can occur although, since the total spin is one, three orientations would be expected. This is a special property of quantum electrodynamics with a vanishing photon mass and is a consequence of the Lorentz condition.

No matter how attractive this method looks at first glance, there are certain mathematical difficulties hidden in it.[1] For example, the subsidiary condition (8.2) appears to be incompatible with the commutation relations (7.5). From (7.5) we have

$$\left[\frac{\partial A_\mu(x)}{\partial x_\mu},\ A_\nu(x')\right] = -i\frac{\partial}{\partial x_\nu} D(x'-x)\ , \tag{8.11}$$

and therefore

$$\langle\psi|\left[\frac{\partial A_\mu(x)}{\partial x_\mu},\ A_\nu(x')\right]|\psi\rangle = -i\frac{\partial}{\partial x_\nu} D(x'-x) \neq 0 \tag{8.12}$$

instead of zero, as one would expect from (8.2). This difficulty is resolved by the remark that the vector $|\Phi_k\rangle$ in (8.6), (8.9) is not normalizable and therefore that (8.10) is not satisfied for any finite c. In applying (8.2), we do not obtain zero instead of (8.12), but rather an indefinite expression of the form $0 \cdot \infty$. This must be defined by taking some limit.[2] A useful formalism for this purpose can be constructed if we ascribe a small rest mass μ to the photon and therefore modify[3] the equations of motion (5.8) to

$$(\Box - \mu^2)\, A_\mu(x) = 0\ . \tag{8.13}$$

The commutation relations for the operators $A_\mu(x)$ are only changed by the appearance in (7.5) of the singular functions $\Delta(x'-x)$ with

---

1. See F. J. Belinfante, Phys. Rev. 76, 226 (1949); F. Coester and J. M. Jauch, Phys. Rev. 78, 149 (1950); S. T. Ma, Phys. Rev. 80, 729 (1950).

2. R. Utiyama, T. Imamura, S. Sunakawa and T. Dodo, Progr. Theor. Phys. 6, 587 (1951).

3. I am indebted to Prof. W. Pauli for the suggestion to use a small photon mass for this purpose. F. Coester, Phys. Rev. 83, 798 (1951) and R. J. Glauber, Progr. Theor. Phys. 9, 295 (1953) have given formulations of quantum electrodynamics as the limiting case of a theory with a non-zero mass.

mass μ rather than the functions $D(x'-x)$. Now let us introduce a scalar field $B(x)$ and a new vector field $U_\mu(x)$ by the definitions

$$B(x) \doteq -\frac{1}{\mu}\frac{\partial A_\mu(x)}{\partial x_\mu}, \tag{8.14}$$

$$U_\mu(x) = A_\mu(x) + \frac{1}{\mu}\frac{\partial B(x)}{\partial x_\mu}. \tag{8.15}$$

These satisfy the equation of motion (8.13) and, in addition, the field $U_\mu(x)$ identically satisfies the operator equation

$$\frac{\partial U_\mu(x)}{\partial x_\mu} = -\mu B(x) + \frac{1}{\mu}\Box B(x) = 0. \tag{8.16}$$

Because of this, only three components of $U_\mu(x)$ can be taken as dynamically independent. Decomposing the new field into plane waves, we write

$$\left.\begin{aligned} U_k(x) = \frac{1}{\sqrt{V}}\sum_{\boldsymbol{k}}\frac{1}{\sqrt{2\omega}}\Big\{e^{ikx}\Big[\sum_{\lambda=1}^{2} e_k^{(\lambda)}\, u^{(\lambda)}(\boldsymbol{k}) + \frac{\omega}{\mu} e_k^{(3)}\, u^{(3)}(\boldsymbol{k})\Big] + \\ + e^{-ikx}\Big[\sum_{\lambda=1}^{2} e_k^{(\lambda)}\, u^{*(\lambda)}(\boldsymbol{k}) + \frac{\omega}{\mu} e_k^{(3)}\, u^{*(3)}(\boldsymbol{k})\Big]\Big\}, \end{aligned}\right\} \tag{8.17}$$

$$U_0(x) = -i\,U_4(x) = \frac{1}{\sqrt{V}}\sum_{\boldsymbol{k}}\frac{1}{\sqrt{2\omega}}\frac{|\boldsymbol{k}|}{\mu}\left[e^{ikx}\, u^{(3)}(\boldsymbol{k}) + e^{-ikx}\, u^{*(3)}(\boldsymbol{k})\right], \tag{8.18}$$

$$B(x) = \frac{1}{\sqrt{V}}\sum_{\boldsymbol{k}}\frac{1}{\sqrt{2\omega}}\left[e^{ikx}\, b(\boldsymbol{k}) + e^{-ikx}\, b^*(\boldsymbol{k})\right], \tag{8.19}$$

with

$$k^2 + \mu^2 = \boldsymbol{k}^2 - \omega^2 + \mu^2 = 0, \qquad \omega > 0. \tag{8.20}$$

In x-space the new fields have the commutation relations

$$[B(x), B(x')] = \frac{1}{\mu^2}\frac{\partial^2}{\partial x_\mu \partial x'_\nu}[A_\mu(x), A_\nu(x')] = i\Delta(x'-x), \tag{8.21}$$

$$[U_\mu(x), B(x')] = -\frac{1}{\mu}\left[A_\mu(x), \frac{\partial A_\nu(x')}{\partial x'_\nu}\right] + \frac{i}{\mu}\frac{\partial}{\partial x_\mu}\Delta(x'-x) = 0, \tag{8.22}$$

$$[U_\mu(x), U_\nu(x')] = [U_\mu(x), A_\nu(x')] = -i\left(\delta_{\mu\nu} - \frac{1}{\mu^2}\frac{\partial^2}{\partial x_\mu \partial x_\nu}\right)\Delta(x'-x). \tag{8.23}$$

From this, we find for the operators $u^{(\lambda)}(\boldsymbol{k})$ and $b(\boldsymbol{k})$,

$$[u^{(\lambda)}(\boldsymbol{k}), u^{*(\lambda')}(\boldsymbol{k}')] = \delta_{\lambda\lambda'}\,\delta_{\boldsymbol{k}\boldsymbol{k}'}, \tag{8.24}$$

$$[b(\boldsymbol{k}), b^*(\boldsymbol{k}')] = -\delta_{\boldsymbol{k}\boldsymbol{k}'}. \tag{8.25}$$

All other commutators vanish identically. We see that the roles of the creation and annihilation operators for the B-field are reversed, just as they were for $A_4(x)$ above.

We shall now discuss what restrictions on the state vector are necessary to enable us to take the limit $\mu\to 0$ and recover the usual Maxwell theory. For this it is necessary that the expectation value of all products of quantities $\frac{\partial A_\nu(x)}{\partial x_\nu}$ vanish in the limit. This we ensure by requiring that a physically realizable state contain no " B-particles". Hence it is exactly true that

$$\langle\psi|\frac{\partial A_\nu(x)}{\partial x_\nu}|\psi\rangle = -\mu\langle\psi|\,B(x)\,|\psi\rangle = 0. \tag{8.26}$$

Moreover, in the limit $\mu\to 0$, we have

$$\left.\begin{aligned}\lim_{\mu\to 0}\langle\psi|\frac{\partial A_\mu(x)}{\partial x_\mu}\,\frac{\partial A_\nu(x')}{\partial x'_\nu}|\psi\rangle &= \lim_{\mu\to 0}\mu^2\langle\psi|\,B(x)\,B(x')\,|\psi\rangle = \\ &= i\lim_{\mu\to 0}\mu^2\Delta^{(+)}(x'-x) = 0,\ \text{etc.}\end{aligned}\right\} \tag{8.27}$$

We shall now impose the additional requirement[1] that the physical state contain no longitudinal " U-particles", i.e., that

$$u^{(3)}(\boldsymbol{k})\,|\psi\rangle = 0\,, \tag{8.28}$$

$$b^*(\boldsymbol{k})\,|\psi\rangle = 0. \tag{8.29}$$

From (8.14) and (8.15) and from the expansions (5.28), (8.17) to (8.19), we obtain

$$b(\boldsymbol{k}) = -\frac{i}{\mu}\left(|\boldsymbol{k}|\,a^{(3)}(\boldsymbol{k}) + i\,\omega\,a^{(4)}(\boldsymbol{k})\right), \tag{8.30a}$$

$$u^{(\lambda)}(\boldsymbol{k}) = a^{(\lambda)}(\boldsymbol{k})\,,\quad \lambda = 1, 2\,, \tag{8.30b}$$

$$\frac{\omega}{\mu}\,u^{(3)}(\boldsymbol{k}) = a^{(3)}(\boldsymbol{k}) + \frac{i}{\mu}\,|\boldsymbol{k}|\,b(\boldsymbol{k}) = \frac{\omega}{\mu^2}\left[\omega\,a^{(3)}(\boldsymbol{k}) + i\,|\boldsymbol{k}|\,a^{(4)}(\boldsymbol{k})\right]. \tag{8.30c}$$

The vector $|\Phi_{\boldsymbol{k}}\rangle$ in (8.5) therefore satisfies

$$[\omega\,a^{(3)} + i\,|\boldsymbol{k}|\,a^{(4)}]\,|\Phi_{\boldsymbol{k}}\rangle = 0\,, \tag{8.31}$$

$$[|\boldsymbol{k}|\,a^{*(3)} + i\,\omega\,a^{*(4)}]\,|\Phi_{\boldsymbol{k}}\rangle = 0. \tag{8.32}$$

1. This requirement is certainly not necessitated by our previous postulates. If longitudinal U-particles are allowed, it can be shown that they have no interaction with the Dirac field in the limit $\mu\to 0$. See F. Coester, Phys. Rev. 83, 798 (1951) and F.J. Belinfante, Phys. Rev. 75, 1321 (1949). In later applications to coupled fields, these particles can play no part; therefore we shall ignore them from now on.

In the limit $\mu \to 0$, (8.31) and (8.32) obviously go over into (8.3) and (8.4). With a non-zero mass the solution of the equations is normalizable and can be written in the following form:

$$|\Phi_{\boldsymbol{k}}\rangle = \frac{\mu}{\omega} \sum_{n^{(3)}, n^{(4)}} \left(-i \frac{|\boldsymbol{k}|}{\omega}\right)^{n^{(3)}} \delta_{n^{(3)}, n^{(4)}} \,|\, n^{(3)}, n^{(4)}\rangle . \tag{8.33}$$

In this way we get an expression for the expectation value of the commutators (8.11) which has the "correct" value (8.12) in the limit. Such a calculation with unnormalizable state vectors is not especially attractive, but it is scarcely any worse than many other things which we will have to do later. The fact that in this way one obtains an infinite result for the expectation value of the anti-commutators (7.9) is a little more serious. If we restrict ourselves to the state $|\psi_0\rangle$ where no transverse photons are present, we have

$$\langle\psi_0|\{U_\mu(x), U_\nu(x')\}|\psi_0\rangle = \left(\delta_{\mu\nu} - \frac{1}{\mu^2}\frac{\partial^2}{\partial x_\mu \partial x_\nu}\right) \Delta^{(1)}(x'-x)\,, \tag{8.34}$$

$$\langle\psi_0|\{B(x), B(x')\}|\psi_0\rangle = \Delta^{(1)}(x'-x)\,, \tag{8.35}$$

$$\langle\psi_0|\{U_\mu(x), B(x')\}|\psi_0\rangle = 0\,, \tag{8.36}$$

and so

$$\langle\psi_0|\{A_\mu(x), A_\nu(x')\}|\psi_0\rangle = \left(\delta_{\mu\nu} - \frac{2}{\mu^2}\frac{\partial^2}{\partial x_\mu \partial x_\nu}\right) \Delta^{(1)}(x'-x)\,. \tag{8.37}$$

For $\mu \to 0$, the right side of (8.37) becomes infinite. We see, however, that this singularity occurs only in a "gradient". Consider, for example, a gauge invariant expression such as the integral

$$\iint dx\, dx'\, F_{\mu\nu}(x, x') \langle\psi_0|\{A_\mu(x), A_\nu(x')\}|\psi_0\rangle \tag{8.38}$$

with

$$\frac{\partial F_{\mu\nu}(x, x')}{\partial x_\mu} = \frac{\partial F_{\mu\nu}(x, x')}{\partial x'_\nu} = 0\,. \tag{8.39}$$

We can integrate by parts before taking the limit $\mu \to 0$, and the last term in (8.37) then contributes nothing.[1] With these requirements we can compute as if the limit of the expectation value of the anticommutator were given by

$$\langle\psi_0|\{A_\mu(x), A_\nu(x')\}|\psi_0\rangle = \delta_{\mu\nu} D^{(1)}(x'-x)\,. \tag{8.40}$$

We shall derive this result in Sec. 9 by another, more systematic, method.

We have seen that we may work with only the transverse photons for gauge invariant expressions and that the longitudinal and scalar degrees of freedom of the electromagnetic field may be completely ignored. Obviously the gauge invariance of the theory

1. F. J. Dyson, Phys. Rev. 77, 420 (1950).

has been lost by the special choice of the subsidiary condition, and clearly an infinite gauge function was chosen in (8.37). At best, this is an inelegant point in the theory, and it would be preferable to have a formulation where there is always the possibility of different gauges and where we could avoid the transition to unnormalizable state vectors and infinite gauge functions.

9. The Subsidiary Condition: Second Method

The procedure given above for treating the Lorentz condition (5.1) in the quantized theory is not the only possible one. Many other methods have been suggested by different authors.[1] In this section we shall concentrate on a method which has been developed by Gupta and Bleuler.[2]

The difficulties with unnormalizable state vectors arise because both creation and annihilation operators are present in (8.3) and (8.4). A condition like (8.2) can be satisfied without difficulty for an annihilation operator; however, if creation operators are also present, one is led to a system of coupled equations like (8.7) and (8.8). The solution of such a system necessarily entails state vectors with infinitely many particles, and these are not always normalizable, as was explicitly seen above. It is clear that the vanishing of the expectation value

$$\langle\psi| \frac{\partial A_\nu(x)}{\partial x_\nu} |\psi\rangle = 0 \tag{9.1}$$

is ensured by (8.2); however, that equation is not necessary for the vanishing of (9.1). We wish to modify the subsidiary condition (8.2) in such a way that only annihilation operators are involved. We cannot do this simply by throwing out (8.4) and retaining (8.3) unchanged. Certainly this would ensure the vanishing of (9.1), but since the roles of creation and annihilation operators are interchanged for $\lambda = 4$, Eq. (8.3) also contains creation operators. If we do not want unduly complicated transformation properties for the theory, we must choose the following representation for $a^{(4)}$ and $a^{*(4)}$:

$$\langle n+1 | a^{*(4)} | n\rangle = \langle n | a^{(4)} | n+1\rangle = \sqrt{n+1}\,, \tag{9.2}$$

instead of (6.14). By this choice, all $a^{*(\lambda)}$ are made Hermitian conjugates of $a^{(\lambda)}$, and hence all the operators $A_\mu(x)$ are made self-adjoint. Certainly this contradicts the reality requirements for the classical electromagnetic potentials, but the introduction[3] of

1. See, for example, W. Heisenberg and W. Pauli, Z. Physik, 56, 1 (1929) and 59, 168 (1930); F. J. Belinfante, Phys. Rev. 84, 644 (1951); J. G. Valatin, Dan. Mat. Fys. Medd. 26, No. 13 (1951).

2. S. Gupta, Proc. Phys. Soc. Lond. A63, 681 (1950) and 64, 850 (1951); K. Bleuler, Helv. Phys. Acta. 23, 567 (1950). See also W. Heitler, Quantum Theory of Radiation, Third Ed., Oxford, 1954, pp. 90-103.

3. Such an operator was introduced by P. A. M. Dirac in a quite different connection. See Proc. Roy. Soc. Lond. A180, 1 (1942).

a "metric operator" η enables us to get around this objection. With the use of the metric, the norm of a state vector is defined as

$$\text{Norm}\,|\psi\rangle = \langle\psi|\,\eta\,|\psi\rangle. \tag{9.3}$$

In order that the norm always be real, η must be Hermitian:

$$\eta^* = \eta. \tag{9.4}$$

So that we can ignore some trivial numerical factors, we also require

$$\eta^2 = \eta\,\eta^* = 1. \tag{9.5}$$

The expectation value of an operator F is now defined by

$$\bar{F} = \langle\psi|\eta F|\psi\rangle. \tag{9.6}$$

As a consequence of this, a Hermitian operator does not necessarily have a real expectation value, and this is just what we want.

With these new definitions, the norm of a state vector is not always positive. We can divide the state vectors into three classes. The first class contains those vectors of positive norm, and these can be normalized to 1. The usual probability interpretation of the quantum theory can be given for these vectors. The second class contains vectors of negative norm, which we can take as -1, and the third class contains the null vectors. For the last two classes of state vectors there is no probability interpretation, and we must therefore require that every physically realizable state belong to the first class. Fortunately, it will be seen later that our form of the subsidiary condition is compatible with this requirement.

In order that the expectation values for $A_k(x)$ be real and the corresponding quantities for $A_4(x)$ be pure imaginary, we require

$$\langle\psi|\eta\,A_k(x)|\psi\rangle = \langle\psi|A_k^*(x)\,\eta^*|\psi\rangle = \langle\psi|A_k(x)\,\eta|\psi\rangle, \tag{9.7}$$

or

$$[A_k(x), \eta] = 0\,, \tag{9.8}$$

and

$$\langle\psi|\eta\,A_4(x)|\psi\rangle = -\langle\psi|A_4(x)\,\eta|\psi\rangle\,, \tag{9.7a}$$

or

$$\{A_4(x), \eta\} = 0. \tag{9.8a}$$

Going over to momentum space, from (9.8) and (9.8a) we have

$$[a^{(\lambda)}(\boldsymbol{k}), \eta] = 0\,, \quad \lambda \neq 4\,, \tag{9.9}$$

$$\{a^{(4)}(\boldsymbol{k}), \eta\} = 0. \tag{9.10}$$

From (9.9) it follows immediately that η is diagonal in the quantum numbers $n^{(\lambda)}(\boldsymbol{k})$ for $\lambda \neq 4$, and has matrix elements 1. In the representation (9.2), Eq. (9.10) is satisfied by

$$\langle n^{(4)} | \eta | n'^{(4)} \rangle = \delta_{n^{(4)} n'^{(4)}} (-1)^{n^{(4)}}. \tag{9.11}$$

For η we therefore have the solution

$$\langle a | \eta | b \rangle = (-1)^{n_a^{(4)}} \delta_{ab}, \tag{9.12}$$

where the quantity $n_a^{(4)}$ is equal to the sum of all the scalar photons in the state $|a\rangle$:

$$n_a^{(4)} = \sum_{\boldsymbol{k}} n_a^{(4)}(\boldsymbol{k}). \tag{9.13}$$

In this metric, a few of the previous formulas are changed a little because we now have to put $N^{(4)} = a^{*(4)} a^{(4)}$. For example, instead of (6.7) and (6.19), we obtain

$$H = \sum_{\boldsymbol{k}} \omega \sum_{\lambda=1}^{4} N^{(\lambda)}(\boldsymbol{k}), \tag{9.14}$$

$$P_k = \sum_{\boldsymbol{k}} k_k \sum_{\lambda=1}^{4} N^{(\lambda)}(\boldsymbol{k}). \tag{9.15}$$

We now return to the subsidiary condition. In momentum space it has the form

$$[a^{(3)}(\boldsymbol{k}) + i\, a^{(4)}(\boldsymbol{k})] \,|\psi\rangle = 0. \tag{9.16}$$

In x-space (9.16) becomes

$$\frac{\partial A_\nu^{(+)}(x)}{\partial x_\nu} |\psi\rangle = 0. \tag{9.17}$$

In general the symbol $F^{(+)}(x)$ denotes that part of the operator $F(x)$ which has only positive frequencies, i.e., with x-dependence given by e^{ikx}. With the use of (9.8) and (9.8a) in (9.17), we have

$$\langle\psi| \left(\frac{\partial A_k^{(-)}(x)}{\partial x_k} - \frac{\partial A_4^{(-)}(x)}{\partial x_4} \right) \eta = \langle\psi| \eta \frac{\partial A_\nu^{(-)}(x)}{\partial x_\nu} = 0, \tag{9.18}$$

and therefore

$$\langle\psi| \eta \frac{\partial A_\nu(x)}{\partial x_\nu} |\psi\rangle = \langle\psi| \eta \frac{\partial A_\nu^{(-)}(x)}{\partial x_\nu} \times |\psi\rangle + \langle\psi| \eta \times \frac{\partial A_\nu^{(+)}(x)}{\partial x_\nu} |\psi\rangle = 0. \tag{9.19}$$

In our new formalism (9.17) replaces the condition (8.2). In this way (9.1) is satisfied, and the connection with classical theory ensured.[1]

---

1. Actually, it must also be shown that several factors $\frac{\partial A_\nu(x)}{\partial x_\nu}$ have the expectation value zero. It is simple to make the necessary revisions. For example, for two factors,

$$\langle\psi| \eta \frac{\partial A_\nu(x)}{\partial x_\nu} \frac{\partial A_\lambda(x')}{\partial x'_\lambda} |\psi\rangle = \langle\psi| \eta \frac{\partial A_\nu^{(+)}(x)}{\partial x_\nu} \frac{\partial A_\lambda^{(-)}(x')}{\partial x'_\lambda} |\psi\rangle =$$

$$= \langle\psi| \eta \left[\frac{\partial A_\nu^{(+)}(x)}{\partial x_\nu}, \frac{\partial A_\lambda^{(-)}(x')}{\partial x'_\lambda} \right] |\psi\rangle = i \langle\psi|\eta|\psi\rangle \,\Box\, D^{(-)}(x'-x) = 0, \text{ etc.}$$

Equation (8.2) had only the single solution (8.9), but the new subsidiary condition (9.16) has many independent solutions. As before, we write an arbitrary state vector as

$$|\psi\rangle = |\psi_T\rangle \prod_{\boldsymbol{k}} |\Phi_{\boldsymbol{k}}\rangle\,, \tag{9.20}$$

and consequently Eq. (9.16) is only a condition [as (8.2) was previously] on the quantities $|\Phi_{\boldsymbol{k}}\rangle$. As a fundamental system of solutions, we can choose

$$|\Phi^{(0)}\rangle = |0,0\rangle\,, \tag{9.21a}$$

$$|\Phi^{(1)}\rangle = |1,0\rangle + i\,|0,1\rangle\,, \tag{9.21b}$$

$$\vdots$$

$$|\Phi^{(n)}\rangle = \sum_{r=0}^{n} (i)^r \sqrt{\binom{n}{r}}\,|n-r, r\rangle\,. \tag{9.21c}$$

$$\vdots$$

Obviously all the vectors (9.21) are mutually orthogonal, i.e.,

$$\langle\Phi^{(n)}|\,\eta\,|\Phi^{(n')}\rangle = 0\,. \tag{9.22}$$

More important is the norm of the vectors $|\Phi^{(n)}\rangle$:

$$\langle\Phi^{(n)}|\,\eta\,|\,\Phi^{(n)}\rangle = \sum_{r=0}^{n} (-1)^r \binom{n}{r} = \delta_{n0}\,. \tag{9.23}$$

None of these norms is negative, and only the vector $|\Phi^{(0)}\rangle$ has a non-zero norm. Because we require that a physically realizable state satisfy (9.16) and have norm 1, the allowed vectors are of the form

$$|\Phi_{\boldsymbol{k}}\rangle = |\Phi_{\boldsymbol{k}}^{(0)}\rangle + \sum_{n\neq 0} c^{(n)}(\boldsymbol{k})\,|\Phi^{(n)}(\boldsymbol{k})\rangle\,, \tag{9.24}$$

with arbitrary coefficients $c^{(n)}(\boldsymbol{k})$.

If we now calculate the expectation value for the electromagnetic potentials in the states (9.24), we obtain (assuming for simplicity that there are no transverse photons present)

$$\left.\begin{aligned} \langle\psi|\,\eta\,A_\mu(x)|\,\psi\rangle = \frac{1}{\sqrt{V}}\sum_{\boldsymbol{k}}\frac{1}{\sqrt{2\omega}}\{e^{ikx}\,[e_\mu^{(3)}\langle\Phi_{\boldsymbol{k}}|\,\eta\,a^{(3)}(\boldsymbol{k})\,|\Phi_{\boldsymbol{k}}\rangle + \\ + e_\mu^{(4)}\langle\Phi_{\boldsymbol{k}}|\,\eta\,a^{(4)}(\boldsymbol{k})\,|\Phi_{\boldsymbol{k}}\rangle] + e^{-ikx}\,[e_\mu^{(3)}\langle\Phi_{\boldsymbol{k}}|\,\eta\,a^{*(3)}(\boldsymbol{k})\,|\Phi_{\boldsymbol{k}}\rangle + \\ + e_\mu^{(4)}\langle\Phi_{\boldsymbol{k}}|\,\eta\,a^{*(4)}(\boldsymbol{k})\,|\Phi_{\boldsymbol{k}}\rangle]\}. \end{aligned}\right\} \tag{9.25}$$

From the equations

$$a^{(3)}\,|\Phi^{(n)}\rangle = \sqrt{n}\,|\Phi^{(n-1)}\rangle\,, \tag{9.26}$$

$$a^{(4)}\,|\Phi^{(n)}\rangle = i\,\sqrt{n}\,|\Phi^{(n-1)}\rangle\,, \tag{9.27}$$

which follow readily from (9.21), we find

$$\left.\begin{aligned}\langle\Phi_k|\eta\, a^{(3)}(k)\,|\Phi_k\rangle &= \sum_{n\neq 0}\langle\Phi^{(0)}|\,\eta\,|\Phi^{(n-1)}\rangle\sqrt{n}\,c^{(n)}+\\ &+\sum_{\substack{n\neq 0\\ n'\neq 0}} c^{*(n')}\langle\Phi^{(n')}|\,\eta\,|\Phi^{(n-1)}\rangle\sqrt{n}\,c^{(n)} = c^{(1)}(k)\,,\end{aligned}\right\}\quad (9.28)$$

$$\langle\Phi_k|\,\eta\, a^{(4)}(k)\,|\Phi_k\rangle = i\,c^{(1)}(k)\,. \quad (9.29)$$

Using this result in (9.25) gives

$$\langle\psi|\,\eta\, A_\mu(x)\,|\psi\rangle = \frac{\partial\Lambda(x)}{\partial x_\mu}\,, \quad (9.30)$$

with

$$\Lambda(x) = \frac{i}{\sqrt{V}}\sum_k \frac{1}{\sqrt{2\omega^3}}\left[c^{*(1)}(k)\,e^{-ikx} - c^{(1)}(k)\,e^{ikx}\right]. \quad (9.31)$$

Thus we see that with the appropriate choice of the coefficients c in (9.24), we can obtain every gauge function $\Lambda(x)$ which satisfies the wave equation (5.7). This possibility did not exist in the previous method where a definite gauge function was chosen. On the other hand, it follows that the different state vectors (9.24) correspond only to different choices of gauge, and it is expected that physically significant results are independent of the "mixing in" of longitudinal and scalar photons in (9.24). Although only the coefficients $c^{(1)}(k)$ enter the result (9.31), if the expectation value of a product of potentials is calculated, it turns out that the other coefficients play a similar role.

The states (9.24) are not eigenstates of the Hamiltonian if the quantities c do not all vanish. However, if we calculate the expectation value of the energy of such a state, we get

$$\left.\begin{aligned}\langle\psi|\,\eta\, H\,|\psi\rangle = \sum_k\Big(\langle\Phi_k^{(0)}| + \sum_{n\neq 0} c^{*(n)}(k)\,\langle\Phi_k^{(n)}|\Big)\,\eta \sum_{n'\neq 0} n'\,\omega\, c^{(n')}(k)\,|\Phi_k^{(n')}\rangle +\\ + \sum_k \omega\,(n_k^{(1)} + n_k^{(2)}) = \sum_k \omega\,(n_k^{(1)} + n_k^{(2)})\,.\end{aligned}\right\}\quad (9.32)$$

This expectation value is equal to the energy of the transverse photons alone and is independent of the mixing in of longitudinal and scalar photons. If we regard the expectation value as the observable quantity, then all the states (9.24) are equivalent. In many applications it will be useful to arrange our states so that they are actual eigenstates of the Hamiltonian. For example, we must then define the vacuum as that state where no transverse photons are present and where no other particles are mixed in, i.e., where $|\Phi_k\rangle$ is given by (9.21a). The vacuum is therefore the state where no particles at all are present and which we have previously denoted by $|0\rangle$. By this, a special gauge has again been chosen, and it is that gauge where

$$\langle 0|\eta A_\mu(x)|0\rangle = \langle 0|A_\mu(x)|0\rangle = 0. \tag{9.33}$$

With this definition of the vacuum, we immediately obtain

$$\langle 0|\eta\{A_\mu(x), A_\nu(x')\}|0\rangle = \langle 0|\{A_\mu(x), A_\nu(x')\}|0\rangle = \delta_{\mu\nu}D^{(1)}(x'-x). \tag{9.34}$$

This is the same equation as (8.40), and this result differs from (7.13) because the representation for $a^{(4)}$ and $a^{*(4)}$ is changed.

We obtain corresponding results for the spatial displacement operators P_k and for the (spatial) angular momentum components J_{ik}. Likewise, the expectation values are independent of the particular mixture taken in (9.24); however, the state vectors are eigenstates of the operators only if all coefficients c are zero. For the angular momentum, we again find that the component in the direction $\boldsymbol{k}$ can have only the two possible values +1 and -1 for a physical (i.e., transverse) photon.

We must note that the states (9.21) do not form a complete set. This is obviously impossible because otherwise the condition (9.17) would be an operator equation. If we wish to do a sum over "intermediate states" as, for example, in a matrix product like

$$\langle a|A B|b\rangle = \sum_{|z\rangle}\langle a|A|z\rangle\langle z|B|b\rangle, \tag{9.35}$$

then we must include <u>all states (6.12)</u> in the sum over $|z\rangle$. This remark will turn out to be essential for a theory with interactions.

We should give a detailed discussion of the Lorentz invariance of the new method. On a formal level, this can be done in a quite satisfactory manner.[1] For us it is sufficient to state the two most important consequences of the formalism: the definition of the vacuum and the expectation value of the anticommutator are obviously covariant. By comparison, the gauge function (9.31) depends upon the $c(\boldsymbol{k})$ and can therefore depend upon the coordinate system if the c are not very carefully chosen. This is not a serious objection, since the final results are independent of the gauge function. We will be content for now with these simple remarks.

10. The Problem of the Measurement of the Electric and Magnetic Field Strengths

In the usual quantum mechanics of (point) particles, the commutator of p and q is non-vanishing:

$$[p, q] = -i\,. \tag{10.1}$$

This has the well known consequence that one cannot simultaneously measure p and q with arbitrary accuracy. This is the

1. See, for example, F. J. Belinfante, Phys. Rev. 96, 780 (1954).

content of the Heisenberg uncertainty principle, which says[1] that the uncertainties Δp and Δq must satisfy the inequality

$$\Delta p \cdot \Delta q \gtrsim 1 . \tag{10.2}$$

Since the mathematical formalism in quantum electrodynamics is just the same, we should expect that a similar uncertainty relation would follow from (7.5), (7.38). Because the electromagnetic potentials themselves do not have a direct physical significance (only the field strengths are observable quantities), we must differentiate these equations. This gives the uncertainty relation between the field strengths:

$$\left.\begin{aligned}[F_{\mu\nu}(x), F_{\lambda\varrho}(x')] = -i\Big[\delta_{\nu\varrho}\frac{\partial^2}{\partial x_\mu \partial x'_\lambda} - \delta_{\nu\lambda}\frac{\partial^2}{\partial x_\mu \partial x'_\varrho} - \delta_{\mu\varrho}\frac{\partial^2}{\partial x_\nu \partial x'_\lambda} + \\ + \delta_{\mu\lambda}\frac{\partial^2}{\partial x_\nu \partial x'_\varrho}\Big] D(x'-x).\end{aligned}\right\} \tag{10.3}$$

Introducing the usual E and H in (10.3) according to (5.2) gives

$$[\mathsf{E}_x(x), \mathsf{E}_x(x')] = [\mathsf{H}_x(x), \mathsf{H}_x(x')] = i\,[A_{xx}^{(xx')} - A_{xx}^{(x'x)}] , \tag{10.4}$$

$$[\mathsf{E}_x(x), \mathsf{E}_y(x')] = [\mathsf{H}_x(x), \mathsf{H}_y(x')] = i\,[A_{xy}^{(xx')} - A_{xy}^{(x'x)}] , \tag{10.5}$$

$$[\mathsf{E}_x(x), \mathsf{H}_x(x')] = 0, \tag{10.6}$$

$$[\mathsf{E}_x(x), \mathsf{H}_y(x')] = -[\mathsf{H}_x(x), \mathsf{E}_y(x')] = i\,[B_{xy}^{(xx')} - B_{xy}^{(x'x)}] , \tag{10.7}$$

and cyclic permutations of these equations for the other field components. Here we are using the notation

$$A_{xx}^{(xx')} = -\frac{1}{4\pi}\left(\frac{\partial^2}{\partial x_1 \partial x'_1} - \frac{\partial^2}{\partial x_0 \partial x'_0}\right)\left[\frac{1}{r}\,\delta(x'_0 - x_0 - r)\right], \tag{10.8}$$

$$A_{xy}^{(xx')} = -\frac{1}{4\pi}\frac{\partial^2}{\partial x_1 \partial x'_2}\left[\frac{1}{r}\delta(x'_0 - x_0 - r)\right], \tag{10.9}$$

$$B_{xy}^{(xx')} = -\frac{1}{4\pi}\frac{\partial^2}{\partial x_0 \partial x'_3}\left[\frac{1}{r}\,\delta(x'_0 - x_0 - r)\right], \tag{10.10}$$

$$r = |\boldsymbol{x}' - \boldsymbol{x}| . \tag{10.11}$$

The right sides of these equations contain quantities which are as singular as the Dirac delta function and its derivatives. Unlike (10.2), it is clear that these equations do not apply directly to experimentally attainable situations. Instead of field strengths at a single point, we shall consider their <u>average</u> <u>values</u> over a finite region of space and time. This enables us to integrate both sides of (10.4) and (10.7) and get meaningful expressions. We have developed a formalism, and it is essential to understand that only such mean values are involved in the interpretation of this for-

1. Actually, (10.1) implies the sharper inequality $\Delta p \cdot \Delta q \gtrsim 1/2$, if a suitable definition of the uncertainties is used. This factor of 2 is not significant for us.

malism. As an example, we can show from Eqs. (10.4) and (10.8) that

$$[\bar{E}_x(x), \bar{E}_x(x')] = i\,[\overline{A^{(xx')}_{xx}} - \overline{A^{(x'x)}_{xx}}] \,, \tag{10.12}$$

$$\bar{E}_x(x) = \frac{1}{V_1 T_1} \int\limits_{V_1} d^3x \int\limits_{T_1} dx_0 \, E_x(x) \,, \tag{10.13}$$

$$\overline{A^{(xx')}_{xx}} = \frac{1}{V_1 V_2 T_1 T_2} \int\limits_{V_1} d^3x \int\limits_{V_2} d^3x' \int\limits_{T_1} dx_0 \int\limits_{T_2} dx'_0 \, A^{(xx')}_{xx} . \tag{10.14}$$

From (10.12) we obtain the uncertainty relation

$$\Delta\, \overline{E_x(x)} \cdot \Delta\, \overline{E_x(x')} \gtrsim |\overline{A^{(xx')}_{xx}} - \overline{A^{(x'x)}_{xx}}| \,. \tag{10.15}$$

This result puts no restriction at all on the measurement of a single field strength at one point. The minus sign on the right side of (10.15) is crucial. From this, it follows that two components of the field strength can be measured to arbitrary accuracy if the two regions (V_1, T_1) and (V_2, T_2) coincide. This is the important difference from the usual quantum mechanics, where two quantities with the same time coordinates do not always commute. Similar results can be proved for the other components of the field strengths by using Eqs. (10.5) through (10.7). From these equations we also see that the mutual interference of two measurements can only occur when the two regions can be connected by light signals. It is therefore expected that the uncertainty (10.15) comes about as follows: An electromagnetic disturbance, produced by a measurement, propagates with the velocity of light to the other region of measurement. There it influences this other measurement in some uncontrollable way. The first term on the right side can be interpreted as the disturbance of region (2) by region (1) and vice versa for the last term. It is the difference of these two terms which appears, rather than their sum, as one might at first expect.

These considerations are quite formal, and we ought to invent gedanken experiments which explicitly exhibit this measurability. Although this is not a simple problem, it has been exhaustively discussed by Bohr and Rosenfeld.[1] Limitations of space prevent us from going into the details of the measuring apparatus, so we must refer the reader to the original article. We shall remark that the discussion of Bohr and Rosenfeld has completely confirmed the equations given here. In order to make the best measurement, it has been shown that the test particles, which are used to define the field strengths, should not be elementary particles. Rather,

---

1. N. Bohr and L. Rosenfeld, Dan. Mat. Fys. Medd. 12, No. 8 (1933), and Phys. Rev. 78, 794 (1950). A short summary is given by W. Heitler, op. cit., pp. 76-86. See also L. Rosenfeld, Physica, Haag 19, 859 (1953) and E. Corinaldesi, Nuovo Cim. Suppl. 10, 83 (1953).

they must be macroscopic bodies with large, uniformly distributed charges and masses.[1] One should think of them as a large number of elementary particles, held together. Moreover, the test bodies must be rigid; i.e., the whole body must be set in motion at the same time and must not bend. Because all forces have a finite velocity of propagation, this is no trivial problem! Bohr and Rosenfeld have shown that, in principle, one can construct such objects. The solution of the problem is not particularly simple and can scarcely lead to anything practical. The essential point is straightforward: The present formulation of the quantum theory cannot forbid the existence of such test bodies. On the other hand, this makes it clear that quantum electrodynamics cannot attack the problem of the elementary particles because macroscopic bodies are essential for the whole interpretation of the theory. For this reason, we do not expect that the value of the charge of the electron, $\frac{e^2}{4\pi} \approx \frac{1}{137}$, can be determined from the theory. When we later study the interaction of the electron and the electromagnetic field, we shall regard the charge as an arbitrary parameter and even allow a time variation of the charge. For the reasons stated above, this formal procedure does not contradict the basis of the theory and will turn out to be quite useful in going to the limit discussed in connection with Eq. (7.18).

11. The Electromagnetic Field in Interaction with Classical Currents

As preparation for our later discussion of the interaction between photons and electrons, we shall now consider the much simpler problem[2] of the interaction between photons and a given classical current density $j_\mu(x)$. The Lagrangian density for this problem is

$$\mathscr{L} = -\frac{1}{4} F_{\mu\nu} F_{\mu\nu} - \frac{1}{2} \frac{\partial A_\nu}{\partial x_\nu} \frac{\partial A_\lambda}{\partial x_\lambda} + A_\mu j_\mu . \tag{11.1}$$

For the potentials, we obtain the equations of motion

$$\Box A_\mu(x) = -j_\mu(x) \tag{11.2}$$

and the Hamiltonian operator is

1. In an earlier work of Landau and Peierls the measurements are discussed in terms of elementary particles (as test bodies), and a contradiction is found to the equations given here. See Z. Physik 69, 56 (1931).

2. A similar problem was first considered by F. Bloch and A. Nordsieck, Phys. Rev. 52, 54 (1937). Similar questions were later considered by many authors. See, for example, W. Pauli and M. Fierz, Nuovo Cim. 15, 167 (1938); W. Thirring and B. Touschek, Phil. Mag. 42, 244 (1951); R. J. Glauber, Phys. Rev. 84, 395 (1951); J. M. Jauch and F. Rohrlich, Helv. Phys. Acta 27, 613 (1954).

$$H = \frac{1}{2}\int d^3x \left[\frac{\partial A_\mu}{\partial x_0}\frac{\partial A_\mu}{\partial x_0} + \frac{\partial A_\mu}{\partial x_k}\frac{\partial A_\mu}{\partial x_k}\right] - \int d^3x\, j_\mu(x)\, A_\mu(x). \tag{11.3}$$

In general the components $j_\mu(x)$ will be time-dependent. Because the Hamiltonian is not invariant under time translations, there will be no conservation law for the total energy. We cannot expect stationary states of the Hamiltonian. If we require that the currents vanish for $x_0 \to -\infty$ at a rate fast enough so that the integral

$$\int_{-\infty}^{x_0} D(x-x')\, j_\mu(x')\, dx' \tag{11.4}$$

converges, then a solution of (11.2) is found to be

$$A_\mu(x) = A_\mu^{(0)}(x) + \int D_R(x-x')\, j_\mu(x')\, dx'. \tag{11.5}$$

The operators $A_\mu^{(0)}(x)$ satisfy the homogeneous wave equation

$$\Box A_\mu^{(0)}(x) = 0\,, \tag{11.5a}$$

and have the same commutation relations as the total potentials $A_\mu(x)$, because the last term in (11.5) is a c-number. They are therefore identical with the operators $A_\mu(x)$ for the free fields which were introduced earlier. Using the simplifications of the method of the indefinite metric, we have

$$[A_\mu^{(0)}(x), A_\nu^{(0)}(x')] = -i\,\delta_{\mu\nu} D(x'-x)\,, \tag{11.6}$$

$$\langle 0|\{A_\mu^{(0)}(x), A_\nu^{(0)}(x')\}|0\rangle = \delta_{\mu\nu} D^{(1)}(x'-x). \tag{11.7}$$

The Eqs. (11.5) through (11.7) give the solution of the equation of motion (11.2) and the canonical commutation relations. As can be seen from (11.5), the operators $A_\mu^{(0)}(x)$ correspond to the initial values of the complete operators $A_\mu(x)$ for $x_0 \to -\infty$. They are frequently called the "in-fields".

If we now compute the energy for this solution, we obtain, after integration by parts,

$$\left.\begin{aligned} H = {} & H^{(0)}(A_\mu^{(0)}) + E(x_0) + \\ & + \int d^3x \int dx' \left[\frac{\partial A_\mu^{(0)}(x)}{\partial x_0} D_R(x-x') - A_\mu^{(0)}(x)\frac{\partial D_R(x-x')}{\partial x_0}\right]\frac{\partial j_\mu(x')}{\partial x_0'}\,, \end{aligned}\right\} \tag{11.8}$$

$$H^{(0)}(A_\mu^{(0)}) = \frac{1}{2}\int d^3x \left[\frac{\partial A_\mu^{(0)}(x)}{\partial x_0}\frac{\partial A_\mu^{(0)}(x)}{\partial x_0} + \frac{\partial A_\mu^{(0)}(x)}{\partial x_k}\frac{\partial A_\mu^{(0)}(x)}{\partial x_k}\right], \tag{11.9}$$

$$\left.\begin{aligned} E(x_0) = {} & \frac{1}{2}\iint dx'\, dx'' \int d^3x\, D_R(x-x')\, D_R(x-x'')\left[\frac{\partial j_\mu(x')}{\partial x_0'}\frac{\partial j_\mu(x'')}{\partial x_0''} + \right. \\ & \left. + \frac{\partial j_\mu(x')}{\partial x_k'}\frac{\partial j_\mu(x'')}{\partial x_k''}\right] - \int d^3x \int dx'\, j_\mu(x)\, D_R(x-x')\, j_\mu(x'). \end{aligned}\right\} \tag{11.10}$$

The term (11.9) is obviously a constant of the motion, although the time derivative of the other terms of (11.8) does not vanish in general. It is easy to show that

$$\frac{dH}{dx_0} = -\int d^3x \, \frac{\partial j_\mu(x)}{\partial x_0} \left[A_\mu^{(0)}(x) + \int D_R(x - x') j_\mu(x') \, dx' \right]. \tag{11.11}$$

We now require that the current vanish for $x_0 \to +\infty$, as well as $x_0 \to -\infty$. From (11.11) it then follows that the expectation value of the energy tends to a finite limit for $x_0 \to \pm\infty$ for time-independent state vectors. As is seen from (11.8), the energy operator for $x_0 \to -\infty$ is simply

$$\lim_{x_0 \to -\infty} H = H^{(0)}(A_\mu^{(0)}). \tag{11.12}$$

If we choose a representation where the operator (11.12) is diagonal, we can characterize each state vector by the particle number operators, which are constructed from the operators $A_\mu^{(0)}(x)$ by the method of Sec. 9. The operators $A_\mu(x)$ coincide with the $A_\mu^{(0)}(x)$ for $x_0 \to -\infty$ under the requirements we have imposed here. In this limit the energy is given by (11.12). The particles constructed in this way are therefore to be interpreted as those which are present in the system as $x_0 \to -\infty$. They are called the "incoming particles" or the "in-states".

It is obviously quite possible to study other solutions of (11.2). For example, we can consider

$$A_\mu(x) = \tilde{A}_\mu^{(0)}(x) + \int D_A(x - x') j_\mu(x') \, dx'. \tag{11.13}$$

The operators $\tilde{A}_\mu^{(0)}(x)$ also satisfy the homogeneous wave equation

$$\Box \tilde{A}_\mu^{(0)}(x) = 0, \tag{11.14}$$

and have the commutation relations

$$[\tilde{A}_\mu^{(0)}(x), \tilde{A}_\nu^{(0)}(x')] = -i \, \delta_{\mu\nu} D(x' - x). \tag{11.15}$$

For the energy, there are expressions similar to Eqs. (11.8) through (11.10), except that all the retarded D-functions are replaced by advanced ones, and the fields $A_\mu^{(0)}(x)$ are replaced by $\tilde{A}_\mu^{(0)}(x)$. In this case, for $x_0 \to +\infty$, we have

$$\lim_{x_0 \to +\infty} A_\mu(x) = \tilde{A}_\mu^{(0)}(x), \tag{11.16}$$

$$\lim_{x_0 \to +\infty} H = H^{(0)}(\tilde{A}_\mu^{(0)}). \tag{11.17}$$

If we choose a representation in which the operator (11.17) is diagonal, then it is quite clear that the particles we have constructed are those which are present in the system as $x_0 \to +\infty$. Subsequently, we shall call them the "outgoing particles" or the "out-states" and $\tilde{A}_\mu^{(0)}(x)$ the "out-fields". In general, the states constructed by this procedure are different from the "in-states" which we introduced before. For example, if we take the state $|\tilde{0}\rangle$ where there are no out-particles (the vacuum of the outgoing particles), then obviously we have

$$\langle\tilde{0}|\{\tilde{A}_\mu^{(0)}(x), \tilde{A}_\nu^{(0)}(x')\}|\tilde{0}\rangle = \delta_{\mu\nu} D^{(1)}(x'-x). \tag{11.18}$$

Note that this state is different from the vacuum state of (11.7).

From (11.5) and (11.13) we obtain the relation between the in- and out-fields:

$$A_\mu^{(0)}(x) - \tilde{A}_\mu^{(0)}(x) = \int D(x-x') j_\mu(x')\, dx'. \tag{11.19}$$

We now go over to p-space and write

$$A_\mu^{(0)}(x) = \frac{1}{\sqrt{V}} \sum_{\boldsymbol{k},\lambda} \frac{e_\mu^{(\lambda)}(\boldsymbol{k})}{\sqrt{2\omega}} [e^{ikx} a^{(\lambda)}(\boldsymbol{k}) + e^{-ikx} a^{*(\lambda)}(\boldsymbol{k})], \tag{11.20}$$

$$\tilde{A}_\mu^{(0)}(x) = \frac{1}{\sqrt{V}} \sum_{\boldsymbol{k},\lambda} \frac{e_\mu^{(\lambda)}(\boldsymbol{k})}{\sqrt{2\omega}} [e^{ikx} \tilde{a}^{(\lambda)}(\boldsymbol{k}) + e^{-ikx} \tilde{a}^{*(\lambda)}(\boldsymbol{k})]. \tag{11.21}$$

From these equations and (11.19) we have

$$a^{(\lambda)}(\boldsymbol{k}) - \tilde{a}^{(\lambda)}(\boldsymbol{k}) = \frac{-i}{\sqrt{2\omega}} j^{(\lambda)}(\boldsymbol{k}, \omega)\, i^{\delta_{\lambda 4}}. \tag{11.22}$$

Here we have introduced the Fourier components of the current density by setting

$$j_\mu(x) = \frac{1}{2\pi} \int_{k_0>0} dk_0 \frac{1}{\sqrt{V}} \sum_{\boldsymbol{k},\lambda} e_\mu^{(\lambda)}(\boldsymbol{k}) [e^{ikx} j^{(\lambda)}(\boldsymbol{k}, k_0) + e^{-ikx} j^{*(\lambda)}(\boldsymbol{k}, k_0)]\, i^{\delta_{\lambda 4}}. \tag{11.23}$$

The complex conjugate of $j^{(\lambda)}(\boldsymbol{k}, k_0)$ is always $j^{*(\lambda)}(\boldsymbol{k}, k_0)$, and the factor $i^{\delta_{\lambda 4}}$ ensures the correct reality properties for the $j_\mu(x)$.

Similarly, for the particle number operators we find

$$\left.\begin{aligned} \tilde{N}^{(\lambda)}(\boldsymbol{k}) = N^{(\lambda)}(\boldsymbol{k}) &+ \frac{(i)^{1+\delta_{\lambda 4}}}{\sqrt{2\omega}} j^{(\lambda)}(\boldsymbol{k}, \omega)\, a^{*(\lambda)}(\boldsymbol{k}) + \\ &+ \frac{(-i)^{1+\delta_{\lambda 4}}}{\sqrt{2\omega}} j^{*(\lambda)}(\boldsymbol{k}, \omega)\, a^{(\lambda)}(\boldsymbol{k}) + \frac{1}{2\omega} |j^{(\lambda)}(\boldsymbol{k}, \omega)|^2, \end{aligned}\right\} \tag{11.24}$$

with

$$N^{(\lambda)}(\boldsymbol{k}) = a^{*(\lambda)}(\boldsymbol{k})\, a^{(\lambda)}(\boldsymbol{k}), \tag{11.25}$$

and

$$\tilde{N}^{(\lambda)}(\boldsymbol{k}) = \tilde{a}^{*(\lambda)}(\boldsymbol{k})\, \tilde{a}^{(\lambda)}(\boldsymbol{k}). \tag{11.26}$$

As an example, we shall consider the state which has no in-particles. This is the state which we denote as the in-vacuum $|0\rangle$. A question of physical interest is the calculation of the number of outgoing particles of given momentum vector $\boldsymbol{k}$ and given polarization λ. We shall confine ourselves to a single momentum

$\boldsymbol{k}$ and a single polarization λ, for simplicity. In our notation an in-state of n particles is $|n\rangle$, and one of n out-particles is $|\tilde{n}\rangle$. The probability that n such particles are emitted by our system is clearly

$$w_n^{(\lambda)}(\boldsymbol{k}) = |\langle\tilde{n}|0\rangle|^2. \tag{11.27}$$

From the equation

$$|\tilde{n}\rangle = \frac{1}{\sqrt{n!}}(\tilde{a}^*)^n|\tilde{0}\rangle, \tag{11.28}$$

we find

$$\left.\begin{aligned}\langle\tilde{n}|0\rangle &= \frac{1}{\sqrt{n!}}\langle\tilde{0}|[\tilde{a}]^n|0\rangle = \frac{1}{\sqrt{n!}}\sum_{r=0}^{n}\binom{n}{r}\langle\tilde{0}|\,a^r|0\rangle\left(\frac{i^{1+\delta_{\lambda 4}}\,j}{\sqrt{2\omega}}\right)^{n-r} = \\ &= \frac{1}{\sqrt{n!}}\left(\frac{i^{1+\delta_{\lambda 4}}\,j}{\sqrt{2\omega}}\right)^n\langle\tilde{0}|0\rangle.\end{aligned}\right\} \tag{11.29}$$

From (11.24) the expectation value of $\tilde{N}$ is

$$\bar{n} = \langle 0|\,\tilde{N}\,|0\rangle = \frac{1}{2\omega}|j|^2. \tag{11.30}$$

In this notation and using (11.27) and (11.29), we have

$$w_n^{(\lambda)}(\boldsymbol{k}) = \frac{1}{n!}[\bar{n}^{(\lambda)}(\boldsymbol{k})]^n|\langle\tilde{0}|0\rangle|^2. \tag{11.31}$$

The last factor in (11.31) does not depend on n and is determined from the normalization condition

$$\sum_{n=0}^{\infty} w_n^{(\lambda)}(\boldsymbol{k}) = 1 \tag{11.32}$$

to be

$$|\langle\tilde{0}|0\rangle|^2 = e^{-\bar{n}}. \tag{11.33}$$

The emitted photons therefore have a Poisson distribution:

$$w_n = \frac{1}{n!}(\bar{n})^n e^{-\bar{n}}. \tag{11.34}$$

This says that the emission of a single photon is statistically independent of the emission of any other. From (11.8) through (11.10), the total energy of the emitted photons is found to be

$$\left.\begin{aligned}\lim_{x_0\to\infty}\langle 0|H|0\rangle &= \frac{1}{2}\iint dx'\,dx''\int d^3x\,D(x-x')\,D(x-x'')\times \\ &\times\left[\frac{\partial j_\mu(x')}{\partial x_0'}\frac{\partial j_\mu(x'')}{\partial x_0''} + \frac{\partial j_\mu(x')}{\partial x_k'}\frac{\partial j_\mu(x'')}{\partial x_k''}\right] = \sum_{\boldsymbol{k}}\sum_{\lambda=1}^{2}\omega\,\bar{n}^{(\lambda)}(\boldsymbol{k}).\end{aligned}\right\} \tag{11.35}$$

This was expected because of (11.30). The continuity equation for the current $j_\mu(x)$,

$$\frac{\partial j_\mu(x)}{\partial x_\mu} = 0, \tag{11.36}$$

has been employed in obtaining (11.35). That is, from (11.36) and (11.23) we have

$$|j^{(3)}(\boldsymbol{k},\omega)|^2 = |j^{(4)}(\boldsymbol{k},\omega)|^2 \,, \tag{11.37}$$

and from (11.30),

$$\bar{n}^{(3)}(\boldsymbol{k}) = \bar{n}^{(4)}(\boldsymbol{k}) \,. \tag{11.38}$$

Therefore the term in (11.35) with scalar photons and a minus sign is just cancelled by the term involving the longitudinal photons.

In a similar way the distribution of emitted photons can be found for an arbitrary initial state. Because this result does not contain anything essentially different, we shall omit the explicit calculation.

An interesting special case of this general problem is provided by a current distribution which is time independent in some special coordinate system. It appears that the results would be uninteresting because it is quite obvious that no photons are emitted. The probabilities w_n vanish by (11.34), since the factor $\bar{n}$ must vanish for time-independent currents. This follows from (11.30) for all ω different from zero. In this case the interesting quantity is the Hamiltonian, which is given by (11.8) through (11.10):

$$H = H^{(0)}(A_\mu^{(0)}) + E \,, \tag{11.39}$$

$$\left.\begin{aligned} E = \frac{1}{2}\iint dx'\,dx''\int d^3x\,D_R(x-x')\,D_R(x-x'')\,\frac{\partial j_\mu(\boldsymbol{x}')}{\partial x'_k}\,\frac{\partial j_\mu(\boldsymbol{x}'')}{\partial x''_k} - \\ - \iint d^3x\,dx'\,D_R(x-x')\,j_\mu(\boldsymbol{x})\,j_\mu(\boldsymbol{x}'). \end{aligned}\right\} \tag{11.40}$$

By the use of

$$D_R(x-x') = \frac{1}{4\pi\,r_{xx'}}\,\delta(r_{xx'} - x_0 + x'_0) \,, \tag{11.41}$$

$$r_{xx'} = |\boldsymbol{x} - \boldsymbol{x}'| \,, \tag{11.42}$$

the term E in (11.39) can be written

$$\begin{aligned} E &= \frac{1}{2}\int\frac{d^3x}{(4\pi)^2}\iint\frac{d^3x'\,d^3x''}{r_{xx'}\,r_{xx''}}\,\frac{\partial j_\mu(\boldsymbol{x}')}{\partial x'_k}\,\frac{\partial j_\mu(\boldsymbol{x}'')}{\partial x''_k} - \frac{1}{4\pi}\iint\frac{d^3x\,d^3x'}{r_{xx'}}\,j_\mu(\boldsymbol{x})\,j_\mu(\boldsymbol{x}') \\ &= -\frac{1}{2}\int\frac{d^3x}{(4\pi)^2}\iint\frac{d^3x'\,d^3x''}{r_{xx'}}\,j_\mu(\boldsymbol{x}')\,j_\mu(\boldsymbol{x}'')\,\frac{\partial^2}{\partial x_k\,\partial x_k}\left(\frac{1}{r_{xx''}}\right) - \\ &\qquad\qquad - \frac{1}{4\pi}\iint\frac{d^3x'\,d^3x''}{r_{x'x''}}\,j_\mu(\boldsymbol{x}')\,j_\mu(\boldsymbol{x}'') \\ &= -\frac{1}{8\pi}\iint\frac{d^3x'\,d^3x''}{r_{x'x''}}\,j_\mu(\boldsymbol{x}')\,j_\mu(\boldsymbol{x}''). \end{aligned} \tag{11.43}$$

This quantity corresponds to the electromagnetic self-energy of the stationary current distribution. The total energy (11.39) is the sum of the term (11.43) and the energy of the free particles. This is intuitively the correct result, although the static current distribution obviously contradicts our assumption of a vanishing current for $|x_0| \to \infty$. Despite this, the integral (11.4) is convergent, since the infinite time integral over the function (11.41) certainly converges. In a theory with a non-vanishing particle mass, this would not hold because the integration would be over the inside of the retarded light cone with a non-vanishing density function $\Delta_R(x-x')$. With a view toward later applications, we shall give another proof for Eqs. (11.39) through (11.43). We consider an almost stationary current distribution

$$j_\mu(x) = e^{-\alpha|x_0|} \cdot J_\mu(\boldsymbol{x}). \tag{11.44}$$

This vanishes for $|x_0| \to \infty$ if α is non-zero, but for finite times it can be regarded as "almost constant" if α is taken sufficiently small. We will describe this by saying that the current is "switched off" for $|x_0| \to \infty$, and in the limit $\alpha \to 0$, the switching-off is done adiabatically. In this process E is not constant in time as in (11.43), but vanishes for $x_0 \to -\infty$, as is obvious from (11.10). Rather than (11.43), E now becomes

$$\left.\begin{aligned} E = \frac{1}{2} \iint \frac{d^3 x'\, d^3 x''}{(4\pi)^2} \int \frac{d^3 x}{r_{x x'}\, r_{x x''}} \left[\alpha^2 + \frac{\partial}{\partial x'_k} \frac{\partial}{\partial x''_k}\right] e^{-\alpha\{|x_0 - r_{x x'}| + |x_0 - r_{x x''}|\}} \times \\ \times J_\mu(\boldsymbol{x}')\, J_\mu(\boldsymbol{x}'') - \iint \frac{d^3 x\, d^3 x'}{4\pi\, r_{x x'}} e^{-\alpha\{|x_0| + |x_0 - r_{x x'}|\}} J_\mu(\boldsymbol{x})\, J_\mu(\boldsymbol{x}'). \end{aligned}\right\} \tag{11.45}$$

The first term in brackets is of the order of magnitude α (and not α^2!) and vanishes if the switching-on is adiabatic. The other terms go over to (11.43). In taking the limit, x_0 is to be held constant as α goes to zero. For the time-dependent current (11.44), the total energy (11.8) contains one more term of the following form:

$$\left.\begin{aligned} \iint d^3x\, d^3x' \int_{-\infty}^{+\infty} \frac{dx'_0}{4\pi\, r_{x x'}} \left[\frac{\partial A_\mu^{(0)}(x)}{\partial x_0} \delta(r_{x x'} - x_0 + x'_0) + \right. \\ \left. + A_\mu^{(0)}(x) \frac{\partial}{\partial x'_0} \delta(r_{x x'} - x_0 + x'_0)\right] \alpha\, e^{-\alpha|x'_0|} J_\mu(\boldsymbol{x}') = \\ = \frac{\alpha}{4\pi} \iint \frac{d^3 x\, d^3 x'}{r_{x x'}} \left[\frac{\partial A_\mu^{(0)}(x)}{\partial x_0} + \alpha\, A_\mu^{(0)}(x)\right] e^{-\alpha|x_0 - r_{x x'}|} J_\mu(\boldsymbol{x}'). \end{aligned}\right\} \tag{11.46}$$

The integrals appearing here converge for $\alpha=0$; therefore the whole term vanishes in the adiabatic limit. By this explicit calculation, we have verified a special case of the adiabatic theorem of quantum mechanics. We can summarize the most important result of this example as follows: If a state is an eigenstate of the Hamiltonian and if a parameter in the Hamiltonian is adiabatically changed (in this case the current), then the same state is also an eigenstate, after the Hamiltonian is changed, but with a different eigenvalue. If the currents vanish, then the Hamiltonian is just $H^{(0)}(A_\mu^{(0)})$ with well known eigenstates which can be characterized by the particle numbers. From Eq. (11.46) it follows that the non-diagonal terms in the Hamiltonian vanish in this representation, and therefore that these same states are also eigenstates of the complete energy operator $H(A_\mu)$. If the eigenvalue of the operator $H^{(0)}(A_\mu^{(0)})$ is $\Sigma n\omega$ for a state, then, by (11.39), the eigenvalue of the operator $H(A_\mu)$ is $\sum n\omega + E$ for the same state. Here E is given by (11.43).

In conclusion, we should note that there are other ways to introduce free particles into this problem. For example, we can introduce a free field $A_\mu^{(0)}(x, T)$ which coincides with the complete field $A_\mu(x)$ at some arbitrary time T. The field $A_\mu^{(0)}(x, T)$ is defined by

$$A_\mu^{(0)}(x, T) = -\int\limits_{x_0' = T} d^3x' \left[D(x - x')\frac{\partial A_\mu(x')}{\partial x_0'} + \frac{\partial D(x - x')}{\partial x_0} A_\mu(x')\right]. \quad (11.47)$$

This field satisfies the homogeneous wave equation

$$\Box A_\mu^{(0)}(x, T) = 0. \quad (11.48)$$

In this way particles have been constructed which would appear if the current $j_\mu(x)$ were suddenly (not adiabatically) switched off at $x_0 = T$. In general, they are not of much interest. If we were to use a Schroedinger picture which coincided with the Heisenberg picture for $x_0 = T$ then, in a sense, these particles would be the "natural" ones. For simplicity, we take $T = 0$ and write the Schroedinger operators $A_\mu(\boldsymbol{x}, 0)$ and $\frac{\partial A_\mu(\boldsymbol{x}, 0)}{\partial x_0}$ in their Fourier representations

$$A_\mu(\boldsymbol{x}, 0) = \frac{1}{\sqrt{V}} \sum_{\boldsymbol{k}, \lambda} \frac{e_\mu^{(\lambda)}(\boldsymbol{k})}{\sqrt{2\omega}} \left[b^{(\lambda)}(\boldsymbol{k})\, e^{i\boldsymbol{k}\boldsymbol{x}} + b^{*(\lambda)}(\boldsymbol{k})\, e^{-i\boldsymbol{k}\boldsymbol{x}}\right], \quad (11.49)$$

$$\left.\frac{\partial A_\mu(x)}{\partial x_0}\right|_{x_0 = 0} = \frac{-i}{\sqrt{V}} \sum_{\boldsymbol{k}, \lambda} e_\mu^{(\lambda)}(\boldsymbol{k}) \sqrt{\frac{\omega}{2}} \left[b^{(\lambda)}(\boldsymbol{k})\, e^{i\boldsymbol{k}\boldsymbol{x}} - b^{*(\lambda)}(\boldsymbol{k})\, e^{-i\boldsymbol{k}\boldsymbol{x}}\right]. \quad (11.50)$$

For the operators $b^{(\lambda)}(\boldsymbol{k})$ in (11.49) and (11.50), we can use the matrices (6.13) and (9.2). These satisfy the commutation relations in the Schroedinger picture. The Hamiltonian is not diagonal. By means of a simple calculation, we find that the state of lowest energy (the vacuum) has the expansion

$$|0\rangle = c\left[|0^{(b)}\rangle + \sum_{n, \boldsymbol{k}, \lambda} \frac{1}{\sqrt{n!}} \left[\frac{j^{(\lambda)}(\boldsymbol{k})}{\sqrt{2\omega^3}}\right]^n i^{n \cdot \delta_{\lambda 4}} |n^{(b)(\lambda)}(\boldsymbol{k})\rangle\right], \quad (11.51)$$

where

$$|n^{(b)}\rangle = \frac{1}{\sqrt{n!}} [b]^n |0^{(b)}\rangle. \quad (11.52)$$

From the condition $\langle 0|0\rangle = 1$, the normalization constant c is found to be

$$|c| = e^{-\frac{1}{2}\sum\limits_{\boldsymbol{k},\lambda}\frac{|j^{(\lambda)}(\boldsymbol{k})|^2}{2\omega^3}} \tag{11.53}$$

The physical vacuum is therefore a mixture of "particles at time zero" given by (11.51). Particularly in the older literature, this result is often described by saying that the physical vacuum contains a mixture of "free" particles. With the conventions we employ, we must be more cautious with this term because we will often make use of both the incoming and outgoing free fields. According to the results found above, the in-particles are identical with the physical ones, yet because they are described by a free field, they can well be called "free" particles.

In particular, if the current is a point source, then $j^{(\lambda)}(\boldsymbol{k})$ is independent of $\boldsymbol{k}$, and therefore the sum in (11.53) diverges. In this case the physical states cannot be expanded in terms of the states of free particles at time zero.[1] Nevertheless, the two kinds of states exist simultaneously but cannot be expanded in terms of one another. In a strictly mathematical sense, they do not belong to the same Hilbert space. This shows only that no physical meaning is to be attached to the free particles at time zero, at least not in our model with point sources.

The discussion given here of the adiabatic theorem and the various possibilities for introducing free particles has been specific to a model with an electromagnetic field interacting with given classical currents. The adiabatic theorem is very general and holds under quite weak conditions in the ordinary quantum mechanics of point particles.[2] A general proof for a system with infinitely many degrees of freedom has not been given yet. For our model we have verified the hypothesis by an explicit calculation. Later we shall often make use of the adiabatic hypothesis.

Although the various free fields have been constructed only for a special example here, the discussion can be taken over, almost unchanged, to the general case of two interacting fields.

---

1. This point has been particularly emphasized by van Hove, Physica, Haag 18, 145 (1952). In this connection, he has introduced the word "orthogonality". (If the constant c is equal to zero, then every term in (11.51) vanishes and, according to van Hove, the vector $|0\rangle$ is therefore "orthogonal" to every vector $|n^{(b)}\rangle$!) See also R. Haag, Dan. Mat. Fys. Medd. 29, No. 12 (1955) as well as A. S. Wightman and S. S. Schweber, Phys. Rev. 98, 812 (1955).

2. M. Born and V. Fock, Z. Physik 51, 165 (1928).

CHAPTER III

THE FREE DIRAC FIELD*

12. Equations of Motion, Lagrange Function, and an Attempt at a Canonical Quantization

The Dirac equation for a free electron is

$$\left(\gamma \frac{\partial}{\partial x} + m\right) \psi(x) = 0. \tag{12.1}$$

Here ψ designates a quantity with four components $\psi_1(x) \ldots \psi_4(x)$, and the terms γ are matrices $(\gamma_\mu)_{\alpha\beta}$, which obey the anticommutation relations

$$\{\gamma_\mu, \gamma_\nu\} = 2\,\delta_{\mu\nu}. \tag{12.2}$$

If we write out explicitly all the matrix multiplications in (12.1) and (12.2), we obtain

$$\sum_{\beta=1}^{4} \left\{ \sum_{\mu=1}^{4} (\gamma_\mu)_{\alpha\beta} \frac{\partial}{\partial x_\mu} + m\,\delta_{\alpha\beta} \right\} \psi_\beta(x) = 0, \tag{12.3}$$

and

$$\sum_{\beta=1}^{4} [(\gamma_\mu)_{\alpha\beta} (\gamma_\nu)_{\beta\delta} + (\gamma_\nu)_{\alpha\beta} (\gamma_\mu)_{\beta\delta}] = 2\delta_{\mu\nu}\,\delta_{\alpha\delta}. \tag{12.4}$$

Usually the short notation of (12.1) and (12.2) is most convenient and leads to no confusion. An explicit representation for the matrices γ is not necessary for most calculations; however, one can be constructed in the following manner:

$$\gamma_k = \begin{pmatrix} 0 & -i\sigma_k \\ i\sigma_k & 0 \end{pmatrix}, \tag{12.5}$$

$$\gamma_4 = \begin{pmatrix} I & 0 \\ 0 & -I \end{pmatrix}. \tag{12.6}$$

Here 0 is the two-row zero matrix, I is the two-row unit matrix, and the quantities σ_k are the Pauli spin matrices:

$$\sigma_x = \begin{pmatrix} 0 & 1 \\ 1 & 0 \end{pmatrix}, \quad \sigma_y = \begin{pmatrix} 0 & -i \\ i & 0 \end{pmatrix}, \quad \sigma_z = \begin{pmatrix} 1 & 0 \\ 0 & -1 \end{pmatrix}. \tag{12.7}$$

---

* See also Chap. B of the article by W. Pauli in the Handbuch der Physik, edited by S. Flügge, Springer-Verlag, Heidelberg, Vol. V, part I.

The matrices γ are therefore four by four, and one readily verifies that they satisfy Eq. (12.2).

With the use of this representation for the γ, we can attempt to find the plane wave solutions of (12.1). It is straightforward to show that for every wave

$$\psi_\alpha(x) = u_\alpha(\boldsymbol{q}) \, e^{i(\boldsymbol{q}\,\boldsymbol{x} - q_0\,x_0)} \tag{12.8}$$

of given spatial momentum $\boldsymbol{q}$, there are two possible values of q_0:

$$q_0 = \pm E \equiv \pm\sqrt{\boldsymbol{q}^2 + m^2}\,, \tag{12.9}$$

and that for each q_0 there are two independent solutions $u_\alpha^{(r)}(\boldsymbol{q})$. We shall enumerate these solutions in the following table:

$$
\begin{array}{c|cccc}
{}_\alpha\!\diagdown^{r} & 1 & 2 & 3 & 4 \\
\hline
1 & 1 & 0 & -\dfrac{q_z}{m+E} & \dfrac{-q_x+iq_y}{m+E} \\
2 & 0 & 1 & -\dfrac{q_x+iq_y}{m+E} & \dfrac{q_z}{m+E} \\
3 & \dfrac{q_z}{m+E} & \dfrac{q_x-iq_y}{m+E} & 1 & 0 \\
4 & \dfrac{q_x+iq_y}{m+E} & -\dfrac{q_z}{m+E} & 0 & 1
\end{array}
\times \sqrt{\dfrac{m+E}{2E}}\,.
$$

The two solutions with $r=1$ and $r=2$ go with the value $q_0 = E$, and the two others go with the value $q_0 = -E$. These solutions are normalized so that

$$\sum_{\alpha=1}^{4} u_\alpha^{*\,(r)}(\boldsymbol{q})\, u_\alpha^{(s)}(\boldsymbol{q}) = \delta_{rs}\,. \tag{12.10}$$

Moreover, one can show from this table that

$$\sum_{r=1}^{4} u_\alpha^{*\,(r)}(\boldsymbol{q})\, u_\beta^{(r)}(\boldsymbol{q}) = \delta_{\alpha\beta}\,, \tag{12.11}$$

$$\sum_{r=1}^{2} \bar{u}_\alpha^{(r)}(\boldsymbol{q})\, u_\beta^{(r)}(\boldsymbol{q}) = -\frac{1}{2E}\,(i\gamma\,q^{(+)} - m)_{\beta\alpha}\,, \tag{12.12}$$

$$\sum_{r=3}^{4} \bar{u}_\alpha^{(r)}(\boldsymbol{q})\, u_\beta^{(r)}(\boldsymbol{q}) = \frac{1}{2E}\,(i\gamma\,q^{(-)} - m)_{\beta\alpha}\,. \tag{12.13}$$

Here we have used the notation

$$\bar{u}(\boldsymbol{q}) = u^*(\boldsymbol{q}) \cdot \gamma_4\,, \tag{12.14}$$

and

$$q^{(+)} = (\boldsymbol{q},\, iE)\,, \quad q^{(-)} = (\boldsymbol{q},\, -iE)\,. \tag{12.15}$$

Equation (12.10) is the orthogonality relation for the solutions of the wave equation, and (12.11) shows the completeness of the set of solutions. Physically the two different solutions for each value of q_0 correspond to the two possible orientations of the electron spin. Obviously the total spin is then one-half.

The transformation properties of the Dirac theory have been discussed in various textbooks.[1] We shall not enter into a discussion of them here except to note that

$$\overline{\psi}(x)\,\psi(x); \quad \overline{\psi}(x) = \psi^*(x)\,\gamma_4 ,$$

is an invariant and that

$$i\,\overline{\psi}(x)\,\gamma_\mu\,\psi(x)$$

has the transformation properties of a four-vector. In particular, $\psi^*(x)\,\psi(x)$ is not an invariant but is the time component of a vector.

Formally the equation of motion can be obtained from the Lagrangian

$$\mathscr{L} = -\overline{\psi}(x)\left(\gamma\frac{\partial}{\partial x} + m\right)\psi(x) \tag{12.16}$$

by allowing variations of $\psi(x)$ and $\overline{\psi}(x)$ as if they were independent fields. The canonical momenta can be obtained directly from (12.16):

$$\pi_\psi(x) = i\,\overline{\psi}(x)\,\gamma_4 = i\,\psi^*(x), \tag{12.17}$$

$$\pi_{\overline{\psi}}(x) = 0. \tag{12.18}$$

The momentum conjugate to $\overline{\psi}(x)$ vanishes identically, and the time derivatives of these functions do not enter (12.17) and (12.18). Thus it is impossible to express the time derivatives as functions of the momenta. Despite this, we can construct a Hamiltonian which is a function of $\psi(x)$, its spatial derivatives, and the momentum $\pi_\psi(x)$. This Hamiltonian is

$$\left.\begin{aligned} \mathscr{H} = i\,\overline{\psi}(x)\,\gamma_4\frac{\partial\psi(x)}{\partial x_0} + \overline{\psi}(x)\left(\gamma\frac{\partial}{\partial x} + m\right)\psi(x) = \\ = -i\,\pi_\psi(x)\,\gamma_4\left(\gamma_k\frac{\partial}{\partial x_k} + m\right)\psi(x). \end{aligned}\right\} \tag{12.19}$$

In this form the function $\overline{\psi}(x)$ has been eliminated, and the Hamiltonian contains only the independent field $\psi(x)$.

The current density for the Dirac electron is well known:

$$j_\mu(x) = i\,e\,\overline{\psi}(x)\,\gamma_\mu\,\psi(x). \tag{12.20}$$

These quantities transform like a vector under Lorentz transformations and, as a consequence of (12.1), satisfy the continuity equation

$$\left.\begin{aligned} \frac{\partial j_\mu(x)}{\partial x_\mu} = i\,e\left[\overline{\psi}(x)\,\gamma_\mu\frac{\partial\psi(x)}{\partial x_\mu} + \frac{\partial\overline{\psi}(x)}{\partial x_\mu}\gamma_\mu\psi(x)\right] = \\ = i\,e\,[-m\,\overline{\psi}(x)\,\psi(x) + m\,\overline{\psi}(x)\,\psi(x)] = 0. \end{aligned}\right\} \tag{12.21}$$

1. See P. A. M. Dirac, The Principles of Quantum Mechanics, Third Ed., Oxford, 1947, p. 257.

Previously we regarded the electromagnetic potentials as operators which satisfied the canonical commutation relations. Now we shall attempt to interpret the field $\psi(x)$ as an operator. This procedure is often called the "second quantization" of the electron field. In this language, "the ordinary Dirac equation" should be the first quantization, but we shall refer instead to the "classical Dirac theory". In this theory the field $\psi(x)$ is a classical quantity which is also the state vector of the theory. After the second quantization, $\psi(x)$ is not a state vector but is the dynamical variable of the theory. In particular, $\psi^*(x)\,\psi(x)$ can no longer be interpreted as a probability density.

Previously we expanded the electromagnetic field in plane waves. Now we do the same for the field $\psi(x)$:

$$\psi_a(x)=\frac{1}{\sqrt{V}}\sum_{\boldsymbol{q}}\left\{e^{i(\boldsymbol{q}\boldsymbol{x}-E x_0)}\sum_{r=1}^{2}u_\alpha^{(r)}(\boldsymbol{q})\,a^{(r)}(\boldsymbol{q})+e^{i(\boldsymbol{q}\boldsymbol{x}+E x_0)}\sum_{r=3}^{4}u_\alpha^{(r)}(\boldsymbol{q})\,a^{(r)}(\boldsymbol{q})\right\}. \quad (12.22)$$

The quantities $a^{(r)}(\boldsymbol{q})$ are therefore operators which we now regard as the dynamical variables. The Hamiltonian can be expanded in terms of them:

$$H=\int d^3x\,\mathscr{H}(x)=\sum_{\boldsymbol{q}}E\left[\sum_{r=1}^{2}a^{*(r)}(\boldsymbol{q})\,a^{(r)}(\boldsymbol{q})-\sum_{r=3}^{4}a^{*(r)}(\boldsymbol{q})\,a^{(r)}(\boldsymbol{q})\right], \quad (12.23)$$

as can the charge:

$$Q=-i\int j_4(x)\,d^3x=e\sum_{\boldsymbol{q}}\sum_{r=1}^{4}a^{*(r)}(\boldsymbol{q})\,a^{(r)}(\boldsymbol{q}). \quad (12.24)$$

Here we could attempt to use the representation (6.13) for the matrices $a^{(r)}$. This is equivalent to the requirement

$$[\pi_\psi(x),\psi(x')]_{x_0=x_0'}=-i\,\delta(\boldsymbol{x}-\boldsymbol{x}'). \quad (12.25)$$

This does bring the energy into diagonal form; however, this approach is scarcely satisfactory. According to (12.23) the eigenvalues of the energy can take negative as well as positive values. Unlike the electromagnetic field, there is no subsidiary condition here which excludes the states of negative energy. Furthermore, the number of electrons in the same state is not limited, whereas it is known experimentally that electrons obey the exclusion principle. Hence the orthodox method of canonical quantization cannot be employed, and we have to develop another method in Sec. 13 for quantizing the Dirac field.

13. Quantization of the Dirac Field by Anticommutators

In our modified quantization method, we shall require that the commutator of H and an arbitrary field operator be equal to $-i$ times the time derivative of the operator. This clearly holds if we require this property only for the operators $\psi(x)$, since any other operator in the theory can be expressed in terms of $\psi(x)$. From (12.22) and (12.23) it follows that a sufficient prescription for quantization is

$$[a^{*(r)}(\mathbf{q})\, a^{(r)}(\mathbf{q}),\, a^{(s)}(\mathbf{q}')] = -\, a^{(r)}(\mathbf{q})\, \delta_{rs}\, \delta_{\mathbf{q}\mathbf{q}'}\,, \tag{13.1}$$

$$[a^{*(r)}(\mathbf{q})\, a^{(r)}(\mathbf{q}),\, a^{*(s)}(\mathbf{q}')] = a^{*(r)}(\mathbf{q})\, \delta_{rs}\, \delta_{\mathbf{q}\mathbf{q}'}. \tag{13.2}$$

The Eqs. (13.1) and (13.2) have many solutions, of which the canonical quantization is only one. They are obviously also satisfied if we require[1]

$$\{a^{*(r)}(\mathbf{q}),\, a^{(s)}(\mathbf{q}')\} = \delta_{rs}\, \delta_{\mathbf{q}\mathbf{q}'}\,, \tag{13.3}$$

$$\{a^{*(r)}(\mathbf{q}),\, a^{*(s)}(\mathbf{q}')\} = \{a^{(r)}(\mathbf{q}),\, a^{(s)}(\mathbf{q}')\} = 0. \tag{13.4}$$

These relations are completely symmetrical in a and a^*, and we have even more freedom here in the definition of particle number operators than we had for the electromagnetic field. Since a and a^* now anticommute, we can rewrite (12.23) as

$$H = \sum_{\mathbf{q}} E \left\{ \sum_{r=1}^{2} a^{*(r)}(\mathbf{q})\, a^{(r)}(\mathbf{q}) + \sum_{r=3}^{4} a^{(r)}(\mathbf{q})\, a^{*(r)}(\mathbf{q}) - 2 \right\}. \tag{13.5}$$

We now define two new operators $b^{(r)}$ by

$$b^{(1)}(\mathbf{q}) = a^{*(4)}(-\,\mathbf{q})\,, \tag{13.6}$$

$$b^{(2)}(\mathbf{q}) = a^{*(3)}(-\,\mathbf{q})\,. \tag{13.7}$$

Thus $b^{(r)}$ and $b^{*(r)}$ also satisfy the anticommutation relations (13.3) and (13.4). The total energy can be expressed in terms of "particle numbers" N^+ and N^-:

$$H = \sum_{\mathbf{q}} E \sum_{r=1}^{2} \left(N^{+(r)}(\mathbf{q}) + N^{-(r)}(\mathbf{q})\right), \tag{13.8}$$

$$N^{+(r)}(\mathbf{q}) = a^{*(r)}(\mathbf{q})\, a^{(r)}(\mathbf{q})\,, \tag{13.9}$$

$$N^{-(r)}(\mathbf{q}) = b^{*(r)}(\mathbf{q})\, b^{(r)}(\mathbf{q}). \tag{13.10}$$

The last term in (13.5) has been omitted from (13.8) since it is only a c-number and obviously corresponds to the zero-point energy of the field. The quantities N^+ and N^- in (13.9) and (13.10) satisfy

$$N^+(1 - N^+) = N^-(1 - N^-) = 0\,, \tag{13.11}$$

as can readily be shown from (13.4). They therefore have only the eigenvalues 0 and 1 and no others. We shall interpret them as the number of particles of given momentum and given spin. By the choice of the anticommutators in (13.3) and (13.4) we have ensured that there can be only one electron in a given state. This is the exclusion principle which we find here as a consequence of the quantization with anticommutators. It is one of the most important results of the theory of quantized fields that quantization with un-

1. P. Jordan and E. Wigner, Z. Physik 47, 631 (1928).

bounded eigenvalues for the particle numbers leads to unphysical consequences for particles with spin one-half. In particular, the energy is not positive definite. In a similar way it can be shown that quantization of particles of integral spin cannot be carried out with the incorporation of the exclusion principle.[1]

If we drop a "zero-point term" in (12.24), we obtain for the charge,

$$Q = e \sum_{q, r} [N^{+(r)}(q) - N^{-(r)}(q)]. \tag{13.12}$$

This quantity is not positive definite. Rather than being troublesome, this is a most satisfactory feature of the theory. We have now found that the solutions with "negative energy" in (12.8) and (12.9) appear in the quantized theory with positive energy but with reversed charge. This corresponds to the "hole theory" of Dirac; however, it is obtained here as a consequence of our quantization, without any additional assumptions. If we wish to have the particles N^+ represent electrons, then we must take a negative value for e in (13.12). The particles N^- then correspond to positrons.

Dropping the zero-point charge in (13.12) can be formulated quite simply in x-space. Instead of defining the current by (12.20), we take[2]

$$j_\mu(x) = \frac{ie}{2} [\overline{\psi}(x)\gamma_\mu, \psi(x)]. \tag{13.13}$$

Here the notation is

$$[\overline{\psi}(x)\gamma_\mu, \psi(x)] = [\overline{\psi}(x), \gamma_\mu \psi(x)] = \sum_{\alpha,\beta} (\gamma_\mu)_{\alpha\beta} (\overline{\psi}_\alpha(x)\psi_\beta(x) - \psi_\beta(x)\overline{\psi}_\alpha(x)). \tag{13.14}$$

If we introduce the series (12.22) into (13.13), it is a straightforward calculation to obtain the expression (13.12) for the charge Q. In a similar way, the vacuum expectation values of all the components of the current vanish:

$$\langle 0 | j_\mu(x) | 0 \rangle = 0. \tag{13.15}$$

It is not possible to give an analogous simple prescription in x-space for subtracting the zero-point energy.

For the spatial components of the momentum P_k, by using (3.12) we obtain

$$\left.\begin{aligned} P_k = -i\int d^3x\,\overline{\psi}(x)\gamma_4 \frac{\partial \psi(x)}{\partial x_k} &= \sum_q q_k \left\{ \sum_{r=1}^{2} (N^{+(r)}(q) - N^{-(r)}(-q)) + 2 \right\} = \\ &= \sum_{q,r} q_k (N^{+(r)}(q) + N^{-(r)}(q)). \end{aligned}\right\} \tag{13.16}$$

The particles N^-, which were taken to represent positrons, have

1. W. Pauli, Progr. Theor. Phys. 5, 526 (1950). References to the older literature are given here.

2. W. Heisenberg, Z. Physik, 90, 209 (1934).

a momentum $\boldsymbol{q}$ according to the definitions (13.6), (13.7), and (13.10). This is in complete agreement with the general result (3.21), since the last term in (12.22) is a creation operator for positrons according to (13.6) and (13.7). Because of this, the x-dependence of this term must be e^{-ipx} if p is the energy-momentum of the one-positron state.

As a final topic, we compute the angular momentum for these particles. The symmetrical energy-momentum tensor can be constructed by the method given in Sec. 4. The result is

$$\left.\begin{aligned} T_{\mu\nu}(x) = \overline{\psi}(x)\,\gamma_\mu \frac{\partial\psi(x)}{\partial x_\nu} + \frac{1}{8}\frac{\partial}{\partial x_\lambda}\{\overline{\psi}(x)\,(\gamma_\lambda[\gamma_\mu,\gamma_\nu] + \gamma_\mu[\gamma_\nu,\gamma_\lambda] + \\ + \gamma_\nu[\gamma_\mu,\gamma_\lambda])\,\psi(x)\}. \end{aligned}\right\} \tag{13.17}$$

For the spatial components of the angular momentum, after partial integrations, we get

$$J_{ij} = J_{ij}^{(0)} + J_{ij}^{(1)}\,, \tag{13.18}$$

$$J_{ij}^{(0)} = \frac{i}{2}\int d^3x\left(\left[\overline{\psi}(x),\gamma_4\frac{\partial\psi(x)}{\partial x_i}\right]x_j - \left[\overline{\psi}(x),\gamma_4\frac{\partial\psi(x)}{\partial x_j}\right]x_i\right), \tag{13.19}$$

$$\left.\begin{aligned} J_{ij}^{(1)} = \frac{i}{8}\int d^3x\,([\overline{\psi}(x),\gamma_4\gamma_j\gamma_i\psi(x)] - [\overline{\psi}(x),\gamma_4\gamma_i\gamma_j\psi(x)]) = \\ = \frac{1}{4}\int d^3x\,[\psi^*(x),\sigma_{ij}\psi(x)]\,, \end{aligned}\right\} \tag{13.20}$$

with

$$\sigma_{\mu\nu} = -\frac{i}{2}(\gamma_\mu\gamma_\nu - \gamma_\nu\gamma_\mu). \tag{13.20a}$$

In (13.19) and (13.20) the product of $\overline{\psi}(x)$ and $\psi(x)$ has been replaced by the commutator $\frac{1}{2}[\overline{\psi}(x),\psi(x)]$ so that the vacuum expectation value of the angular momentum will vanish. In momentum space the term $J_{ij}^{(0)}$ contains a factor $u_\alpha^{*(r)}(\boldsymbol{q})\,u_\alpha^{(s)}(\boldsymbol{q}) = \delta_{rs}$ and is consequently independent of the state of polarization of the particles. We shall therefore interpret this as the orbital angular momentum. We now consider the component of the other term $J^{(1)}$ in the direction of propagation of a particle. For a single electron state, if we take the direction of propagation as the z-axis, from (13.20) and (12.22) we readily obtain

$$J_{12}^{(1)}\,|q\rangle = \tfrac{1}{2}(-1)^{r+1}\,|q\rangle\,. \tag{13.21}$$

In a similar fashion, for a single positron state we have

$$J_{12}^{(1)}\,|q'\rangle = \tfrac{1}{2}(-1)^{r+1}|q'\rangle. \tag{13.22}$$

The complete similarity of (13.21) and (13.22) has been obtained by the choice of index in (13.6) and (13.7). In working out (13.21) and (13.22) we have used the matrix representation for σ_{ij} which is given in (12.5) and (12.6):

$$\sigma_{ij} = \begin{pmatrix} \sigma_k & 0 \\ 0 & \sigma_k \end{pmatrix}, \text{ and cyclic permutations.} \quad (13.23)$$

14. The Charge Symmetry of the Theory

There is a certain arbitrariness in the interpretation of the theory, as has already been noted in connection with Eq. (13.12). Either we take the particles N^+ as electrons, N^- as positrons, and give a negative value to the quantity e of (13.12) and (13.13), or we interchange the roles of electrons and positrons and then e must be taken positive. More precisely, the theory is invariant under the transformations

$$N^+ \rightleftarrows N^-, \quad (14.1)$$

$$e \leftarrow -e. \quad (14.2)$$

This invariance property can also be expressed in x-space. We recall that the operator $\psi(x)$ contains annihilation operators for the electrons and creation operators for the positrons, while the reverse is true for the operator $\overline{\psi}(x)$ or $\psi^*(x)$. It is clear that the transformation $N^+ \rightleftarrows N^-$ must correspond essentially to an interchange of $\psi(x)$ and $\overline{\psi}(x)$. In order to obtain the complete symmetry, it is not sufficient just to interchange $\psi(x)$ and $\overline{\psi}(x)$, but rather we must investigate the slightly more complicated transformations[1]

$$\psi_\alpha(x) \to \psi'_\alpha(x) = C_{\alpha\beta}\overline{\psi}_\beta(x). \quad (14.3)$$

If it is possible to find a matrix C which has the properties

$$C_{\alpha\beta} = -C_{\beta\alpha}, \quad (14.4)$$

$$(C^{-1})_{\alpha\beta} = (C_{\beta\alpha})^*, \quad (14.5)$$

$$(C^{-1}\gamma_\mu C)_{\alpha\beta} = -(\gamma_\mu)_{\beta\alpha}, \quad (14.6)$$

then the operator $\psi'(x)$ will satisfy the same equation of motion as $\psi(x)$:

$$\left.\begin{aligned} \left(\gamma\frac{\partial}{\partial x} + m\right)\psi'(x) &= \left(\gamma C\frac{\partial}{\partial x} + mC\right)\overline{\psi}(x) = C\left[C^{-1}\gamma C\frac{\partial}{\partial x} + m\right]\overline{\psi}(x) \\ &= C\left[-\frac{\partial\overline{\psi}(x)}{\partial x}\gamma + m\overline{\psi}(x)\right] = 0. \end{aligned}\right\} \quad (14.7)$$

In these transformations, C is a matrix of the same type as the γ-matrices and does not affect the "particle number indices" of the operators $\psi(x)$. If we introduce an expansion for $\psi'(x)$ analogous to (12.22) and interpret the states with positive q_0 as electrons, we see that the Hilbert space of the electrons has simply

1. W. Pauli, Ann. Inst. H. Poincaré 6, 109 (1936). The matrix C of Pauli is a little different from our C. See J. Schwinger, Phys. Rev. 74, 1439 (1948).

been interchanged with that of the positrons.

Moreover, we can show that the entire theory is actually invariant under the simultaneous transformations (14.2) and (14.3). Thus the new current becomes

$$\left.\begin{aligned} j'_\mu(x) &= -\frac{ie}{2}[\overline{\psi}'(x), \gamma_\mu \psi'(x)] = -\frac{ie}{2}[C^{-1}\psi(x), \gamma_\mu C\overline{\psi}(x)] = \\ &= \frac{ie}{2}[\psi(x), C^{-1}\gamma_\mu C\overline{\psi}(x)] = \frac{ie}{2}[\overline{\psi}(x), \gamma_\mu\psi(x)] = j_\mu(x), \end{aligned}\right\} \quad (14.8)$$

Here use has been made of the following result, which is obtained from (14.3) through (14.6):

$$\left.\begin{aligned} \overline{\psi}'_\alpha(x) &= \psi'^*_\beta(x)(\gamma_4)_{\beta\alpha} = -(C^{-1}\gamma_4\psi(x))_\beta(\gamma_4)_{\beta\alpha} = \\ &= -(C^{-1}\gamma_4 C)_{\beta\delta}(C^{-1}\psi(x))_\delta(\gamma_4)_{\beta\alpha} = (C^{-1})_{\alpha\beta}\psi_\beta(x). \end{aligned}\right\} \quad (14.9)$$

The Lagrangian (12.16) is not invariant under this "charge conjugation", even if it is "symmetrized" and written as

$$\mathscr{L}_1 = -\frac{1}{2}\left[\overline{\psi}(x), \left(\gamma\frac{\partial}{\partial x} + m\right)\psi(x)\right]. \quad (14.10)$$

We find that

$$\mathscr{L}_2 = -\frac{1}{2}\left[\overline{\psi}'(x), \left(\gamma\frac{\partial}{\partial x} + m\right)\psi'(x)\right] = -\frac{1}{2}\left[-\frac{\partial\overline{\psi}(x)}{\partial x}\gamma + m\overline{\psi}(x), \psi(x)\right] \neq \mathscr{L}_1. \quad (14.11)$$

Because both of these functions give the correct equations of motion, we can take the completely symmetric expression

$$\left.\begin{aligned} \mathscr{L} = \frac{1}{2}(\mathscr{L}_1 + \mathscr{L}_2) &= -\frac{1}{4}\left[\overline{\psi}(x), \left(\gamma\frac{\partial}{\partial x} + m\right)\psi(x)\right] - \\ &- \frac{1}{4}\left[-\frac{\partial\overline{\psi}(x)}{\partial x}\gamma + m\overline{\psi}(x), \psi(x)\right] \end{aligned}\right\} \quad (14.12)$$

as the Lagrangian. In (14.10) through (14.12), we have given up our original assumption that the fields in the Lagrangian are to be regarded as classical functions for which the commutators are identically zero. The quantities which are varied, $\psi(x)$ and $\overline{\psi}(x)$, are to be regarded as operators although the variations themselves may be regarded as independent c-numbers. Then the varied field operators will not obey the correct anticommutation relations; however, this does not cause trouble since the correct equations of motion are obtained. The energy-momentum density $T_{\mu\nu}$ obtained from the Lagrangian (14.12) is a priori charge symmetric.

We must now show that there actually exists a matrix C which satisfies Eqs. (14.4) through (14.6). This is most simply done by explicitly giving the matrix. Using the representation of (12.5) and (12.6), it is clear that

$$(\gamma_\mu)_{\alpha\beta} = (\gamma_\mu)_{\beta\alpha} \quad \text{for } \mu = 4 \text{ or } 2, \quad (14.13)$$

$$(\gamma_\mu)_{\alpha\beta} = -(\gamma_\mu)_{\beta\alpha} \quad \text{for } \mu = 1 \text{ or } 3. \quad (14.14)$$

If we choose

$$C = \gamma_2 \gamma_4 \,, \tag{14.15}$$

then the matrix C is unitary and antisymmetric, i.e., (14.4) and (14.5) are satisfied. Furthermore,

$$C^{-1}\gamma_\mu C = \gamma_4\gamma_2\gamma_\mu\gamma_2\gamma_4 = -\gamma_\mu \quad \text{for } \mu = 2,4, \tag{14.16}$$

$$C^{-1}\gamma_\mu C = \gamma_4\gamma_2\gamma_\mu\gamma_2\gamma_4 = \gamma_\mu \quad \text{for } \mu = 1,3. \tag{14.17}$$

As a final topic, we shall give a new proof of (13.15) in which only the charge symmetry of the theory is used and not the explicit form of the expansion (12.22). Under the interchange of the two Hilbert spaces, the vacuum is clearly invariant. Therefore as a consequence of charge symmetry, we have

$$\langle 0 | [\overline{\psi}'(x), \gamma_\mu \psi'(x)] | 0 \rangle = \langle 0 | [\overline{\psi}(x), \gamma_\mu \psi(x)] | 0 \rangle . \tag{14.18}$$

But from (14.8) we have the operator identity that

$$[\overline{\psi}'(x), \gamma_\mu \psi'(x)] = -[\overline{\psi}(x), \gamma_\mu \psi(x)]. \tag{14.19}$$

By combining (14.18) and (14.19), we obtain the vacuum expectation value of the vector $j_\mu(x)$:

$$\langle 0 | j_\mu(x) | 0 \rangle = 0. \tag{14.20}$$

In a similar way, we can show that the vacuum expectation value of any odd number of current operators must vanish:

$$\langle 0 | j_{\nu_1}(x_1) j_{\nu_2}(x_2) \ldots j_{\nu_{2n+1}}(x_{2n+1}) | 0 \rangle = 0. \tag{14.21}$$

These proofs are not completely valid since they involve manipulation with unbounded quantities. Physically, at least, the result (14.20) is certainly correct and we shall see later that Eqs. (14.20) and (14.21) can be obtained by a suitable limiting process.

15. Anticommutators and Commutators in x-Space. The S-Functions

In analogy with the treatment of the electromagnetic field, we can work out the commutator and anticommutator of $\overline{\psi}(x)$ and $\psi(x')$. From (13.3) and (13.4), we can expect a simple expression for the anticommutator, which is now a c-number. By (12.22) and (13.3), (13.4), we have

$$\left.\begin{aligned} &\{\overline{\psi}_\alpha(x), \psi_\beta(x')\} = \\ &= \frac{1}{V}\sum_{\boldsymbol{q},\boldsymbol{q}'}\sum_{r,s}[\overline{u}_\alpha^{(r)}(\boldsymbol{q})\, u_\beta^{(s)}(\boldsymbol{q}')\, e^{-iq^{(+)}x + iq'^{(+)}x'}\{a^{*(r)}(\boldsymbol{q}), a^{(s)}(\boldsymbol{q}')\} + \cdots] = \\ &= \frac{1}{V}\sum_{\boldsymbol{q}}\left\{\sum_{r=1}^{2}\overline{u}_\alpha^{(r)}(\boldsymbol{q})\, u_\beta^{(r)}(\boldsymbol{q})\, e^{iq^{(+)}(x'-x)} + \sum_{r=3}^{4}\overline{u}_\alpha^{(r)}(\boldsymbol{q})\, u_\beta^{(r)}(\boldsymbol{q})\, e^{iq^{(-)}(x'-x)}\right\}. \end{aligned}\right\} \tag{15.1}$$

With the use of (12.12) and (12.13) we can perform the summation over r in (15.1):

$$\left.\begin{aligned}
&\{\bar{\psi}_\alpha(x), \psi_\beta(x')\} = \\
&= \frac{-1}{V}\sum_{q}\frac{1}{2E}\{(i\gamma q^{(+)} - m)_{\beta\alpha} e^{iq^{(+)}(x'-x)} - (i\gamma q^{(-)} - m)_{\beta\alpha} e^{iq^{(-)}(x'-x)}\} \rightarrow \\
&\rightarrow \frac{-1}{(2\pi)^3}\int\frac{d^3q}{2E}[(i\gamma q^{(+)} - m)_{\beta\alpha} e^{iq^{(+)}(x'-x)} - (i\gamma q^{(-)} - m)_{\beta\alpha} e^{iq^{(-)}(x'-x)}] \\
&= \frac{-1}{(2\pi)^3}\int dq\, e^{iq(x'-x)}(i\gamma q - m)_{\beta\alpha}\,\delta(q^2+m^2)\,\varepsilon(q).
\end{aligned}\right\} \quad (15.2)$$

The right side of (15.2) will appear quite frequently in what follows, and it will be convenient to have a special notation for it. Accordingly, we shall write

$$\{\bar{\psi}_\alpha(x), \psi_\beta(x')\} = -iS_{\beta\alpha}(x'-x), \quad (15.3)$$

where

$$S_{\alpha\beta}(x) = \frac{-i}{(2\pi)^3}\int dq\, e^{iqx}(i\gamma q - m)_{\alpha\beta}\,\delta(q^2+m^2)\,\varepsilon(q). \quad (15.4)$$

This function $S(x)$ is obviously quite closely related to the functions $\Delta(x)$ which we studied earlier. From the remark following Eq. (7.37) and from (15.4), (7.6), we have

$$S(x) = \left(\gamma\frac{\partial}{\partial x} - m\right)\Delta(x). \quad (15.5)$$

In particular, for $x_0 = 0$, $S(x)$ becomes

$$S(x)|_{x_0=0} = i\gamma_4\,\delta(\boldsymbol{x}). \quad (15.6)$$

Equation (15.6) can be shown directly from the integral representation (15.4). Alternatively, it follows from (15.5) by using (7.35).

Just as the function $\Delta(x)$ can be used to solve the wave equation

$$(\Box - m^2)\,u(x) = 0 \quad (15.7)$$

with given initial conditions, the S-function can be used to solve the Dirac equation

$$\left(\gamma\frac{\partial}{\partial x} + m\right)\psi(x) = 0 \quad (15.8)$$

with the initial condition

$$\psi(x) = u(\boldsymbol{x}) \quad \text{for} \quad x_0 = T. \quad (15.9)$$

From (15.4) or (15.5) it follows that the S-function satisfies the equation

$$\left(\gamma\frac{\partial}{\partial x} + m\right)S(x) = 0, \quad (15.10)$$

and thus that

$$\psi(x) = -i \int_{x_0'=T} d^3x'\, S(x-x')\gamma_4 u(\mathbf{x}') \tag{15.11}$$

is always a solution of the Dirac equation (15.8). Furthermore, according to (15.6) this solution satisfies the initial conditions (15.9) for $x_0 = T$, and hence is the solution of the specified problem.

In a similar way, we can also solve the inhomogeneous equation

$$\left(\gamma \frac{\partial}{\partial x} + m\right)\psi(x) = f(x), \tag{15.12}$$

with the same initial condition as above by the use of

$$\psi(x) = \int_{x_0'=T}^{x_0'=x_0} S(x-x') f(x')\, dx' - i \int_{x_0=T} d^3x'\, S(x-x')\gamma_4 u(\mathbf{x}'). \tag{15.13}$$

As in Sec. 7 we can go over to the limit $T \to -\infty$. With the assumption that the last term tends to the limit $\psi^{(0)}(x)$, we can rewrite (15.13) as

$$\psi(x) = \psi^{(0)}(x) - \int S_R(x-x') f(x')\, dx'. \tag{15.14}$$

Here the retarded S-function $S_R(x)$ is defined by

$$S_R(x) = \begin{cases} -S(x) & \text{for} \quad x_0 > 0, \\ 0 & \text{for} \quad x_0 < 0. \end{cases} \tag{15.15}$$

Just as for the functions $D(x)$ and $\Delta(x)$, we can also introduce an advanced S-function $S_A(x)$ and a function $\overline{S}(x)$ by

$$S_A(x) = \begin{cases} 0 & \text{for } x_0 > 0, \\ S(x) & \text{for } x_0 < 0, \end{cases} \tag{15.16}$$

$$\overline{S}(x) = -\tfrac{1}{2}\varepsilon(x)\, S(x). \tag{15.17}$$

These functions are used in a manner similar to the corresponding Δ-functions. They satisfy the differential equations

$$\left(\gamma\frac{\partial}{\partial x} + m\right)\overline{S}(x) = \left(\gamma\frac{\partial}{\partial x} + m\right) S_R(x) = \left(\gamma\frac{\partial}{\partial x} + m\right) S_A(x) = -\delta(x). \tag{15.18}$$

and have the integral representations[1]

$$\overline{S}(x) = \frac{1}{(2\pi)^4} P \int dp\, e^{ipx} \frac{i\gamma p - m}{p^2 + m^2}, \tag{15.19}$$

$$S_R(x) = \frac{1}{(2\pi)^4}\int dp\, e^{ipx} (i\gamma p - m)\left[P\frac{1}{p^2+m^2} + i\pi\,\varepsilon(p)\,\delta(p^2+m^2)\right], \tag{15.20}$$

1. As in Eq. (7.24), the symbol P refers to the principal value.

$$S_A(x)=\frac{1}{(2\pi)^4}\int dp\, e^{ipx}(i\gamma p-m)\left[P\frac{1}{p^2+m^2}-i\pi\,\varepsilon(p)\,\delta(p^2+m^2)\right]. \quad (15.21)$$

The proof of these equations is not trivial, but requires some care because of the differentiation of the function $\varepsilon(x)$. As an example, we give the calculation for the integral representation (15.19):

$$\overline{S}(x)=-\frac{1}{2}\varepsilon(x)\left(\gamma\frac{\partial}{\partial x}-m\right)\Delta(x)=\left(\gamma\frac{\partial}{\partial x}-m\right)\overline{\Delta}(x)+\frac{i}{2}\gamma_4\Delta(x)\frac{\partial\varepsilon(x)}{\partial x_0}. \quad (15.22)$$

The time derivative of the function $\varepsilon(x)$ is just $2\cdot\delta(x_0)$, and so the last term in (15.22) contains the function $\Delta(x)$ on the surface $x_0=0$. From (7.35) we therefore find

$$\overline{S}(x)=\left(\gamma\frac{\partial}{\partial x}-m\right)\overline{\Delta}(x). \quad (15.23)$$

From this, (15.19) follows by a simple calculation, using the integral representation of $\overline{\Delta}(x)$. The other equations (15.20) and (15.21) can either be obtained in a similar way or can be proved from the relation

$$S_{R,A}(x)=\overline{S}(x)\mp\tfrac{1}{2}S(x). \quad (15.24)$$

The differential equations (15.18) follow most easily from the integral representations.

We now return to the anticommutator (15.3). From (15.6) we see that if the two times are equal, the anticommutator takes the value

$$\{\overline{\psi}_\alpha(x),\psi_\beta(x')\}_{x_0=x_0'}=(\gamma_4)_{\beta\alpha}\,\delta(\boldsymbol{x}-\boldsymbol{x}'). \quad (15.25)$$

This equation can be written in the following form:

$$\{\pi_\psi(x),\psi(x')\}_{x_0=x_0'}=i\,\delta(\boldsymbol{x}-\boldsymbol{x}'). \quad (15.26)$$

The above equation has a certain formal similarity to the canonical commutation relations, the sole difference being the sign.

In a similar way, from (12.22), (13.3), and (13.4), we have

$$\{\psi(x),\psi(x')\}=\{\overline{\psi}(x),\overline{\psi}(x')\}=0. \quad (15.27)$$

In (15.27) it is <u>not</u> assumed that the two points x and x' have space-like separation.

The commutator

$$[\overline{\psi}_\alpha(x),\psi_\beta(x')] \quad (15.28)$$

is not a c-number but we can take the vacuum expectation value of (15.28), just as we did previously for the anticommutator of the electromagnetic potentials. After a straightforward calculation we obtain

$$\langle 0|[\overline{\psi}_\alpha(x),\psi_\beta(x')]|0\rangle=S^{(1)}_{\beta\alpha}(x'-x) \quad (15.29)$$

$$S^{(1)}(x) = \left(\gamma \frac{\partial}{\partial x} - m\right) \Delta^{(1)}(x) = \frac{1}{(2\pi)^3} \int dp\, e^{ipx} (i\gamma p - m)\, \delta(p^2 + m^2). \quad (15.30)$$

The function $S^{(1)}(x)$ does not vanish for space-like x, since the function $\Delta^{(1)}(x)$ does not vanish there. If we attempted a discussion of the measurement of the Dirac field in a manner similar to that which we previously gave for the electromagnetic field, we would find that disturbances propagate with velocity greater than that of light. This would be in contradiction to the basic postulates of relativity and therefore is not admissible. There are no experiments, however--even gedanken experiments--by means of which we can study the field $\psi(x)$ directly. Consequently the field $\psi(x)$ cannot be used to transmit a signal between two observers. By itself, Eq. (15.29) is not in contradiction to relativity. We do have to check that the commutator of two components of the current, which are observables, vanishes for space-like separations. For these quantities we have

$$\left.\begin{aligned} [j_\mu(x), j_\nu(x')] &= -\frac{e^2}{4} [[\overline{\psi}(x), \gamma_\mu \psi(x)], [\overline{\psi}(x'), \gamma_\nu \psi(x')]] = \\ &= -e^2 [\overline{\psi}(x)\, \gamma_\mu \psi(x), \overline{\psi}(x')\, \gamma_\nu \psi(x')] = \\ &= e^2 (\overline{\psi}(x')\, \gamma_\nu \psi(x')\, \overline{\psi}(x)\, \gamma_\mu \psi(x) - \overline{\psi}(x)\, \gamma_\mu \psi(x)\, \overline{\psi}(x')\, \gamma_\nu \psi(x')) . \end{aligned}\right\} \quad (15.31)$$

The first transformation in (15.31) is verified by noting that the difference between $\frac{1}{2}[\overline{\psi}_\alpha(x), \psi_\beta(x')]$ and $\overline{\psi}_\alpha(x)\, \psi_\beta(x')$ is a c-number which vanishes in the commutator. By means of the equation

$$\psi_\alpha(x)\, \overline{\psi}_\beta(x') = -i\, S_{\alpha\beta}(x - x') - \overline{\psi}_\beta(x')\, \psi_\alpha(x) , \quad (15.32)$$

we can rewrite (15.31) in the following way:

$$\left.\begin{aligned} [j_\mu(x), j_\nu(x')] &= i\, e^2 (\overline{\psi}(x)\, \gamma_\mu\, S(x - x')\, \gamma_\nu \psi(x') - \overline{\psi}(x')\, \gamma_\nu\, S(x' - x)\, \gamma_\mu \psi(x)) + \\ &+ e^2 (\overline{\psi}_\alpha(x)\, \overline{\psi}_\beta(x') (\gamma_\mu \psi(x))_\alpha (\gamma_\nu \psi(x'))_\beta - \overline{\psi}_\beta(x')\, \overline{\psi}_\alpha(x) (\gamma_\nu \psi(x'))_\beta (\gamma_\mu \psi(x))_\alpha) . \end{aligned}\right\} \quad (15.33)$$

Because of (15.27), the last two terms in (15.33) cancel, and upon using

$$\overline{\psi}_\alpha(x)\, \psi_\beta(x') = \frac{1}{2} [\overline{\psi}_\alpha(x), \psi_\beta(x')] - \frac{i}{2} S_{\beta\alpha}(x' - x) , \quad (15.34)$$

we obtain the result

$$\left.\begin{aligned} [j_\mu(x), j_\nu(x')] &= \frac{i\, e^2}{2} \{[\overline{\psi}(x), \gamma_\mu S(x - x')\, \gamma_\nu \psi(x')] - [\overline{\psi}(x'), \gamma_\nu S(x' - x)\, \gamma_\mu \psi(x)]\} \\ &+ \frac{e^2}{2} \{\mathrm{Sp}\, [\gamma_\mu\, S(x - x')\, \gamma_\nu\, S(x' - x)] - \mathrm{Sp}\, [\gamma_\nu\, S(x' - x)\, \gamma_\mu\, S(x - x')]\} \\ &= \frac{i\, e^2}{2} \{[\overline{\psi}(x), \gamma_\mu\, S(x - x')\, \gamma_\nu \psi(x')] - [\overline{\psi}(x'), \gamma_\nu\, S(x' - x)\, \gamma_\mu \psi(x)]\}. \end{aligned}\right\} \quad (15.35)$$

Since a factor $S(x' - x)$ appears in each term in (15.35), the commutator vanishes for all non-zero, space-like separations.

16. The Dirac Equation with a Time-Independent, External Electromagnetic Field

The classical Dirac equation for an electron in an external electromagnetic field is

$$\left[\gamma\left(\frac{\partial}{\partial x} - i e A\right) + m\right]\psi(x) = 0. \tag{16.1}$$

If the external field is time-independent, we can expect solutions of the form

$$\psi_\alpha(x) = u_\alpha(\boldsymbol{x})\, e^{-i p_0 x_0}. \tag{16.2}$$

The function $u(\boldsymbol{x})$ satisfies the eigenvalue equation

$$\mathscr{H} u(\boldsymbol{x}) = p_0 u(\boldsymbol{x}), \tag{16.3}$$

with the Hermitian operator $\mathscr{H}$,

$$\mathscr{H} = \alpha_k\left(-i\frac{\partial}{\partial x_k} - e A_k(\boldsymbol{x})\right) + m\gamma_4 + e A_0(\boldsymbol{x}), \tag{16.4}$$

and the Hermitian matrices α_k,

$$\alpha_k = i\gamma_4\gamma_k = \begin{pmatrix} 0 & \sigma_k \\ \sigma_k & 0 \end{pmatrix}. \tag{16.5}$$

Therefore we can assume that the solutions $u_\alpha^{(n)}(\boldsymbol{x})$ of (16.3) are orthonormal and complete:

$$\int d^3x\, u_\alpha^{*(n)}(\boldsymbol{x})\, u_\alpha^{(n')}(\boldsymbol{x}) = \delta_{nn'}, \tag{16.6}$$

$$\sum_n u_\alpha^{*(n)}(\boldsymbol{x})\, u_\beta^{(n)}(\boldsymbol{x}') = \delta_{\alpha\beta}\,\delta(\boldsymbol{x} - \boldsymbol{x}'). \tag{16.7}$$

The eigenvalue p_0 in (16.3) can be either positive or negative. From the charge symmetry of the theory, we see immediately that if $p_0 = E$, $E > 0$, is an eigenvalue, and with the corresponding eigenfunction denoted by $u_\alpha(\boldsymbol{x}, E)$, then the function $u'_\alpha(\boldsymbol{x}, E) = C_{\alpha\beta}\bar{u}_\beta(\boldsymbol{x}, -E)$ is an eigenfunction[1] of $\mathscr{H}$ with the eigenvalue $p_0 = -E$. Here the matrix C is the one used in Sec. 14.

In the quantized theory we write

$$\psi_\alpha(x) = \sum_{E_n > 0} \{u_\alpha^{(n)}(\boldsymbol{x})\, e^{-iE_n x_0} a^{(n)} + u_\alpha'^{(n)}(\boldsymbol{x})\, e^{iE_n x_0} b^{*(n)}\}, \tag{16.8}$$

and, as we did previously, we require

$$\{\bar{\psi}_\alpha(x), \psi_\beta(x')\}_{x_0 = x_0'} = (\gamma_4)_{\beta\alpha}\,\delta(\boldsymbol{x} - \boldsymbol{x}'), \tag{16.9}$$

---

1. (Translator's note) In general, the function u' is an eigenfunction (with negative energy) of a Hamiltonian equal to that of Eq. (16.4), except that the sign of the four-vector potential must be reversed. Consideration of the problem of an electron in a Coulomb field makes this clear. Only if there are additional symmetries can the change of sign of A be ignored.

$$\{\bar{\psi}_\alpha(x), \bar{\psi}_\beta(x')\}_{x_0=x_0'} = \{\psi_\alpha(x), \psi_\beta(x')\}_{x_0=x_0'} = 0. \tag{16.10}$$

From Eqs. (16.6) through (16.10) it follows directly that

$$\{a^{*(n)}, a^{(n')}\} = \{b^{*(n)}, b^{(n')}\} = \delta_{nn'}, \tag{16.11}$$

$$\{a^{(n)}, a^{(n')}\} = \{a^{(n)}, b^{(n')}\} = \cdots = 0. \tag{16.12}$$

Instead of (16.9), we obtain for two arbitrary points x and x',

$$\{\bar{\psi}_\alpha(x), \psi_\beta(x')\} = -i\, S_{\beta\alpha}(x', x), \tag{16.13}$$

$$S_{\beta\alpha}(x', x) = i \sum_{E_n > 0} [\bar{u}_\alpha^{(n)}(\boldsymbol{x})\, u_\beta^{(n)}(\boldsymbol{x}')\, e^{-iE_n(x_0'-x_0)} + \bar{u}_\alpha'^{(n)}(\boldsymbol{x})\, u_\beta'^{(n)}(\boldsymbol{x}')\, e^{iE_n(x_0'-x_0)}]. \tag{16.14}$$

Clearly the function $S(x, x')$ satisfies the differential equation

$$\left[\gamma\left(\frac{\partial}{\partial x} - i e A(\boldsymbol{x})\right) + m\right] S(x, x') = 0, \tag{16.15}$$

with the initial condition

$$S(x, x') = i\gamma_4 \delta(\boldsymbol{x} - \boldsymbol{x}') \quad \text{for} \quad x_0 = x_0'. \tag{16.16}$$

In fact, the function $S(x, x')$ can be taken as defined by (16.15) and (16.16). This S-function differs from the corresponding function for the free field in several respects. In particular, it depends upon two quantities x and x' and is not just a function of the difference $x - x'$.

In analogy to (15.5), let us try to express $S(x, x')$ as

$$S(x, x') = \left\{\gamma\left[\frac{\partial}{\partial x} - i e A(\boldsymbol{x})\right] - m\right\} \Delta(x, x'). \tag{16.17}$$

The function $\Delta(x, x')$ then satisfies the differential equation

$$\left\{\left[\frac{\partial}{\partial x} - i e A(\boldsymbol{x})\right]^2 - m^2 - \frac{e}{2}\sigma_{\mu\nu} F_{\mu\nu}(\boldsymbol{x})\right\} \Delta(x, x') = 0, \tag{16.18}$$

with the initial conditions

$$\left.\begin{aligned} \Delta(x, x') &= 0 \qquad &(16.19)\\ \frac{\partial \Delta(x, x')}{\partial x_0} &= -\delta(\boldsymbol{x} - \boldsymbol{x}') \qquad &(16.20)\end{aligned}\right\} \quad \text{for} \quad x_0 = x_0'.$$

Actually, these transformations do not really help too much, since the function $\Delta(x, x')$ still contains two spin indices, owing to the last term in (16.18). In general the solution of Eq. (16.18) is almost as involved as the solution of the original Eq. (16.15). As a simple example, let us first consider the case of a constant magnetic field H. In this problem the transformation (16.17) does turn out to be useful. With no loss of generality, we can assume that H is in the direction of the z-axis and that $e\mathrm{H} > 0$. Then we may take the vector potential $\boldsymbol{A}(\boldsymbol{x})$ as

$$\boldsymbol{A}(\boldsymbol{x}) = (0, \mathrm{H}\, x, 0). \tag{16.21}$$

This brings the differential equation (16.18) into the form

$$\left\{\Delta - \frac{\partial^2}{\partial x_0^2} - 2ie\mathsf{H}\,x\frac{\partial}{\partial y} - e^2\mathsf{H}^2x^2 - m^2 + e\sigma_{12}\mathsf{H}\right\}\Delta(x, x') = 0\,. \quad (16.22)$$

First, let us consider the simpler equation

$$\left[\Delta - \frac{\partial^2}{\partial x_0^2} - 2i\,x\frac{\partial}{\partial y} - x^2 - m^2\right]G(x, x') = 0\,, \quad (16.23)$$

and the corresponding eigenvalue problem

$$\left[-\Delta + 2i\,x\frac{\partial}{\partial y} + x^2 + m^2\right]u(\boldsymbol{x}) = E^2u(\boldsymbol{x})\,. \quad (16.24)$$

The eigenfunctions for (16.24) are obviously

$$u_{kln}(\boldsymbol{x}) = \frac{1}{2\pi}e^{i(ly+kz)}\,\mathrm{H}_n(x-l)\,, \quad (16.25)$$

$$E_{kn}^2 = m^2 + k^2 + 2n + 1\,. \quad (16.26)$$

Here the function $\mathrm{H}_n(x)$ in (16.25) is the normalized eigenfunction for a harmonic oscillator with frequency 1. For this function, we shall use the integral representation

$$\mathrm{H}_n(x) = \frac{\sqrt{2^n\cdot n!}}{\sqrt[4]{\pi}}\,e^{-\frac{x^2}{2}}\,\frac{1}{2\pi i}\oint\frac{dp}{p^{n+1}}\,e^{-\frac{p^2}{4}}\,e^{px}\,. \quad (16.27)$$

Thus the function $G(x, x')$ in (16.23) can be written

$$G(x,x') = \sum_{n=0}^{\infty}\iint\frac{dk\,dl}{(2\pi)^2}\,e^{i[l(y-y')+k(z-z')]}\,\mathrm{H}_n(x-l)\,\mathrm{H}_n(x'-l)\frac{\sin[E_{kn}(x_0'-x_0)]}{E_{kn}}\,. \quad (16.28)$$

Both the integration over l and the summation over n can be done explicitly. It is then possible to transform the integration on k so that (16.28) becomes[1]

$$G(x, x') = \frac{\varepsilon(x'-x)}{16\pi^2}\,e^{-\frac{i}{2}(y'-y)(x'+x)}\int_{-\infty}^{+\infty}\frac{d\alpha}{\alpha}\,e^{\frac{i}{2}\left[-\alpha\lambda + \frac{m^2}{\alpha}\right]}\,\frac{e^{\frac{i\sigma}{2}\left(2\alpha - \cot\frac{1}{2\alpha}\right)}}{\sin\frac{1}{2\alpha}}\,, \quad (16.29)$$

$$\lambda = (x'-x)^2 + (y'-y)^2 + (z'-z)^2 - (x_0'-x_0)^2\,, \quad (16.30)$$

$$\sigma = \tfrac{1}{2}\left[(x'-x)^2 + (y'-y)^2\right]. \quad (16.31)$$

From this solution of Eq. (16.23), we can immediately find the corresponding solution of Eq. (16.22). We introduce the notation

$$M = e\sigma_{12}\mathsf{H}\,, \quad (16.32)$$

so that

$$M^2 = e^2\mathsf{H}^2. \quad (16.33)$$

---

1. J. Géhéniau, Physica, Haag 16, 822 (1950). See also J. Géhéniau and M. Demeur, Physica, Haag 17, 71 (1951); M. Demeur, Physica 17, 933 (1951), Mem. Acad. Roy. Belg. 28, No. 5 (1953); Y. Katayama, Progr. Theor. Phys. 6, 309 (1951); J. Schwinger, Phys. Rev. 82, 664 (1951).

We then find

$$\left.\begin{aligned}\Delta(x,x') &= \frac{\varepsilon(x'-x)}{8\pi^2}\, e^{-\frac{ieH}{2}(y'-y)(x'+x)}\int\limits_{-\infty}^{+\infty} d\alpha\, e^{\frac{i}{2}\left[-\alpha\lambda+\frac{m^2-M}{\alpha}\right]}\,\frac{e^{\frac{i\sigma}{2}eH\left(\frac{2\alpha}{eH}-\cot\frac{eH}{2\alpha}\right)}}{\frac{2\alpha}{eH}\sin\frac{eH}{2\alpha}}\\ &= \frac{\varepsilon(x'-x)}{8\pi^2}\,\Phi(x,x')\int\limits_{-\infty}^{+\infty} d\alpha\, e^{\frac{i}{2}\left[-\alpha\lambda+\frac{m^2}{\alpha}\right]}\left[\cos\frac{eH}{2\alpha}-\frac{iM}{eH}\sin\frac{eH}{2\alpha}\right]\frac{e^{\frac{i\sigma}{2}eH\left(\frac{2\alpha}{eH}-\cot\frac{eH}{2\alpha}\right)}}{\frac{2\alpha}{eH}\sin\frac{eH}{2\alpha}}.\end{aligned}\right\} \quad (16.34)$$

Here we have defined

$$\Phi(x,x') = e^{-\frac{ieH}{2}(y'-y)(x'+x)}. \quad (16.35)$$

Just as we did in Sec. 7, we can introduce a retarded, an advanced, and an "barred" Δ-function. These are obtained directly from (16.34) by multiplication by $-\frac{1}{2}[1+\varepsilon(x-x')]$, $\frac{1}{2}[1-\varepsilon(x-x')]$, or $-\frac{1}{2}\varepsilon(x-x')$.

If we let H tend to zero in these functions we obtain, for example for $\bar{\Delta}(x,x')$,

$$\bar{\Delta}(x,x') \to \frac{1}{16\pi^2}\int d\alpha\, e^{\frac{i}{2}\left[-\alpha\lambda+\frac{m^2}{\alpha}\right]}. \quad (16.36)$$

This is the same result which we obtained in Sec. 7, since from

$$P\frac{1}{p^2+m^2} = \frac{1}{2i}\int\limits_{-\infty}^{+\infty} dw\, e^{iw(p^2+m^2)}\frac{w}{|w|} \quad (16.37)$$

it follows that

$$\left.\begin{aligned}\frac{1}{(2\pi)^4}P\int dp\frac{e^{ipx}}{p^2+m^2} &= \frac{1}{2i}\frac{1}{(2\pi)^4}\int dw\frac{w}{|w|}e^{iwm^2-\frac{ix^2}{4w}}\int dp\, e^{iw\left(p+\frac{x}{2w}\right)^2}\\ &= \frac{1}{32\pi^2}\int\frac{dw}{w^2}e^{\frac{i}{2}\left(-\frac{\lambda}{2w}+2wm^2\right)} = \frac{1}{16\pi^2}\int d\alpha\, e^{\frac{i}{2}\left[-\alpha\lambda+\frac{m^2}{\alpha}\right]}.\end{aligned}\right\} \quad (16.38)$$

In this expression, we have used the integral

$$\left.\begin{aligned}\int dp\, e^{iwp^2} &= \left[\int\limits_{-\infty}^{+\infty} dp_x\left(\cos(|w|p_x^2)+i\frac{w}{|w|}\sin(|w|p_x^2)\right)\right]^3\times\\ &\times\left[\int\limits_{-\infty}^{+\infty} dp_0\left(\cos(|w|p_0^2)-i\frac{w}{|w|}\sin(|w|p_0^2)\right)\right] = \frac{i\pi^2}{w|w|}.\end{aligned}\right\} \quad (16.39)$$

In a similar way we can define a function $\Delta^{(1)}(x,x')$ by

$$\langle 0|[\bar{\psi}_\alpha(x),\psi_\beta(x')]|0\rangle = \left\{\gamma\left[\frac{\partial}{\partial x'}-ieA(x')\right]-m\right\}_{\beta\delta}\Delta^{(1)}_{\delta\alpha}(x',x). \quad (16.40)$$

Obviously this function can be written

$$\left.\begin{aligned}\Delta^{(1)}(x, x') = \sum_{n=0}^{\infty} \iint \frac{dk\, dl}{(2\pi)^2}\, e_i^{[l(y-y')+k(z-z')]} \times \\ \times\, \mathrm{H}_n(x-l)\, \mathrm{H}_n(x'-l)\, \frac{\cos\left[E_{kn}(x_0'-x_0)\right]}{E_{kn}},\end{aligned}\right\} \tag{16.41}$$

and the subsequent calculation proceeds along lines similar to that for $\Delta(x, x')$. The result is

$$\left.\begin{aligned}\Delta^{(1)}(x, x') = \frac{i}{8\pi^2}\, \Phi(x, x') \int_{-\infty}^{+\infty} d\alpha\, \frac{\alpha}{|\alpha|}\, e^{\frac{i}{2}\left[-\alpha\lambda + \frac{m^2}{\alpha}\right]} \times \\ \times \left[\cos\frac{e\mathrm{H}}{2\alpha} - \frac{iM}{e\mathrm{H}} \sin\frac{e\mathrm{H}}{2\alpha}\right] \frac{e^{\frac{i\sigma e\mathrm{H}}{2}\left(\frac{2\alpha}{e\mathrm{H}} - \cot\frac{e\mathrm{H}}{2\alpha}\right)}}{\frac{2\alpha}{e\mathrm{H}} \sin\frac{e\mathrm{H}}{2\alpha}}.\end{aligned}\right\} \tag{16.42}$$

At this point it is possible to define a function

$$\Delta_F(x, x') = -2i\,\overline{\Delta}(x, x') + \Delta^{(1)}(x, x'),$$

and so forth.

The factor $\Phi(x, x')$ in (16.34) and (16.42) which was defined by (16.35) can also be written in the following way:

$$\Phi(x, x') = e^{-ie\int_{x'}^{x} A_\nu(\xi)\, d\xi_\nu}. \tag{16.43}$$

The line integral in (16.43) is not independent of the path, at least not if the electromagnetic field is not identically zero, but is defined as the integral along the straight line between x and x'. In our case, with a constant field, the other factor in (16.34) and (16.42) is a function only of $x - x'$. In general, this is not the case: it is possible to show that after splitting off the factor $\Phi(x, x')$ the remainder is independent of the gauge which is used.

If the external field is not constant, but depends upon the spatial coordinates, the calculation is very much more complicated than that of the previous example. As yet, no explicit examples for the singular functions have been given in a spatially varying external field. Recently, however, Wichmann and Kroll[1] have evaluated the expression

$$\varrho(r) = \langle 0|[\overline{\psi}(x), \gamma_4 \psi(x)]|0\rangle - \int dx' \langle 0|[\overline{\psi}(x'), \gamma_4 \psi(x')]|0\rangle \cdot \delta(x) \tag{16.44}$$

for a Coulomb field. This expression is quite important for the so-called "vacuum polarization" (c.f. Sec. 29). Their result is

$$\left.\begin{aligned}Q(p) = \int 4\pi r^2 \varrho(r)\, e^{-pr}\, dr = \frac{q}{\pi Z} \sum_{k=1}^{\infty} \int_0^{\infty} dt \int_0^1 dz \left[\left(\frac{1}{(1-z)(1+uz)} + \frac{1}{1-z+uz}\right)\right. \\ \times \left.\frac{\gamma \cos g}{(1+t^2)^{\frac{3}{2}}} + \frac{2t\, s(k) \sin g}{(1+t^2)(1+uz)} + \frac{2\gamma t^2 \cos g}{(1+t^2)^{\frac{3}{2}}(1+uz)}\right] p^{-s(k)}.\end{aligned}\right\} \tag{16.45}$$

1. E. H. Wichmann and N. M. Kroll, Phys. Rev. 96, 232 (1954); 101, 843 (1956).

Here $q = Ze$, the charge of the external field, and the other abbreviations have the following definitions:

$$p = \frac{(1 + uz)(1 - z(1 - u))}{1 - z}, \qquad (16.46a)$$

$$g = \frac{\gamma t}{\sqrt{1 + t^2}} \log \frac{(1 - z)(1 + uz)}{1 - z(1 - u)}, \qquad (16.46b)$$

$$u = \frac{p}{2\sqrt{1 + t^2}}, \qquad (16.46c)$$

$$s(k) = \sqrt{k^2 - \gamma^2}, \qquad (16.46d)$$

$$\gamma = \frac{Ze^2}{4\pi}. \qquad (16.46e)$$

The summation over k in (16.45) corresponds to a summation over eigenfunctions of various angular momenta. The integrations are obtained by transformations of the integral representations of the radial eigenfunctions.

Equation (16.45) represents an analytic function of γ in the region $|\gamma| < 1$. A power series in γ, which results from the use of ordinary perturbation theory in the external field problem, therefore gives a series which is convergent for $|\gamma| < 1$.

CHAPTER IV

THE DIRAC FIELD AND THE ELECTROMAGNETIC FIELD IN INTERACTION. PERTURBATION THEORY

17. Lagrangian and Equations of Motion

We are now ready to attack our major problem: the interaction of a quantized electron field and a quantized electromagnetic field. As equations of motion, we will continue to use the Dirac equation (16.1), except that $A_\mu(x)$ is now the operator of the quantized electromagnetic field, and not an external field. In certain problems it is necessary to use not only the quantized field, but also to add an external field. Then $A_\mu(x)$ becomes the sum of both these terms. In order not to complicate the formalism too much, we shall initially confine ourselves to the quantized field only.

As a second system of operator equations, we will use Eq. (11.2) with the current given by the Dirac operators of Eq. (13.13). We obtain all of these equations if we choose the Lagrangian to be the following charge-symmetric expression:

$$\mathscr{L} = \mathscr{L}_\psi + \mathscr{L}_A + \mathscr{L}_W , \tag{17.1}$$

$$\mathscr{L}_\psi = -\frac{1}{4}\left[\overline{\psi}(x), \left(\gamma\frac{\partial}{\partial x} + m\right)\psi(x)\right] - \frac{1}{4}\left[-\frac{\partial\overline{\psi}(x)}{\partial x}\gamma + m\overline{\psi}(x), \psi(x)\right], \tag{17.2}$$

$$\mathscr{L}_A = -\frac{1}{4}\left(\frac{\partial A_\nu(x)}{\partial x_\mu} - \frac{\partial A_\mu(x)}{\partial x_\nu}\right)\left(\frac{\partial A_\nu(x)}{\partial x_\mu} - \frac{\partial A_\mu(x)}{\partial x_\nu}\right) - \frac{1}{2}\frac{\partial A_\mu(x)}{\partial x_\mu}\frac{\partial A_\nu(x)}{\partial x_\nu}, \tag{17.3}$$

$$\mathscr{L}_W = \frac{ie}{2}A_\mu(x)\,[\overline{\psi}(x), \gamma_\mu\psi(x)]\,. \tag{17.4}$$

The two terms $\mathscr{L}_\psi$ and $\mathscr{L}_A$ are formally identical with (14.12) and (5.10) and the term $\mathscr{L}_W$ can be written as

$$\mathscr{L}_W = A_\mu(x)\,j_\mu(x)\,. \tag{17.5}$$

In the usual way, we obtain the desired equations of motion from this Lagrangian:

$$\left(\gamma\frac{\partial}{\partial x} + m\right)\psi(x) = ie\gamma A(x)\,\psi(x)\ , \tag{17.6}$$

$$\Box A_\mu(x) = -\frac{ie}{2}[\overline{\psi}(x), \gamma_\mu\psi(x)] = -j_\mu(x)\,. \tag{17.7}$$

The canonical quantization is considerably simplified because the "new" term $\mathscr{L}_W$ does not contain any time derivatives of the field operators. The canonical momenta are therefore the same functions of the dynamical variables as previously, and we can immediately write down the equal time canonical commutation relations:

$$[A_\mu(x), A_\nu(x')]_{x_0=x_0'} = \left[\frac{\partial A_\mu(x)}{\partial x_0}, \frac{\partial A_\nu(x')}{\partial x_0'}\right]_{x_0=x_0'} = 0\,, \tag{17.8}$$

$$\left[\frac{\partial A_\mu(x)}{\partial x_0}, A_\nu(x')\right]_{x_0=x_0'} = -i\,\delta_{\mu\nu}\,\delta(\boldsymbol{x}-\boldsymbol{x}')\,, \tag{17.9}$$

$$\{\psi(x), \psi(x')\}_{x_0=x_0'} = \{\overline{\psi}(x), \overline{\psi}(x')\}_{x_0=x_0'} = 0\,, \tag{17.10}$$

$$\{\overline{\psi}(x), \psi(x')\}_{x_0=x_0'} = \gamma_4\,\delta(\boldsymbol{x}-\boldsymbol{x}')\,, \tag{17.11}$$

$$[A_\mu(x), \psi(x')]_{x_0=x_0'} = \left[\frac{\partial A_\mu(x)}{\partial x_0}, \psi(x')\right]_{x_0=x_0'} = 0\,, \tag{17.12}$$

$$[A_\mu(x), \overline{\psi}(x')]_{x_0=x_0} = \left[\frac{\partial A_\mu(x)}{\partial x_0}, \overline{\psi}(x')\right]_{x_0=x_0'} = 0. \tag{17.13}$$

For free fields we were able to construct the commutators for arbitrary times from those for equal times. This is now impossible, since the commutators do not satisfy simple equations of motion. In particular, for time-like separations, they are not even c-numbers and we are not able to give explicit expressions for them.

In Sec. 11 we were able to solve the equation of motion (11.2) by means of a retarded singular function in (11.5). In the present case, we are not able to solve the equation of motion by this method; however, we can use it to make a transformation to an integral equation. Instead of the differential equations (17.6) and (17.7) we can write

$$\psi(x) = \psi^{(0)}(x) - \int S_R(x-x')\, ie\gamma A(x')\,\psi(x')\,dx'\,, \tag{17.14}$$

$$A_\mu(x) = A_\mu^{(0)}(x) + \int D_R(x-x')\,\frac{ie}{2}\,[\overline{\psi}(x'), \gamma_\mu\psi(x')]\,dx'. \tag{17.15}$$

Here it is necessary to introduce two new operators $\psi^{(0)}(x)$ and $A_\mu^{(0)}(x)$ into the equations of motion. These are obviously solutions of the free-field equations

$$\left(\gamma\frac{\partial}{\partial x} + m\right)\psi^{(0)}(x) = 0\,, \tag{17.16}$$

$$\Box A_\mu^{(0)}(x) = 0. \tag{17.17}$$

Because the integral equations (17.14) and (17.15) contain the retarded functions, $\psi^{(0)}(x)$ and $A_\mu^{(0)}(x)$ are formally the initial values of the Heisenberg operators $\psi(x)$ and $A_\mu(x)$ for $x_0 \to -\infty$. In principle, we can regard these equations of motion as allowing us to calculate the field operators as functions of their initial values.

At first sight this is quite a different approach to the problem from that studied previously (where we looked for the eigenvalues of a Hamiltonian operator). Here the analogous problem would be to find such a representation of the operators $\psi(x)$ and $A_\mu(x)$ that the Hamiltonian operator is diagonal:

$$H(A,\psi) = H^{(0)}(A,\psi) + H^{(1)}(A,\psi) , \tag{17.18}$$

$$H^{(0)}(A,\psi) = H_1^{(0)}(A) + H_2^{(0)}(\psi) , \tag{17.19}$$

$$H_1^{(0)}(A) = \frac{1}{2}\int d^3x \left[\frac{\partial A_\mu(x)}{\partial x_0}\,\frac{\partial A_\mu(x)}{\partial x_0} + \frac{\partial A_\mu(x)}{\partial x_k}\,\frac{\partial A_\mu(x)}{\partial x_k}\right], \tag{17.20}$$

$$H_2^{(0)}(\psi) = \frac{1}{2}\int d^3x \left[\overline{\psi}(x), \left(\gamma_k \frac{\partial}{\partial x_k} + m\right)\psi(x)\right] , \tag{17.21}$$

$$H^{(1)}(A,\psi) = -\frac{ie}{2}\int d^3x\, A_\mu(x)\,[\overline{\psi}(x), \gamma_\mu \psi(x)] . \tag{17.22}$$

This is the "classical" statement of the problem of quantum electrodynamics. If we were to attempt such a calculation in detail, we would soon find difficulties. In fact, it would turn out that the theory formulated here is not well-defined mathematically and contains several infinite quantities. Great progress has been made in recent years with the realization that it is sufficient to regard these infinite quantities as a physically unobservable "renormalization" of the constants m and e. As we shall see later, it is quite important to have the formal relativistic covariance apparent at all stages, for otherwise the infinite quantities cannot be uniquely identified. The diagonalization of the Hamiltonian (17.18) is formally not a covariant problem and it would be quite difficult to arrive at a unique interpretation of the infinite parts of the theory in this way. It is therefore preferable to attack the problem only by the use of the covariant equations of motion (17.14) and (17.15), rather than by means of the Hamiltonian directly.

A previous example has shown that it is possible to diagonalize the complete Hamiltonian (11.8) by solving the equations of motion for the field operators (11.5) with the use of "adiabatic switching". Although the proof of the validity of adiabatic switching given above does not hold here, it is straightforward to attempt a similar method. The justification of the method will presumably follow when we know the results of the calculation. We therefore write the equations of motion (17.14) and (17.15) in the form

$$\psi(x,\alpha) = \psi^{(0)}(x) - ie\int S_R(x-x')\, e^{-\alpha|x_0'|}\,\gamma A(x',\alpha)\,\psi(x',\alpha)\,dx' , \tag{17.23}$$

$$A_\mu(x,\alpha) = A_\mu^{(0)}(x) + \frac{ie}{2}\int D_R(x-x')\, e^{-\alpha|x_0'|}\,[\overline{\psi}(x',\alpha), \gamma_\mu \psi(x',\alpha)]\,dx', \tag{17.24}$$

and regard ψ and A_μ as functions of $\psi^{(0)}$, $A_\mu^{(0)}$, and α. The physically interesting quantities are

$$\psi(x) = \lim_{\alpha\to 0} \psi(x,\alpha) , \tag{17.25}$$

$$A_\mu(x) = \lim_{\alpha\to 0} A_\mu(x,\alpha) . \tag{17.26}$$

We expect that the two limiting values (17.25) and (17.26) exist and diagonalize the Hamiltonian operator (17.18). At present, we cannot give a general proof of these conjectures but must test explicitly whether or not they hold when we find a particular solution.

The operators $\psi^{(0)}(x)$ and $A_\mu^{(0)}(x)$ are therefore to be chosen so that they diagonalize the Hamiltonian operator for $x_0 \to -\infty$; i.e.,

$$H(A,\psi)|_{x_0\to-\infty} = H^{(0)}(A^{(0)},\psi^{(0)}) = H_1^{(0)}(A^{(0)}) + H_2^{(0)}(\psi^{(0)}) \tag{17.27}$$

is to be diagonal. Finding two such operators is just the problem which we have solved in Chap. II and III, so we can use those solutions here. From now on we regard the operators $\psi^{(0)}(x)$ and $A_\mu^{(0)}(x)$ as known. They satisfy

$$\{\overline{\psi}^{(0)}(x), \psi^{(0)}(x')\} = -i\,S(x'-x)\,, \tag{17.28}$$

$$\langle 0|\,[\overline{\psi}^{(0)}(x), \psi^{(0)}(x')]\,|0\rangle = S^{(1)}(x'-x)\,, \tag{17.29}$$

$$[A_\mu^{(0)}(x), A_\nu^{(0)}(x')] = -i\,\delta_{\mu\nu}\,D(x'-x), \tag{17.30}$$

$$\langle 0|\,\{A_\mu^{(0)}(x), A_\nu^{(0)}(x')\}\,|0\rangle = \delta_{\mu\nu}\,D^{(1)}(x'-x). \tag{17.31}$$

In (17.31) it is implied either that we are evaluating only gauge-invariant expressions or that we are using the method of the indefinite metric. Since it will later be useful to discuss quantities which are not formally gauge-invariant, we shall require from now on that the longitudinal and scalar photons be treated by means of the indefinite metric. The metric operator η is prescribed as commuting with $\psi^{(0)}$ and $\overline{\psi}^{(0)}$:

$$[\psi^{(0)}(x), \eta] = [\overline{\psi}^{(0)}(x), \eta] = 0. \tag{17.32}$$

For the complete operator $A_\mu(x)$, we have

$$[A_k(x), \eta] = \{A_4(x), \eta\} = 0\,, \tag{17.33}$$

exactly as for the incoming field. For the complete Dirac field we cannot prescribe the relations corresponding to (17.32), because the coupled equations of motion do not leave these quantities independent of the degrees of freedom of the incoming electromagnetic field. In order that the quantities present in the theory have the correct reality properties, we must define $\overline{\psi}(x)$ by

$$\overline{\psi}(x) = \eta\,\psi^*(x)\,\eta\,\gamma_4\,. \tag{17.34}$$

In the theory with coupled fields, the continuity equation for the current,

$$\frac{\partial j_\mu(x)}{\partial x_\mu} = \frac{ie}{2}\frac{\partial}{\partial x_\mu}[\overline{\psi}(x), \gamma_\mu\psi(x)] = 0\,, \tag{17.35}$$

implies for $A_\mu(x)$:

$$\frac{\partial A_\mu(x)}{\partial x_\mu} = \frac{\partial A_\mu^{(0)}(x)}{\partial x_\mu}. \tag{17.36}$$

Both (17.35) and (17.36) are operator identities. If the Lorentz condition for the incoming field holds in the form

$$\frac{\partial A_\mu^{(0)(+)}(x)}{\partial x_\mu}|\psi\rangle = 0, \tag{17.37}$$

then it also holds for the complete field.

18. Perturbation Calculations in the Heisenberg Picture

For the coupled fields, we cannot exhibit an exact solution to the equations of motion. Rather, we must attempt to find useful methods of approximation. Because of the small value of the charge, $\frac{e^2}{4\pi} \approx \frac{1}{137}$, it is straightforward to consider the right sides of (17.14) and (17.15) or (17.23) and (17.24) to be "small" and to attempt a solution in the form of a power series in e. We therefore write

$$\psi(x) = \psi^{(0)}(x) + e\,\psi^{(1)}(x) + e^2\,\psi^{(2)}(x) + \cdots, \tag{18.1}$$

$$A_\mu(x) = A_\mu^{(0)}(x) + e\,A_\mu^{(1)}(x) + e^2\,A_\mu^{(2)}(x) + \cdots, \tag{18.2}$$

and, upon introducing these series into the equations of motion, we obtain recursion relations for the different orders of approximation:

$$\psi^{(n+1)}(x) = -\frac{i}{2}\int S_R(x-x')\,\gamma_\nu \sum_{m=0}^{n} \{A_\nu^{(m)}(x'), \psi^{(n-m)}(x')\}\,dx', \tag{18.3}$$

$$A_\mu^{(n+1)}(x) = \frac{i}{2}\int D_R(x-x') \sum_{m=0}^{n} [\overline{\psi}^{(m)}(x'), \gamma_\mu\,\psi^{(n-m)}(x')]\,dx'. \tag{18.4}$$

In (18.3) we have symmetrized the right side. Although this is not necessary, it will turn out to be convenient. The symmetrization is certainly allowed, because by (17.12) and (17.13) the operators $A_\mu(x)$ and $\psi(x)$ [or $\overline{\psi}(x)$] commute for equal times.

In the lowest order, we have

$$\psi^{(1)}(x) = -i\int S_R(x-x')\,\gamma A^{(0)}(x')\,\psi^{(0)}(x')\,dx', \tag{18.5}$$

$$\overline{\psi}^{(1)}(x) = -i\int \overline{\psi}^{(0)}(x')\,\gamma A^{(0)}(x')\,S_A(x'-x)\,dx', \tag{18.6}$$

$$A_\mu^{(1)}(x) = \frac{i}{2}\int D_R(x-x')\,[\overline{\psi}^{(0)}(x'), \gamma_\mu\,\psi^{(0)}(x')]\,dx', \tag{18.7}$$

$$\left.\begin{aligned}\psi^{(2)}(x) = \tfrac{1}{4}\iint S_R(x-x')\,\gamma_\nu \{\psi^{(0)}(x'), [\overline{\psi}^{(0)}(x''), \gamma_\nu\,\psi^{(0)}(x'')]\}\times\\ \times D_R(x'-x'')\,dx'\,dx'' - \tfrac{1}{2}\iint S_R(x-x')\,\gamma_{\nu_1}\,S_R(x'-x'')\,\gamma_{\nu_2}\times\\ \times\psi^{(0)}(x'')\,\{A_{\nu_1}^{(0)}(x'), A_{\nu_2}^{(0)}(x'')\}\,dx'\,dx'',\end{aligned}\right\} \tag{18.8}$$

$$\left.\begin{aligned}A_\mu^{(2)}(x) = \tfrac{1}{2}\iint D_R(x-x')\,([\bar\psi^{(0)}(x'),\gamma_\mu S_R(x'-x'')\gamma_\nu\psi^{(0)}(x'')] + \\ + [\bar\psi^{(0)}(x'')\gamma_\nu S_A(x''-x'),\gamma_\mu\psi^{(0)}(x')])\,A_\nu^{(0)}(x'')\,dx'\,dx''.\end{aligned}\right\}\quad(18.9)$$

In principle this process can be continued to arbitrary order without any difficulties. Thus we can obtain the general equations for the n-th order approximation to the field operators. The physically observable quantities, for example, the components of the current, can also be constructed in a similar fashion. Because we shall later study the current operator, we give the lowest approximations here:

$$j_\mu^{(0)}(x) = \frac{i}{2}[\bar\psi^{(0)}(x),\gamma_\mu\psi^{(0)}(x)]\,,\qquad(18.10)$$

$$\left.\begin{aligned}j_\mu^{(1)}(x) = \tfrac{1}{2}\int dx'\,([\bar\psi^{(0)}(x),\gamma_\mu S_R(x-x')\gamma_\nu\psi^{(0)}(x')] + \\ + [\bar\psi^{(0)}(x')\gamma_\nu S_A(x'-x),\gamma_\mu\psi^{(0)}(x)])\,A_\nu^{(0)}(x')\,,\end{aligned}\right\}\quad(18.11)$$

$$\left.\begin{aligned}j_\mu^{(2)}(x) = \frac{i}{8}\iint dx'\,dx''\,[\bar\psi^{(0)}(x),\gamma_\mu S_R(x-x')\gamma_\nu\{\psi^{(0)}(x'),[\bar\psi^{(0)}(x''),\gamma_\nu\psi^{(0)}(x'')]\}]\times \\ \times D_R(x'-x'') - \\ -\frac{i}{4}\iint dx'\,dx''\,[\bar\psi^{(0)}(x),\gamma_\mu S_R(x-x')\gamma_{\nu_1} S_R(x'-x'')\gamma_{\nu_2}\psi^{(0)}(x'')]\times \\ \times\{A_{\nu_1}^{(0)}(x'),A_{\nu_2}^{(0)}(x'')\} - \\ -\frac{i}{2}\iint dx'\,dx''\,[\bar\psi^{(0)}(x')\gamma A^{(0)}(x')\,S_A(x'-x),\gamma_\mu S_R(x-x'')\times \\ \times\gamma A^{(0)}(x'')\,\psi^{(0)}(x'')] + \\ +\frac{i}{8}\iint dx'\,dx''\,[\{[\bar\psi^{(0)}(x''),\gamma_\nu\psi^{(0)}(x'')],\psi^{(0)}(x')\}\gamma_\nu S_A(x'-x)\gamma_\mu,\psi^{(0)}(x)]\times \\ \times D_R(x'-x'') - \\ -\frac{i}{4}\iint dx'\,dx''\,[\bar\psi^{(0)}(x'')\gamma_{\nu_2} S_A(x''-x')\gamma_{\nu_1} S_A(x'-x),\gamma_\mu\psi^{(0)}(x)]\times \\ \times\{A_{\nu_1}^{(0)}(x'),A_{\nu_2}^{(0)}(x'')\}.\end{aligned}\right\}\quad(18.12)$$

As is already apparent from (18.12), the higher approximations to the operators rapidly become very complicated. We shall refrain from giving any further explicit examples,[1] and instead turn to a detailed discussion of the properties of the lowest-order approximations.

In Eq. (18.6) an advanced singular function appears for the first time. This does not mean that the operators $\bar\psi(x)$ become $\bar\psi^{(0)}(x)$ as $x_0\to+\infty$, but rather it is only a consequence of the "reversed" order of writing the difference of coordinates $x'-x$. The same holds for the advanced functions appearing in (18.9), (18.11), and (18.12).

As can be seen from (18.5) through (18.12), the solutions found in this way contain products of operators $\psi^{(0)}(x)$, $A_\mu^{(0)}(x)$ and singular functions. This means, for example, that the complete operator $\psi(x)$ has a non-zero matrix element between the vacuum and a state with an incoming electron and incoming photon. Even the order $\psi^{(1)}(x)$ contains the product $A_\mu^{(0)}(x')\cdot\psi^{(0)}(x')$, which allows

1. G. Källén, Ark. Fysik 2, 187 (1950); 2, 371 (1950).

this transition:

$$\langle 0|\psi(x)|q,k\rangle = -ie\int S_R(x-x')\gamma_\nu \langle 0|A_\nu^{(0)}(x')|k\rangle \langle 0|\psi^{(0)}(x')|q\rangle\, dx' + \cdots. \tag{18.13}$$

Here $|q,k\rangle$ is the state mentioned above; $|q\rangle$ is a state with only the electron and $|k\rangle$ a state with only the photon. The same term also gives a contribution to the matrix element $\langle k|\psi(x)|q\rangle$:

$$\langle k|\psi(x)|q\rangle = -ie\int S_R(x\;\, x')\gamma_\nu \langle k|A_\nu^{(0)}(x')|0\rangle \langle 0|\psi^{(0)}(x')|q\rangle\, dx' + \cdots. \tag{18.14}$$

The right sides of (18.13) and (18.14) contain integrals which are not actually convergent. We recall that these integrals are to be taken as limiting values, according to (17.23) and (17.24), so that for (18.13), for example,

$$\left.\begin{aligned}\langle 0|\psi(x)|q,k\rangle &= -ie\lim_{\alpha\to 0}\int dx'\int \frac{dp}{(2\pi)^4}\frac{e^{ip(x-x')}}{p^2+m^2}(i\gamma p - m)\,\gamma\, e^{(\lambda)}\times\\ &\quad\times\frac{e^{ikx'}}{\sqrt{2V\omega}}\,\frac{u(q)}{\sqrt{V}}\,e^{iqx'}\,e^{-\alpha|x_0'|}+\cdots=\\ &= -ie\frac{e^{i(q+k)x}}{(q+k)^2+m^2}\,[i\gamma(q+k)-m]\,\gamma\, e^{(\lambda)}\,u(q)\,\frac{1}{V\cdot\sqrt{2\omega}}+\cdots.\end{aligned}\right\} \tag{18.15}$$

In (18.15) we have used the following expressions:

$$\langle 0|A_\nu^{(0)}(x)|k\rangle = \frac{e_\nu^{(\lambda)}}{\sqrt{2V\omega}}\,e^{ikx}, \tag{18.16a}$$

[c.f. (5.28)] and

$$\langle 0|\psi^{(0)}(x)|q\rangle = \frac{u^{(r)}(q)}{\sqrt{V}}\,e^{iqx}, \tag{18.16b}$$

[c.f. (12.22)]. Moreover, the fact that $(q+k)^2+m^2=2qk$ is not zero for all photons (with $\omega\neq 0$) has been used.

In the theory of free fields, we have seen that the field operators create states with only one particle if they operate on the vacuum. From the calculation above it follows that this is no longer true for the theory with coupled fields. For example, if we operate with $\bar{\psi}$ on the vacuum, we obtain, among other states, states with two incoming particles (one electron and one photon), according to (18.15). On the other hand, we also obtain the one-particle states in this manner. The matrix elements for these transitions are different from the corresponding matrix elements for the free field. From (18.8) we have

$$\left.\begin{aligned}\langle 0|\psi(x)|q\rangle &= \langle 0|\psi^{(0)}(x)|q\rangle + \frac{e^2}{4}\iint S_R(x-x')\,\gamma_\nu\times\\ &\quad\times\langle 0|\{\psi^{(0)}(x'),[\bar\psi^{(0)}(x''),\gamma_\nu\psi^{(0)}(x'')]\}|q\rangle\, D_R(x'-x'')\,dx'\,dx''-\\ &\quad-\frac{e^2}{2}\iint S_R(x-x')\,\gamma_{\nu_1}\,S_R(x'-x'')\,\gamma_{\nu_2}\times\\ &\quad\times\langle 0|\psi^{(0)}(x'')\{A_{\nu_1}^{(0)}(x'),A_{\nu_2}^{(0)}(x'')\}|q\rangle\,dx'\,dx''+\cdots.\end{aligned}\right\} \tag{18.17}$$

The expressions appearing in (18.17) can be simplified considerably. Beginning with the second term, we have

$$\left.\begin{aligned}&\langle 0|\,\psi^{(0)}(x'')\,\{A^{(0)}_{\nu_1}(x'),\,A^{(0)}_{\nu_2}(x'')\}\,|q\rangle = \\ &\quad = \sum_{|Z\rangle} \langle 0|\,\psi^{(0)}(x'')\,|Z\rangle\langle Z|\,\{A^{(0)}_{\nu_1}(x'),\,A^{(0)}_{\nu_2}(x'')\}\,|q\rangle.\end{aligned}\right\} \quad (18.18)$$

The second factor of (18.18) can only give transitions in which the number of photons is changed: Either two photons are produced or the same photon is first produced and then destroyed. The sum over intermediate states in (18.18) therefore contains states of either an electron, or an electron and two photons. For the last class, however, the first factor is zero, and we have

$$\left.\begin{aligned}&\langle 0|\,\psi^{(0)}(x'')\,\{A^{(0)}_{\nu_1}(x'),\,A^{(0)}_{\nu_2}(x'')\}\,|q\rangle = \\ &\quad = \langle 0|\,\psi^{(0)}(x'')\,|q\rangle\,\langle q|\,\{A^{(0)}_{\nu_1}(x'),\,A^{(0)}_{\nu_2}(x'')\}\,|q\rangle = \\ &\quad = \langle 0|\,\psi^{(0)}(x'')\,|q\rangle\,\langle 0|\,\{A^{(0)}_{\nu_1}(x'),\,A^{(0)}_{\nu_2}(x'')\}\,|0\rangle = \\ &\quad = \delta_{\nu_1\nu_2}\,D^{(1)}(x'-x'')\,\langle 0|\,\psi^{(0)}(x'')\,|q\rangle.\end{aligned}\right\} \quad (18.19)$$

In a similar way, we obtain

$$\left.\begin{aligned}&\langle 0|\,\{\psi^{(0)}(x'),\,[\overline{\psi}^{(0)}(x''),\,\gamma_\nu\,\psi^{(0)}(x'')]\}\,|q\rangle = \\ &\qquad\qquad = -\,2\,S^{(1)}(x'-x'')\,\gamma_\nu\,\langle 0|\,\psi^{(0)}(x'')\,|q\rangle.\end{aligned}\right\} \quad (18.20)$$

In working out (18.20), we have taken

$$\langle q|\,[\overline{\psi}^{(0)}(x''),\,\psi^{(0)}(x')]\,|q\rangle = \langle 0|\,[\overline{\psi}^{(0)}(x''),\,\psi^{(0)}(x')]\,|0\rangle = S^{(1)}(x'-x''), \quad (18.20a)$$

although the state $|q\rangle$ is already occupied and should not occur in the sum because of the exclusion principle. However, the contribution of this state is proportional to V^{-1} and it can therefore be neglected if V becomes infinite.

Collecting our results, we have

$$\left.\begin{aligned}&\langle 0|\,\psi(x)\,|q\rangle = \langle 0|\,\psi^{(0)}(x)\,|q\rangle - \frac{e^2}{2}\iint S_R(x-x')\,\gamma_\nu \times \\ &\quad\times [S^{(1)}(x'-x'')\,D_R(x'-x'') + S_R(x'-x'')\,D^{(1)}(x'-x'')] \times \\ &\quad\times \gamma_\nu\,\langle 0|\,\psi^{(0)}(x'')\,|q\rangle\,dx'\,dx'' + \cdots.\end{aligned}\right\} \quad (18.21)$$

The integrals in (18.21) can be done and we shall work them out in Sec. 31. For the moment, let us avoid this and consider (18.21) only as an example showing how matrix elements for simple transitions can be developed from complicated expressions like (18.8).

It remains to establish that the series obtained here satisfy the postulates (17.25), (17.26), and that they diagonalize the Hamiltonian. Regarding the diagonalization of the Hamiltonian, it is an immediate result that the conservation law (4.22) for the energy-momentum tensor is no longer valid, because the Lagrangian contains the coordinates x_μ explicitly. They are contained in the charge. By the equations of motion or by means of considerations similar to those of (4.26), it follows that

$$\frac{\partial T_{\mu\nu}}{\partial x_\mu} = \frac{\partial \mathscr{L}}{\partial x_\nu}, \tag{18.22}$$

rather than (4.22). The right side of (18.22) is to be understood as the derivative with respect to the coordinates with the field operators held fixed. In the present case of a variable charge, we have

$$\frac{\partial T_{\mu\nu}}{\partial x_\mu} = \frac{\partial \mathscr{L}}{\partial e}\frac{\partial e}{\partial x_\nu}. \tag{18.23}$$

We consider the time derivative of the Hamiltonian. From (18.23) we have

$$\frac{\partial H}{\partial x_0} = -\int d^3x \frac{\partial T_{44}}{\partial x_0} = -i\int d^3x \frac{\partial T_{\mu 4}}{\partial x_\mu} = -\int d^3x \frac{\partial \mathscr{L}}{\partial e}\cdot\frac{\partial e}{\partial x_0}. \tag{18.24}$$

The change in the Hamiltonian during a time interval (x_0, x_0') can therefore be calculated from the implicit equation

$$H(A(x), \psi(x)) = H(A(x'), \psi(x')) - \int_{x'}^{x} dx'' \frac{\partial \mathscr{L}(A(x''), \psi(x''))}{\partial e(x'')}\frac{\partial e(x'')}{\partial x_0''}. \tag{18.25}$$

Up to now, the detailed nature of the dependence of the Lagrangian upon e has not been used, and Eq. (18.25) is completely general. For our Lagrangian (17.1) through (17.4) we obtain

$$\frac{\partial \mathscr{L}}{\partial e} = \frac{i}{2}[\overline{\psi}(x), \gamma_\mu \psi(x)]\, A_\mu(x), \tag{18.26}$$

and therefore we have

$$H(A(x), \psi(x)) = H(A(x'), \psi(x')) - \frac{i}{2}\int_{x'}^{x} dx''\, [\overline{\psi}(x''), \gamma_\mu \psi(x'')]\, A_\mu(x'')\frac{\partial e(x'')}{\partial x_0''}. \tag{18.27}$$

In Eq. (18.27) we can allow x_0' to go to $-\infty$ and, if we introduce the time derivative of the charge explicitly, we find

$$\left.\begin{aligned} H(A(x), \psi(x)) = H^{(0)}(A^{(0)}(x), \psi^{(0)}(x)) - \alpha\frac{ie}{2}\int_{-\infty}^{x} dx'\, e^{-\alpha|x_0'|} \times \\ \times [\overline{\psi}(x'), \gamma_\mu \psi(x')]\, A_\mu(x'). \end{aligned}\right\} \tag{18.28}$$

In (18.28) it must be pointed out that the operator $H^{(0)}(A^{(0)}, \psi^{(0)})$ is time-independent. The last term in (18.28) contains a factor α, but without further calculation it cannot be concluded from this that this term vanishes for $\alpha \to 0$. If the matrix element under consideration tends to a finite limit for $x_0 \to -\infty$, the integral has order of magnitude

$$\alpha \int_{-\infty}^{x} e^{\alpha x_0'} dx_0' = e^{\alpha x_0} = O(1), \tag{18.29}$$

and therefore does not vanish for $\alpha \to 0$. With the use of our solution, we can verify that each term in the series development of the integral has a time dependence of the form

$$\alpha \int_{-\infty}^{x_0} x_0'^{\,n_1-1} e^{(n_2\alpha + i p_0) x_0'} dx_0' = \frac{\alpha}{(n_2\alpha + i p_0)^{n_1}} \int_{-\infty}^{x_0(n_2\alpha+ip_0)} z^{n_1-1} e^{z} dz\,. \tag{18.30}$$

Here p_0 is the difference of the eigenvalues of the operator $H^{(0)}(A^{(0)}, \psi^{(0)})$ for the two states being considered and n_1 and n_2 are positive integers. For the terms in which p_0 is different from zero, the integral (18.30) is therefore of order α. For the states for which $H^{(0)}$ in (18.28) has the same eigenvalue, there are always symmetry operators (parity, momentum, etc.) which commute with all three terms of (18.28) and which have different eigenvalues for the states under consideration. These quantities are constants of the motion, even during the switching, and matrix elements of the last term in (18.28) between two such states must vanish. Therefore, in the limit $\alpha \to 0$ the difference between the two operators $H(A, \psi)$ and $H^{(0)}(A^{(0)}, \psi^{(0)})$ has been expressed in a power series in e in which each term has diagonal form. This shows that our solution of the equations of motion (17.23) and (17.24) actually "diagonalizes" the Hamiltonian and therefore that the search for these solutions is equivalent to the "classical" statement of the problem of quantum electrodynamics.

It is more complicated to discuss the existence of the field operators for $\alpha \to 0$. For example, it is clear that (18.30) does not exist in the limit if $n_1 > 1$, since this expression is of order of magnitude α^{-n_1+1} if p_0 vanishes. A more careful investigation, which we avoid here, shows that the appearance of such terms is connected with the change of the eigenvalue of the energy, and that they are to be understood as series expansions of expressions of the type

$$e^{-i\delta E[x_0 - O(\alpha^{-1})]}. \tag{18.31}$$

If we had fixed the initial conditions for a finite time T, we would have had an expression $x_0 - O(T)$, rather than $x_0 - O(\alpha^{-1})$. Such an infinite phase in the field operators is of no physical significance and can obviously be avoided by a very careful treatment of the boundary conditions. We do not pursue the matter further because we shall later remove the change in energy from our equations (c.f. Chap. VII). Apart from these infinite phases, α is present in our solutions only in integrals like

$$\int_{-\infty}^{+\infty} dx_0\, e^{-n\alpha|x_0| - i p_0 x_0} = \frac{2n\alpha}{n^2\alpha^2 + p_0^2} \to 2\pi\,\delta(p_0), \tag{18.32}$$

or sometimes in integrals of the form

$$\int_{-\infty}^{x_0} dx_0' \, e^{-n\alpha|x_0 - x_0'| - ip_0(x_0 - x_0')} = \frac{1}{n\alpha + ip_0} = -iP\frac{1}{p_0} + \pi\,\delta(p_0). \qquad (18.33)$$

In the limit $\alpha \to 0$ the right sides of (18.32) and (18.33) have singularities no stronger than delta-functions in p_0. Because these expressions must later be integrated over p (actually, only space-time averages of the operators have a physical meaning), the existence of these expressions is certain. We shall content ourselves with these admittedly incomplete remarks about the adiabatic hypothesis.

Actually, we ought to discuss whether or not the series used here are convergent and define a solution, and, if so, whether one is justified in operating termwise with them. We shall later return to these questions and shall even try to carry through the discussion without power series. For the present, we shall not go into this further, but rather study the applications of the theory more closely.

19. The S-Matrix

The field operators which we have considered up to now are not especially suited to practical applications of the theory. In principle, the electromagnetic field strengths are measurable, as we have indicated in Sec. 10, but this is more a question of the method of interpretation of the theory than of the experiments which are actually done. In actual measurements, the determination of interaction cross sections is of great importance. Typically, in these experiments a number of particles of known momenta and energies meet each other. During a rather short time they interact with each other and then they continue again as independent particles with measurable energy-momentum vectors. In general, these new energy-momentum vectors differ from the original ones. Under certain conditions, new particles can be produced in the collision or some of the original ones annihilated. It is clear that our formulation of quantum electrodynamics with incoming and outgoing particles is quite appropriate for the discussion of such collision problems. In order to relate the incoming and outgoing fields, we have to find a generalization of the method used in Sec. 11. We shall denote[1] the incoming fields, or in-fields, by $A_\mu^{(\mathrm{ein})}(x)$ and $\psi^{(\mathrm{ein})}(x)$ and the outgoing fields, or out-fields by $A_\mu^{(\mathrm{aus})}(x)$ and $\psi^{(\mathrm{aus})}(x)$. First of all, we know that these quantities obey the same canonical commutation relations. Ac-

1. (Translator's Note) The usual notation in English for the in-field is $A^{(\mathrm{in})}$ and not $A^{(\mathrm{ein})}$. We have retained Källén's notation only in order to avoid resetting all the equations. Similar remarks apply to $A^{(\mathrm{out})} = A^{(\mathrm{aus})}$.

cording to well-known theorems,[1] there must exist a matrix S with the following properties:

$$\psi^{(\mathrm{aus})}(x) = S^{-1}\psi^{(\mathrm{ein})}(x)\, S\,, \tag{19.1}$$

$$A_\mu^{(\mathrm{aus})}(x) = S^{-1} A_\mu^{(\mathrm{ein})}(x)\, S\,, \tag{19.2}$$

$$S\,S^* = S^*\,S = 1\,. \tag{19.3}$$

By means of this matrix we can express the Hamiltonian for $x_0 \to +\infty$,

$$H^{(0)}(A^{(\mathrm{aus})}, \psi^{(\mathrm{aus})}),$$

as a function of the Hamiltonian for $x_0 \to -\infty$:

$$H^{(0)}(A^{(\mathrm{aus})}, \psi^{(\mathrm{aus})}) = S^{-1} H^{(0)}(A^{(\mathrm{ein})}, \psi^{(\mathrm{ein})})\, S\,. \tag{19.4}$$

Introducing the eigenvectors of the operator $H^{(0)}(A^{(\mathrm{ein})}, \psi^{(\mathrm{ein})})$ and denoting them simply by $|n\rangle$, we find from (19.4),

$$H^{(0)}(A^{(\mathrm{ein})}, \psi^{(\mathrm{ein})})\,|n\rangle = E_n\,|n\rangle\,, \tag{19.5}$$

$$H^{(0)}(A^{(\mathrm{aus})}, \psi^{(\mathrm{aus})})\, S^{-1}|n\rangle = E_n\, S^{-1}|n\rangle\,. \tag{19.6}$$

The states $S^{-1}|n\rangle$ are therefore eigenstates of the energy for very large times and consequently the probability that a state $|n\rangle$ makes the transition to a state $|n'\rangle$ is given by

$$w_{n n'} = |\langle n'|\, S\, |n\rangle|^2\,. \tag{19.7}$$

The general problem of finding the outgoing particles, given those which are incoming, is therefore solved if the matrix S is known.[2]

Equation (19.3) says nothing about the general structure of the matrix S except for its unitarity. It is quite clear that S has to be the unit matrix if $e=0$, and that even if e is non-zero, the matrix elements of S can be non-zero only if the states $|n\rangle$ and $|n'\rangle$ have the same total energy and total momentum. We therefore expect that S has the general form

$$\langle n'|S|n\rangle = \delta_{n'n} + \langle n'|R|n\rangle\, \delta(p'-p)\,. \tag{19.8}$$

1. See, for example, P. A. M. Dirac, The Principles of Quantum Mechanics, Third Ed., Oxford, 1947, p. 106. Here it is explicitly assumed that the states $|n\rangle$ form a complete system which is equivalent to the assumption that no bound states exist. See also A. S. Wightman and S. S. Schweber, Phys. Rev. 98, 812 (1955), especially p. 825.

2. The S-matrix was originally introduced into quantum theory by W. Heisenberg, Z. Physik 120, 513 (1943). The theory of the S-matrix in quantum electrodynamics has been developed, for example, by F. J. Dyson, Phys. Rev. 75, 486, 1736 (1949), as well as by C. N. Yang and D. Feldman, Phys. Rev. 79, 972 (1950).

Here $\langle n'|R|n\rangle$ is a non-singular function of the energy-momentum vectors of the particles considered, and p' and p are the total energy-momentum vectors of the states $|n'\rangle$ and $|n\rangle$. In fact, when we explicitly calculate the S-matrix below, we shall find that these conjectures, implicit in (19.8), are confirmed. If the expression (19.8) is squared in order to obtain the physically interesting transition probabilities (19.7), the delta function enters quadratically: a completely meaningless result! We have to go back a step in our calculations momentarily and recall that with a (finite) periodic boundary condition, the spatial delta function is replaced by the symbol $\delta_{\boldsymbol{p}\boldsymbol{p}'}$ and this quantity can be squared without difficulty. If $\alpha \neq 0$ we do not obtain an exact delta function for the energy in (19.8), but rather expressions of the form

$$\frac{1}{\pi}\frac{\alpha}{\alpha^2+(p_0'-p_0)^2}. \tag{19.9}$$

The expression (19.9) can be squared without difficulty, but the integral

$$\lim_{\alpha\to 0}\frac{1}{\pi^2}\int\frac{F(x)\,\alpha^2\,dx}{(\alpha^2+x^2)^2}, \tag{19.10}$$

where $F(x)$ is a regular function, does not exist. Simple considerations show that the integral

$$\lim_{\alpha\to 0}\frac{1}{\pi^2}\int\frac{F(x)\,\alpha^3}{(\alpha^2+x^2)^2}\,dx=\frac{1}{2\pi}F(0) \tag{19.11}$$

does exist. We can therefore write symbolically

$$w_{nn'}=|\langle n'|R|n\rangle|^2\,\delta_{\boldsymbol{p}\boldsymbol{p}'}\,\frac{\delta(p_0'-p_0)}{2\pi}\,\frac{1}{\alpha}. \tag{19.12}$$

If α goes to zero the transition probability becomes very large. In ordinary quantum mechanics, in treating collision problems, the quantity of interest is not the total transition probability but the transition rate. Recalling that the particles have been interacting with each other for a time of the order of $1/\alpha$, we see from (19.12) that in quantum electrodynamics also, the transition probability per unit time has a finite limit if α goes to zero. As we shall see in Sec. 20, we can write

$$\langle n'|S|n\rangle=\delta_{n'n}+\langle n'|R|n\rangle\,\delta_{\boldsymbol{p}'\boldsymbol{p}}\,\delta(p_0'-p_0), \tag{19.13}$$

rather than (19.8), and obtain for these quantities

$$\frac{\partial w_{nn'}}{\partial t}=|\langle n'|R|n\rangle|^2\,\delta_{\boldsymbol{p}'\boldsymbol{p}}\,\frac{\delta(p_0'-p_0)}{2\pi}. \tag{19.14}$$

From the quantities (19.14), the interaction cross sections, etc., can be obtained in well-known ways. We shall not give general formulas here, because they cannot be written without complicated notation. In later sections we shall often compute particular interaction cross sections from the S-matrix.

In principle, we can determine the S-matrix from the equations

$$(\psi^{(aus)}(x) =)\ S^{-1}\psi^{(0)}(x)\,S = \psi^{(0)}(x) + \int S(x-x')\,ie\gamma A(x')\psi(x')\,dx', \quad (19.15)$$

$$(A_\mu^{(aus)}(x) =)\ S^{-1}A_\mu^{(0)}(x)\,S = A_\mu^{(0)}(x) - \int D(x-x')\frac{ie}{2}[\overline{\psi}(x'),\gamma_\mu\psi(x')]\,dx', \quad (19.16)$$

or

$$[S, \psi^{(0)}(x)] = -S\int S(x-x')\,ie\gamma A(x')\psi(x')\,dx', \quad (19.17)$$

$$[S, A_\mu^{(0)}(x)] = S\int D(x-x')\frac{ie}{2}[\overline{\psi}(x'),\gamma_\mu\psi(x')]\,dx'. \quad (19.18)$$

In (19.15) through (19.18) we have again used the earlier notation $A_\mu^{(0)}(x)$ and $\psi^{(0)}(x)$ for the in-fields. We now develop the matrix S in a power series in e, analogous to those for the field operators:

$$S = 1 + e\,S^{(1)} + \cdots. \quad (19.19)$$

In the first approximation, from (19.17) and (19.18), we find

$$[S^{(1)}, \psi^{(0)}(x)] = -i\int dx'\,S(x-x')\,\gamma A^{(0)}(x')\psi^{(0)}(x'), \quad (19.20)$$

$$[S^{(1)}, A_\mu^{(0)}(x)] = \frac{i}{2}\int dx'\,D(x-x')\,[\overline{\psi}^{(0)}(x'), \gamma_\mu\psi^{(0)}(x')]. \quad (19.21)$$

Since the operator $A_\mu^{(0)}(x)$ commutes with $\psi^{(0)}(x)$ and $\overline{\psi}^{(0)}(x)$, it follows from (19.21) that the first approximation $S^{(1)}$ to the S-matrix must be of the form

$$S^{(1)} = -\tfrac{1}{2}\int dx'\,[\overline{\psi}^{(0)}(x'), \gamma_\mu\psi^{(0)}(x')]\,A_\mu^{(0)}(x') + s^{(1)}. \quad (19.22)$$

In (19.22) the term $s^{(1)}$ is independent of $A_\mu^{(0)}(x)$. From (19.20) we find that $s^{(1)}$ is also independent of the operators for the Dirac field and consequently is a c-number. This c-number obviously cannot be determined from the commutation relations (19.17) and (19.18). From the unitarity of the S-matrix, it follows that $S^{(1)} = -S^{(1)*}$; that is, the number $s^{(1)}$ must be pure imaginary. Without altering the properties (19.1) through (19.3), we can always multiply the S-matrix by a factor $e^{i\delta}$, where the phase δ is a real number. We can therefore set $s^{(1)}$ in (19.22) equal to zero by definition; this only fixes the arbitrary quantity δ. Thus we have as the first approximation to the S-matrix the result

$$S^{(1)} = -\tfrac{1}{2}\int dx'\,[\overline{\psi}^{(0)}(x'), \gamma_\mu\psi^{(0)}(x')]\,A_\mu^{(0)}(x'). \quad (19.23)$$

In a similar way it can be shown that the next approximation is given by

$$\left.\begin{aligned} S^{(2)} = \tfrac{1}{4}\int_{-\infty}^{+\infty} dx' \int_{-\infty}^{x'} dx''\,[\overline{\psi}^{(0)}(x'), \gamma_{\nu_1}\psi^{(0)}(x')]\,A_{\nu_1}^{(0)}(x')\,\times \\ \times\,[\overline{\psi}^{(0)}(x''), \gamma_{\nu_2}\psi^{(0)}(x'')]\,A_{\nu_2}^{(0)}(x''). \end{aligned}\right\} \quad (19.24)$$

In (19.24) an arbitrary c-number has been set equal to zero, as was done before. The method indicated here leads to rather complicated calculations if the higher orders in the S-matrix are considered. Despite this, the results are remarkably simple, and it is to be expected that there is a more direct method to calculate this matrix.

20. Treatment of Quantum Electrodynamics by Means of a Time-Dependent Canonical Transformation

In this section we shall develop another method of solving the differential equations of quantum electrodynamics. This "new" method is the one which has played the greatest role in the development of the modern theories. It is closely connected with the interaction picture given in Sec. 2. From the fact that the in- and out-fields satisfied the same canonical commutation relations, we concluded in Sec. 19 that there must exist a matrix S with the properties (19.1) through (19.3). Actually, for an arbitrary time x_0 the field operators satisfy the same commutation relations as the in-fields, and from this we can conclude that there is a time-dependent matrix $U(x_0)$ which has the following properties:

$$\psi(x) = U^{-1}(x_0)\,\psi^{(0)}(x)\,U(x_0)\,, \tag{20.1}$$

$$A_\mu(x) = U^{-1}(x_0)\,A_\mu^{(0)}(x)\,U(x_0), \tag{20.2}$$

$$U^*(x_0)\,U(x_0) = U(x_0)\,U^*(x_0) = 1\,. \tag{20.3}$$

To determine the matrix $U(x_0)$ we have the two differential equations

$$\left.\begin{aligned}\left(\gamma\frac{\partial}{\partial x}+m\right)\psi(x) &= -i\left[-U^{-1}\frac{\partial U}{\partial x_0}U^{-1}\gamma_4\psi^{(0)}U + U^{-1}\gamma_4\psi^{(0)}\frac{\partial U}{\partial x_0}\right] = \\ &= i\,U^{-1}\left[\frac{\partial U}{\partial x_0}U^{-1},\ \gamma_4\psi^{(0)}\right]U = i\,e\,U^{-1}\gamma\,A^{(0)}\psi^{(0)}\,U,\end{aligned}\right\} \tag{20.4}$$

or

$$\left[\frac{\partial U}{\partial x_0}U^{-1},\ \psi^{(0)}\right] = e\,\gamma_4\gamma\,A^{(0)}\psi^{(0)}, \tag{20.5}$$

and

$$\left.\begin{aligned}\Box A_\mu(x) = U^{-1}\Bigg\{&\left[\frac{\partial}{\partial x_0}\left(\frac{\partial U}{\partial x_0}U^{-1}\right),\ A_\mu^{(0)}\right] + 2\left[\frac{\partial U}{\partial x_0}U^{-1},\ \frac{\partial A_\mu^{(0)}}{\partial x_0}\right] + \\ &+\left[\left[\frac{\partial U}{\partial x_0}U^{-1},\ A_\mu^{(0)}\right],\frac{\partial U}{\partial x_0}U^{-1}\right]\Bigg\}U = -\frac{ie}{2}U^{-1}[\overline{\psi}^{(0)},\gamma_\mu\psi^{(0)}]\,U,\end{aligned}\right\} \tag{20.6}$$

or

$$\left.\begin{aligned}&\left[\frac{\partial}{\partial x_0}\left(\frac{\partial U}{\partial x_0}U^{-1}\right),\ A_\mu^{(0)}\right] + 2\left[\frac{\partial U}{\partial x_0}U^{-1},\frac{\partial A_\mu^{(0)}}{\partial x_0}\right] + \left[\left[\frac{\partial U}{\partial x_0}U^{-1},\ A_\mu^{(0)}\right],\frac{\partial U}{\partial x_0}U^{-1}\right] = \\ &\qquad = -\frac{ie}{2}[\overline{\psi}^{(0)},\gamma_\mu\psi^{(0)}]\,.\end{aligned}\right\} \tag{20.7}$$

From (20.5) and (20.7) it follows that $\frac{\partial U}{\partial x_0}U^{-1}$ must be a function of $A_\mu^{(0)}(x)$ and $\psi^{(0)}(x)$. Moreover, from (20.5) we have that $\frac{\partial U}{\partial x_0}U^{-1}$ must contain a term of the form

$$-\frac{e}{2}\int_{x_0'=x_0} d^3x' \,[\overline{\psi}^{(0)}(x'), \gamma_\mu \psi^{(0)}(x')]\, A_\mu^{(0)}(x'), \tag{20.8}$$

and we readily verify that this term is sufficient to satisfy (20.7) also. The matrix $U(x_0)$ therefore satisfies the differential equation

$$i\frac{\partial U(x_0)}{\partial x_0} = H^{(1)}\left(A^{(0)}(x), \psi^{(0)}(x)\right) U(x_0), \tag{20.9}$$

with the boundary condition

$$U(-\infty) = 1. \tag{20.10}$$

As we did before for the S-matrix, we have set an arbitrary c-number equal to zero in Eq. (20.9). If x_0 becomes large, we clearly have

$$U(+\infty) = S. \tag{20.11}$$

In order that U tend to a well-defined limit for $|x_0| \to \infty$, the charge must be adiabatically switched on and off, as was done previously. With these assumptions, (20.9) and (20.10) can be incorporated into the integral equation

$$U(x_0) = 1 - i\int_{-\infty}^{x_0} dx_0' \, H^{(1)}\left(A^{(0)}(x'), \psi^{(0)}(x')\right) U(x_0'). \tag{20.12}$$

The canonical transformations (20.1) through (20.3) can be regarded as the transition from the Heisenberg picture, $A_\mu(x)$, $\psi(x)$ to another picture, $A_\mu^{(0)}(x)$, $\psi^{(0)}(x)$. In the new picture the operators are known, but the state vectors which are given by $U(x_0)|n\rangle$ are not simple and must be found by solving the "Schroedinger equation" (20.12) or (20.9), (20.10). Here we are using $|n\rangle$ for the state vector in the Heisenberg picture. This new picture is identical with the interaction picture discussed previously, and Eq. (20.9) agrees with Eq. (2.11).

The integral equation (20.12) can be solved formally by a power series:

$$\left.\begin{aligned} U(x_0) &= 1 - i\int_{-\infty}^{x_0} H^{(1)}(x_0')\,dx_0' - \int_{-\infty}^{x_0} dx_0' \int_{-\infty}^{x_0'} dx_0''\, H^{(1)}(x_0') H^{(1)}(x_0'') + \cdots = \\ &= \sum_{n=0}^{\infty} (-i)^n \int_{-\infty}^{x_0} dx_0' \int_{-\infty}^{x_0'} dx_0'' \cdots \int_{-\infty}^{x_0^{(n-1)}} dx_0^{(n)}\, H^{(1)}(x_0') \ldots H^{(1)}(x_0^{(n)}). \end{aligned}\right\} \tag{20.13}$$

Here we have written $H^{(1)}(x_0)$ as a shorthand for $H^{(1)}\left(A^{(0)}(x), \psi^{(0)}(x)\right)$. Clearly, the solution (20.13) can, in principle, be used to calculate the field operators $A_\mu(x)$, $\psi(x)$ as functions of the in-fields. After some complicated transformations, which we shall not go into here,[1] it can be shown that the Heisenberg operators are given by the expressions

1. See, for example, F. J. Dyson, Phys. Rev. 75, 486 (1949).

$$\left.\begin{aligned} A_\mu(x) = A_\mu^{(0)}(x) + \sum_{n=1}^{\infty} i^n \int_{-\infty}^{x_0} dx_0' \ldots \int_{-\infty}^{x_0^{(n-1)}} dx_0^{(n)} \times \\ \times \left[H^{(1)}(x_0^{(n)}), \left[\ldots \left[H^{(1)}(x_0'), A_\mu^{(0)}(x)\right] \ldots\right]\right], \end{aligned}\right\} \quad (20.14a)$$

$$\left.\begin{aligned} \psi(x) = \psi^{(0)}(x) + \sum_{n=1}^{\infty} i^n \int_{-\infty}^{x_0} dx_0' \ldots \int_{-\infty}^{x_0^{(n-1)}} dx_0^{(n)} \times \\ \times \left[H^{(1)}(x_0^{(n)}), \left[\ldots \left[H^{(1)}(x_0'), \psi^{(0)}(x)\right] \ldots\right]\right]. \end{aligned}\right\} \quad (20.14b)$$

These formulas are not especially suited to practical calculations and several rather complicated transformations are necessary before the simple formulas (18.5) through (18.12) result.[1] For the calculation of the field operators, the method of the interaction picture is not especially suitable. The situation is quite different for the S-matrix. Here the series (20.13) essentially contains the result. It is

$$S = 1 + \sum_{n=1}^{\infty} (-i)^n \int_{-\infty}^{+\infty} dx_0' \int_{-\infty}^{x_0'} dx_0'' \ldots \int_{-\infty}^{x_0^{(n-1)}} dx^{(n)} H^{(1)}(x_0') \ldots H^{(1)}(x_0^{(n)}). \quad (20.15)$$

For what follows, it is useful to transform the expression (20.15) somewhat.[2] If $F(x_1 \ldots x_n)$ is symmetric in all variables, it is evident that

$$\int_a^b dx_1 \int_a^{x_1} dx_2 \ldots \int_a^{x_{n-1}} dx_n F(x_1 \ldots x_n) = \frac{1}{n!} \int_a^b dx_1 \ldots \int_a^b dx_n F(x_1 \ldots x_n). \quad (20.16)$$

The product of the interaction operators in (20.15) is certainly not symmetric in the time variables, because these do not commute with each other for different times. Despite this, we can transform the region of integration according to (20.16) if we require that an operator with a greater time always stands to the left of an operator with a smaller time. To do this, we introduce a "time ordering" or simply a "P-symbol" in the following way:

$$P(A(x_0)\, B(x_0')) = \begin{cases} A(x_0)\, B(x_0') & \text{for} \quad x_0 > x_0' \\ B(x_0')\, A(x_0) & \text{for} \quad x_0' > x_0. \end{cases} \quad (20.17)$$

The generalization of the P-symbol to several factors is obvious. The S-matrix (20.15) can now be written as an integral from $-\infty$ to $+\infty$:

$$S = 1 + \sum_{n=1}^{\infty} \frac{(-i)^n}{n!} \int dx_0' \ldots \int dx_0^{(n)}\, P(H^{(1)}(x_0') \ldots H^{(1)}(x_0^{(n)})). \quad (20.18)$$

The expressions (19.23) and (19.24) found earlier for the first ap-

1. See, for example, J. Schwinger, Phys. Rev. 74, 1439 (1948), 75, 651 (1949); or G. Källén, Helv. Phys. Acta 22, 637 (1949).
2. See footnote 1, p. 90.

proximations to the S-matrix are clearly special cases of (20.15). Obviously one can verify[1] directly that the formulas (20.14) and (20.18) satisfy the differential equations (17.14), (17.15), and the relations (19.17), (19.18).

In conclusion, we shall give a proof of the relation (19.14) by means of the methods developed here. We first remark that the series (20.13) contains a factor of the form

$$\langle n' | U(x_0) | n \rangle = \langle n' | R | n \rangle \delta_{\mathbf{p}\mathbf{p}'} \frac{1}{2\pi} \int_{-\infty}^{x_0} e^{i(p_0'-p_0)x_0'} dx_0' \qquad (20.19)$$

in every order of approximation to $U(x_0)$. We shall therefore suppose that all matrix elements of the complete operator U have the form (20.19). If we let x_0 become very large in this equation, we find

$$\langle n' | S | n \rangle = \langle n' | R | n \rangle \delta_{\mathbf{p}\mathbf{p}'} \delta(p_0' - p_0), \quad (|n'\rangle \neq |n\rangle), \qquad (20.20)$$

where the matrix R in (20.20) is to be identified with the matrix R in (19.13). For finite x_0 the probability that the system be in state $|n'\rangle$ at time x_0 is

$$|\langle n' | U(x_0) | n \rangle|^2 = \frac{|\langle n' | R | n \rangle|^2}{4\pi^2} \delta_{\mathbf{p}\mathbf{p}'} \int_{-\infty}^{x_0} dx_0' \int_{-\infty}^{x_0} dx_0'' \, e^{i(p_0'-p_0)(x_0'-x_0'')}. \qquad (20.21)$$

The time derivative of the quantity (20.21) is the desired transition probability per unit time:

$$\left.\begin{aligned} \frac{d}{dx_0} |\langle n' | U(x_0) | n \rangle|^2 &= \frac{|\langle n' | R | n \rangle|^2}{4\pi^2} \delta_{\mathbf{p}\mathbf{p}'} \left[\int_{-\infty}^{x_0} dx_0' \, e^{i(p_0'-p_0)(x_0'-x_0)} + \right. \\ &\left. + \int_{-\infty}^{x_0} dx_0'' \, e^{i(p_0'-p_0)(x_0-x_0'')} \right] = \frac{|\langle n' | R | n \rangle|^2}{4\pi^2} \delta_{\mathbf{p}\mathbf{p}'} \int_{-\infty}^{+\infty} dx_0' \, e^{i(p_0'-p_0)(x_0'-x_0)} = \\ &= \frac{|\langle n' | R | n \rangle|^2}{2\pi} \delta_{\mathbf{p}\mathbf{p}'} \delta(p_0' - p_0). \end{aligned}\right\} \qquad (20.22)$$

The result (20.22) is identical with Eq. (19.14) and is time-independent.

At first sight the time independence of the quantity (20.22) seems surprising. Physically one might expect that the transition probability would slowly fall to zero for increasing time, because of the decrease of the probability of the state $|n\rangle$ due to the decay. Our failure to obtain such a result is clearly dependent on the use of the ansatz (20.19) for the matrix U. Actually (20.19) is only correct up to terms which become very small if V goes to infinity. In many applications of (20.22), the sum of all transition probabilities from a given state $|n\rangle$ is usually proportional to V^{-1}. Thus, the probability that the state $|n\rangle$ is present, even after the collision,

---

1. G. Källén, Ark Fysik 2, 187, 371 (1950).

is practically equal to 1. Physically then, it is clear that the probability that a finite number of particles collide must be proportional to V^{-1} if they are "released" in a very large volume V. Only with this requirement does the concept of an interaction cross section make sense. In this way we can understand why (20.22) is formally time-independent. It is now clear that this method can encounter difficulties for very large times. In particular, if the transition rate is time-independent, then for very large times the total transition probability may be so large that we should not employ our formal solution (20.19). In these circumstances, evidently we shall have to include those terms which were omitted in the perturbation series because of their dependence on V. In particular, only if the interaction time α^{-1} can be chosen so brief that $\frac{1}{\alpha V}$ is very small is (19.12) correct. The limit $\alpha \to 0$ which would give an infinite transition probability is not actually allowed in (19.12). For problems in which there are also bound states present it can happen that the simultaneous limit $\alpha \to 0$ and $V \to \infty$ is impossible. The method of calculating the S-matrix developed here cannot be applied without some changes. In Sec. 28 we shall study a simple example of such a case and use a slightly modified method of integration. In those cases where we obtain sensible interaction cross sections by means of (20.22), we shall use the S-matrix and the concomitant restrictions without hesitation.

21. Calculation of the P-Symbol. The Normal Product

Our next task is to calculate the P-symbol in (20.18), or to calculate matrix elements of the form

$$\left.\begin{aligned}\langle n'|S^{(n)}|n\rangle = \frac{(-e)^n}{2^n \cdot n!}\int dx' \ldots \int dx^{(n)} \langle n'|\,P([\overline{\psi}^{(0)}(x'),\gamma_{\nu_1}\psi^{(0)}(x')]\,A^{(0)}_{\nu_1}(x')\ldots\\ \ldots[\overline{\psi}^{(0)}(x^{(n)}),\gamma_{\nu_n}\psi^{(0)}(x^{(n)})]\,A^{(0)}_{\nu_n}(x^{(n)}))\,|n\rangle\,.\end{aligned}\right\} \quad (21.1)$$

Here the states $|n\rangle$ and $|n'\rangle$ are eigenstates of the Hamiltonian for the incoming particles, i.e., of $H^{(0)}(A^{(0)},\psi^{(0)})$. We can consider these states to be characterized by a given number of electrons and photons and write them as products of the form

$$|n\rangle = |n_k\rangle\,|n_q\rangle. \quad (21.2)$$

Here $|n_k\rangle$ is an eigenvector of $H_1^{(0)}(A^{(0)}_\mu)$ and describes the photons which are present, while $|n_q\rangle$ is an eigenvector of $H_2^{(0)}(\psi^{(0)})$ and describes the electrons [c.f. Eqs. (17.20) and (17.21)]. Because the operators $A^{(0)}_\mu$ and $\psi^{(0)}$ commute with each other, we can write the matrix element (21.1) as a product of two factors:

$$\left.\begin{aligned}\langle n'|S^{(n)}|n\rangle = \frac{(-e)^n}{n!\,2^n}\int dx' \ldots \int dx^{(n)} \langle n_q'|\,P([\overline{\psi}^{(0)}(x'),\gamma_{\nu_1}\psi^{(0)}(x')]\ldots\\ \ldots[\overline{\psi}^{(0)}(x^{(n)}),\gamma_{\nu_n}\psi^{(0)}(x^{(n)})])\,|n_q\rangle\langle n_k'|\,P(A^{(0)}_{\nu_1}(x')\ldots A^{(0)}_{\nu_n}(x^{(n)}))\,|n_k\rangle.\end{aligned}\right\} \quad (21.3)$$

The individual P-symbols appearing in (21.3) can be worked out by the same method which was used earlier in the proof of (18.21). In order to make the procedure more systematic, we shall follow Wick[1] and introduce the so-called "normal product" of the free field operators. According to (5.28), we can write the electromagnetic field as a sum of two parts:

$$A_\mu^{(0)}(x) = A_\mu^{(+)}(x) + A_\mu^{(-)}(x). \tag{21.4}$$

Here the operator $A_\mu^{(+)}(x)$ contains only the annihilation operators:

$$A_\mu^{(+)}(x) = \frac{1}{\sqrt{V}} \sum_{\boldsymbol{k}, \lambda} \frac{e_\mu^{(\lambda)}}{\sqrt{2\omega}} e^{ikx} a^{(\lambda)}(\boldsymbol{k}), \tag{21.5}$$

and the operator $A_\mu^{(-)}(x)$ contains only the creation operators:

$$A_\mu^{(-)}(x) = \frac{1}{\sqrt{V}} \sum_{\boldsymbol{k}, \lambda} \frac{e_\mu^{(\lambda)}}{\sqrt{2\omega}} e^{-ikx} a^{*(\lambda)}(\boldsymbol{k}). \tag{21.6}$$

These quantities satisfy the commutation relations

$$[A_\mu^{(+)}(x), A_\nu^{(+)}(x')] = [A_\mu^{(-)}(x), A_\nu^{(-)}(x')] = 0, \tag{21.7}$$

$$[A_\mu^{(+)}(x), A_\nu^{(-)}(x')] = -i\,\delta_{\mu\nu} D^{(-)}(x'-x) = \langle 0 | A_\mu^{(0)}(x) A_\nu^{(0)}(x') | 0 \rangle. \tag{21.8}$$

In a similar way we write the Dirac field operators as

$$\psi^{(0)}(x) = \psi^{(+)}(x) + \psi^{(-)}(x), \tag{21.9}$$

$$\overline{\psi}^{(0)}(x) = \overline{\psi}^{(+)}(x) + \overline{\psi}^{(-)}(x), \tag{21.10}$$

where the "+ operators" contain only annihilation operators and the "− operators" contain only creation operators. In place of (21.7) and (21.8), we have the commutation relations

$$\{\overline{\psi}^{(+)}(x), \psi^{(-)}(x')\} = -i\,S^{(-)}(x'-x) = \langle 0 | \overline{\psi}^{(0)}(x) \psi^{(0)}(x') | 0 \rangle, \tag{21.11}$$

$$\{\overline{\psi}^{(-)}(x), \psi^{(+)}(x')\} = -i\,S^{(+)}(x'-x) = \langle 0 | \psi^{(0)}(x') \overline{\psi}^{(0)}(x) | 0 \rangle, \tag{21.12}$$

$$\{\psi^{(+)}(x), \psi^{(+)}(x')\} = \{\overline{\psi}^{(+)}(x), \psi^{(+)}(x')\} = \{\psi^{(+)}(x), \psi^{(-)}(x')\} = \cdots = 0. \tag{21.13}$$

Now we define the normal product of two or more factors as an operator product where the <u>creation operators always stand to the left of the annihilation operators</u>. We denote the normal product by : . . . : and we abbreviate $A_{\nu_i}^{(0)}(x^{(i)})$ as $A(i)$. We then have the normal product of operators $A_\mu^{(0)}(x)$,

1. G. C. Wick, Phys. Rev. 80, 268 (1950). These products had already been used by A. Houriet and A. Kind, Helv. Phys. Acta, 22, 319 (1949). The name "normal product" was introduced by F. J. Dyson, Phys. Rev. 82, 428 (1951).

$$\left.\begin{aligned}&:A(1)\dots A(n):=A^{(+)}(1)\dots A^{(+)}(n)+\\&+\sum_{i=1}^{n}A^{(-)}(1)\,A^{(+)}(1)\dots A^{(+)}(i-1)\,A^{(+)}(i+1)\dots A^{(+)}(n)+\\&+\sum_{i<j}A^{(-)}(i)\,A^{(-)}(j)\,A^{(+)}(1)\dots A^{(+)}(n)+\dots+A^{(-)}(1)\dots A^{(-)}(n).\end{aligned}\right\}\tag{21.14}$$

In (21.14) the normal product is clearly completely symmetric in all of its variables. It is not necessary to state exactly the sequence of the $A^{(+)}$ (or $A^{(-)}$) operators with each other since, according to (21.7), they commute. Such a statement is necessary for the ψ-operators, however, because in (21.11) through (21.13) the anticommutators are present. We now give this normal product a well-defined meaning by the conventions

$$\left.\begin{aligned}&:\varphi(1)\dots\varphi(n):=\varphi^{(+)}(1)\dots\varphi^{(+)}(n)+\sum_{i=1}^{n}\delta_P\,\varphi^{(-)}(i)\,\varphi^{(+)}(1)\dots\varphi^{(+)}(n)+\\&+\sum_{i<j}\delta_P\,\varphi^{(-)}(i)\,\varphi^{(-)}(j)\,\varphi^{(+)}(1)\dots\varphi^{(+)}(n)+\dots+\varphi^{(-)}(1)\dots\varphi^{(-)}(n).\end{aligned}\right\}\tag{21.15}$$

In (21.15) $\varphi(i)$ stands for a factor which can be either $\psi^{(0)}(x^{(i)})$ or $\overline{\psi}^{(0)}(x^{(i)})$. The symbol δ_P is +1 if the numbers $(i, j, \dots)$ are an even permutation of the original numbers (1, 2, 3, ...) and −1 if they are an odd permutation. With this definition, the following symmetry condition obviously holds:

$$:\varphi(1)\dots\varphi(n):=\delta_P:\varphi(i)\,\varphi(j)\dots:.\tag{21.16}$$

Thus in a normal product we can always calculate as if the $A_\mu^{(0)}(x)$ were commuting operators and the $\psi^{(0)}(x)$ and $\overline{\psi}^{(0)}(x)$ were anticommuting operators. To transform an ordinary product into a normal product, we have equations of the form

$$A(1)\,A(2)=:A(1)\,A(2):+\langle 0|A(1)\,A(2)|0\rangle,\tag{21.17}$$

$$\psi(1)\,\psi(2)=:\psi(1)\,\psi(2):\equiv:\psi(1)\,\psi(2):+\langle 0|\psi(1)\,\psi(2)|0\rangle,\tag{21.18}$$

$$\overline{\psi}(1)\,\overline{\psi}(2)=:\overline{\psi}(1)\,\overline{\psi}(2):\equiv:\overline{\psi}(1)\,\overline{\psi}(2):+\langle 0|\overline{\psi}(1)\,\overline{\psi}(2)|0\rangle,\tag{21.19}$$

$$\overline{\psi}(1)\,\psi(2)=:\overline{\psi}(1)\,\psi(2):+\langle 0|\overline{\psi}(1)\,\psi(2)|0\rangle,\tag{21.20}$$

$$\psi(1)\,\overline{\psi}(2)=:\psi(1)\,\overline{\psi}(2):+\langle 0|\psi(1)\,\overline{\psi}(2)|0\rangle.\tag{21.21}$$

Equations (21.17) through (21.21) can be proved directly from (21.4) through (21.16). As an example of the application of these equations, we shall write the free field current operator as a normal product:

$$\left.\begin{aligned}\tfrac{1}{2}[\overline{\psi}^{(0)}(x),\gamma_\mu\psi^{(0)}(x)]&=\tfrac{1}{2}\{:\overline{\psi}^{(0)}(x)\gamma_\mu\psi^{(0)}(x):+\langle 0|\overline{\psi}^{(0)}(x)\,\gamma_\mu\psi^{(0)}(x)|0\rangle-\\&\quad-:\psi^{(0)}(x)\,\gamma_\mu^T\overline{\psi}^{(0)}(x):-\langle 0|\psi^{(0)}(x)\,\gamma_\mu^T\overline{\psi}^{(0)}(x)|0\rangle\}=\\&=:\overline{\psi}^{(0)}(x)\,\gamma_\mu\psi^{(0)}(x):+\tfrac{1}{2}\langle 0|[\overline{\psi}^{(0)}(x),\gamma_\mu\psi^{(0)}(x)]|0\rangle=\\&=:\overline{\psi}^{(0)}(x)\,\gamma_\mu\psi^{(0)}(x):.\end{aligned}\right\}\tag{21.22}$$

This result shows clearly the vanishing of the vacuum expectation value of the current operator. It also illustrates the advantage of the normal product over the usual product: For a given transition, where a fixed set of particles is changed, there is one and only one kind of term, among all possible normal products, which contributes. These terms are just those having the correct annihilation and creation operators. This statement does not hold for the ordinary product where a creation and subsequent annihilation can take place in the intermediate states--as we have seen in Sec. 18. It is important to require that no particle in the final state is identical with one in the initial state. If this is not true, there are circumstances where several different normal products can give non-zero contributions to the result. As an example, consider the following matrix element:

$$\langle q, k | F | q, k' \rangle = A \langle k | : A^{(0)} A^{(0)} : | k' \rangle + B \langle q, k | : \overline{\psi}^{(0)} \psi^{(0)} A^{(0)} A^{(0)} : | q, k' \rangle . \tag{21.23}$$

In all such cases it turns out that the "complicated" terms contain additional factors of V^{-n} and therefore that they can be dropped if the volume V is made very large. Therefore we need consider only the first non-vanishing term in expressions such as (21.23). A similar simplification has already been used in (18.20a).

Our major problem is now the transformation of the P-symbol into a sum of terms each of which contains operators only as normal products. If we knew the solution of this problem, the calculation of matrix elements in (21.3) would be straightforward. To make the algebraic manipulations as transparent as possible, we shall introduce two more symbols. We define

$$T\left(\varphi(1)\, \varphi(2) \, \ldots \, \varphi(n)\right) = \delta_P \, \varphi(i) \, \varphi(j) \, \ldots . \tag{21.24}$$

Here the factors on the left side are chronologically ordered, just as in the P-symbol. The two symbols differ by the factor δ_P, which is defined as in (21.15). Moreover, we shall always take this factor to be +1 if the operators $\varphi(x)$ are $A_\mu^{(0)}(x)$. With this convention, the following equations can be used for (21.14) as well as (21.15). We also define

$$: \overset{\vee}{\varphi}(1) \, \overset{\vee}{\varphi}(2) \, \varphi(3) \ldots \varphi(n) : = \langle 0 | \varphi(1) \, \varphi(2) | 0 \rangle : \varphi(3) \ldots \varphi(n) : , \tag{21.25}$$

$$: \varphi(1) \ldots \overset{\vee}{\varphi}(i) \ldots \overset{\vee}{\varphi}(j) \ldots \varphi(n) : = \delta_P : \overset{\vee}{\varphi}(i) \, \overset{\vee}{\varphi}(j) \, \varphi(1) \ldots \varphi(n) : . \tag{21.26}$$

From (21.17) through (21.21), we prove by induction (from n to n +1) that

$$\varphi(1) : \varphi(2) \ldots \varphi(n) : = : \varphi(1) \, \varphi(2) \ldots \varphi(n) : + \sum_{i=2}^{n} : \overset{\vee}{\varphi}(1) \, \varphi(2) \ldots \overset{\vee}{\varphi}(i) \ldots \varphi(n) : , \tag{21.27}$$

and from this (also by induction from n to n+1)

$$\left.\begin{aligned}&\varphi(1)\,\varphi(2)\ldots\varphi(n)=:\varphi(1)\ldots\varphi(n):+\\&+\sum_{i<j}:\varphi(1)\,\varphi(2)\ldots\overset{\mathrm{v}}{\varphi}(i)\ldots\overset{\mathrm{v}}{\varphi}(j)\ldots\varphi(n):+\\&+\sum_{i_1<j_1;\,i_2<j_2}:\varphi(1)\ldots\overset{\mathrm{v}}{\varphi}(i_1)\ldots\overset{\mathrm{w}}{\varphi}(i_2)\ldots\overset{\mathrm{v}}{\varphi}(j_1)\ldots\overset{\mathrm{w}}{\varphi}(j_2)\ldots\varphi(n):+\cdots.\end{aligned}\right\}\quad(21.28)$$

According to Eqs. (21.26) and (21.16), in the "contracted" normal products the sequence of the factors can be changed only if a factor δ_P is added. Care must be taken that two factors which are contracted with each other are never interchanged. For example, we have

$$:\varphi(1)\ldots\overset{\mathrm{v}}{\varphi}(i)\,\varphi(j)\,\overset{\mathrm{v}}{\varphi}(k)\ldots\varphi(n):=-:\varphi(1)\ldots\overset{\mathrm{v}}{\varphi}(i)\,\overset{\mathrm{v}}{\varphi}(k)\,\varphi(j)\ldots\varphi(n):\quad(21.29)$$

but

$$:\varphi(1)\ldots\overset{\mathrm{v}}{\varphi}(i)\,\overset{\mathrm{v}}{\varphi}(j)\ldots\varphi(n):\neq-:\varphi(1)\ldots\overset{\mathrm{v}}{\varphi}(j)\,\overset{\mathrm{v}}{\varphi}(i)\ldots\varphi(n):.\quad(21.30)$$

[For the operators $A_\mu^{(0)}(x)$ the sign factor is not present in (21.29).] Because two contracted factors on the right side of (21.28) will then always have the same order as they do on the left side, it follows from (21.24), (21.28) and the symmetry condition (21.16) that

$$\left.\begin{aligned}&T(\varphi(1)\ldots\varphi(n))=:\varphi(1)\ldots\varphi(n):+\sum_{i<j}:\varphi(1)\ldots\dot{\varphi}(i)\ldots\dot{\varphi}(j)\ldots\varphi(n):+\\&\quad+\sum_{i_1<j_1;\,i_2<j_2}:\varphi(1)\ldots\dot{\varphi}(i_1)\ldots\ddot{\varphi}(i_2)\ldots\dot{\varphi}(j_1)\ldots\ddot{\varphi}(j_2)\ldots\varphi(n):+\cdots.\end{aligned}\right\}\quad(21.31)$$

In (21.31) the new contractions are obviously defined by

$$:\dot{\varphi}(1)\,\dot{\varphi}(2)\,\varphi(3)\ldots\varphi(n):=\langle 0|\,T(\varphi(1)\,\varphi(2))\,|0\rangle:\varphi(3)\ldots\varphi(n):,\quad(21.32)$$

$$:\varphi(1)\ldots\dot{\varphi}(i)\ldots\dot{\varphi}(j)\ldots\varphi(n):=\delta_P:\dot{\varphi}(i)\,\dot{\varphi}(j)\,\varphi(1)\ldots\varphi(n):.\quad(21.33)$$

The vacuum expectation value of the T-product can be calculated immediately. The non-vanishing terms are

$$\left.\begin{aligned}\langle 0|\,T(A_\mu^{(0)}(x)\,A_\nu^{(0)}(x'))\,|0\rangle&=\langle 0|\,P(A_\mu^{(0)}(x)\,A_\nu^{(0)}(x'))\,|0\rangle=\\&=\tfrac{1}{2}\,\delta_{\mu\nu}\,[D^{(1)}(x'-x)-i\,\varepsilon(x-x')\,D(x'-x)]=\\&=\tfrac{1}{2}\,\delta_{\mu\nu}\,[D^{(1)}(x'-x)-2i\,\bar{D}(x'-x)]=\\&=\tfrac{1}{2}\,\delta_{\mu\nu}\,D_F(x'-x)\ ,\end{aligned}\right\}\quad(21.34)$$

Here the function $D_F(x)$ is that defined by (7.29) and

$$\langle 0|\,T(\bar{\psi}^{(0)}(x)\,\psi^{(0)}(x'))\,|0\rangle=\tfrac{1}{2}\,S_F(x'-x)\,,\quad(21.35)$$

$$\langle 0|\,T(\psi^{(0)}(x)\,\bar{\psi}^{(0)}(x'))\,|0\rangle=-\tfrac{1}{2}\,S_F(x-x')=-\langle 0|\,T(\bar{\psi}^{(0)}(x')\,\psi^{(0)}(x)\,|0\rangle.\quad(21.36)$$

Because of the antisymmetry in (21.36) [and the symmetry in (21.34)], we can change the order of two operators which are contracted with each other in (21.33). In particular, we find

$$:\varphi(1)\ldots\dot{\varphi}(i)\ldots\dot{\varphi}(j)\ldots\varphi(n): = \delta_P:\varphi(r)\ldots\dot{\varphi}(j)\ldots\dot{\varphi}(i)\ldots\varphi(s):. \quad (21.37)$$

Equations (21.31) through (21.36) contain the desired result for a T-product; it can now be written as a sum of normal products. The coefficients of the normal products can be constructed from the singular F-functions. Admittedly, in applications of (21.3) we shall have P-products, but always with the same number of operators $\psi^{(0)}(x)$ and $\overline{\psi}^{(0)}(x)$. It is clear that

$$P(:\overline{\psi}(1)\,\psi(1):\ldots:\overline{\psi}(n)\,\psi(n):) = T(:\overline{\psi}(1)\,\psi(1):\ldots:\overline{\psi}(n)\,\psi(n):). \quad (21.38)$$

For the mixed T-product in (21.38) one can readily show that there is an expansion analogous to (21.31). It must be remembered that two factors with the same time are not to be contracted with each other. Therefore we have, for example,

$$\left.\begin{aligned}P(:\overline{\psi}(1)\psi(1)::\overline{\psi}(2)\,\psi(2):) &= :\overline{\psi}(1)\,\psi(1)\,\overline{\psi}(2)\,\psi(2): + :\dot{\overline{\psi}}(1)\,\psi(1)\,\overline{\psi}(2)\,\dot{\psi}(2): +\\ &+ :\overline{\psi}(1)\,\dot{\psi}(1)\,\dot{\overline{\psi}}(2)\,\psi(2): + :\dot{\overline{\psi}}(1)\,\ddot{\psi}(1)\,\ddot{\overline{\psi}}(2)\,\dot{\psi}(2): =\\ &= :\overline{\psi}(1)\,\psi(1)\,\overline{\psi}(2)\,\psi(2): + \tfrac{1}{2}\,S_F(21)\,:\psi(1)\,\overline{\psi}(2): -\\ &- \tfrac{1}{2}\,S_F(12)\,:\overline{\psi}(1)\,\psi(2): - \tfrac{1}{4}\,S_F(21)\,S_F(12).\end{aligned}\right\} \quad (21.39)$$

With these rules of calculation, it is possible to write down immediately the expansion of an arbitrary P-symbol into normal products. All that is lacking is a general discussion of the omen hidden in (21.33) and (21.38). We shall forego this now, since we can continue the discussion much more simply in Sec. 22 by means of a method developed there.

22. A Graphical Representation of the S-Matrix

We now have all necessary means to compute any desired element of the S-matrix to arbitrary order. Even in rather simple problems the expressions which are obtained are very complicated and not easy to understand. There is an extraordinary method for arranging the results systematically which was originally discovered by Feynman.[1] In this method each normal product is represented by means of a graphical figure and there are rules according to which one can immediately draw these "graphs" and convert them into analytical terms. In his paper, Feynman originally introduced these graphs and rules of calculation quite intuitively, and only later were they established by means of the usual formalism.[2] We shall not explore these intuitive considerations further, but rather shall show that our analytical expressions can be represented by means of the graphs. Essentially, we follow the work

1. R. P. Feynman, Phys. Rev. 76, 749, 769 (1949).
2. F. J. Dyson, Phys. Rev. 75, 486, 1736 (1949). See also R. P. Feynman, Phys. Rev. 80, 440 (1950).

of Dyson.[1]

In order to represent a matrix element such as

$$\left.\begin{aligned}\frac{(-e)^n}{n!}\int dx' \ldots \int dx^n \langle n_q' | P\left(:\bar{\psi}(1)\,\gamma_{\nu_1}\psi(1): \ldots :\bar{\psi}(n)\,\gamma_{\nu_n}\psi(n):\right) | n_q\rangle \times \\ \times \langle n_k' | P\left(A(1) \ldots A(n)\right) | n_k\rangle\end{aligned}\right\} \quad (22.1)$$

graphically, we draw n points on the paper and associate each variable x^i with one point. Each variable x^i appears three times in each term in the expansion into normal products. From the operator $A(i)$ there results either a factor in a normal product or a contraction $D_F(ij)$ with another operator $A(j)$. If two variables are present in the same function D_F, we connect the corresponding points in our graph by a dashed line. If an operator $A(i)$ enters a normal product, we draw a dashed line from the point x^i to the edge of the figure. We proceed in a corresponding way for the operators $\psi(i)$ and $\bar{\psi}(j)$. If the two operators are contracted into a function S_F, we draw a solid, directed line from point x^j to point x^i. If these operators are in a normal product, we again draw the line from the point x^i or x^j to the edge. In order to differentiate among the various kinds of operators, the "electron lines" will be solid and the "photon lines" dashed. Moreover, the electron lines are provided with an arrow which is always to show the direction from the point of the operator $\bar{\psi}(j)$ to the point of the operator $\psi(i)$. As an example, we show the diagram of the matrix element

$$\left.\begin{aligned}&\frac{(-e)^2}{2!}\iint dx\,dx' \langle q | P(:\bar{\psi}^{(0)}(x)\,\gamma_\mu\psi^{(0)}(x)::\bar{\psi}^{(0)}(x')\,\gamma_\nu\psi^{(0)}(x'):) | q'\rangle \times \\ &\qquad\qquad \times \langle 0 | P(A_\mu^{(0)}(x)\,A_\nu^{(0)}(x')) | 0\rangle = \\ &= -\frac{e^2}{8}\iint dx\,dx'\,[\langle q | :\bar{\psi}^{(0)}(x)\,\gamma_\lambda\,S_F(x-x')\,\gamma_\lambda\psi^{(0)}(x'): | q'\rangle + \\ &\quad + \langle q | :\bar{\psi}^{(0)}(x')\,\gamma_\lambda\,S_F(x'-x)\,\gamma_\lambda\psi^{(0)}(x): | q'\rangle]\,D_F(x'-x) = \\ &= -\frac{e^2}{4}\iint dx\,dx' \langle q | \bar{\psi}^{(0)}(x) | 0\rangle \gamma_\lambda S_F(x-x')\,\gamma_\lambda \langle 0 | \psi^{(0)}(x') | q'\rangle \times \\ &\qquad\qquad \times D_F(x'-x).\end{aligned}\right\} \quad (22.2)$$

in the corresponding figure. Obviously the diagram (Fig. 2) gives an intuitive picture of the arrival of an electron q', the emission

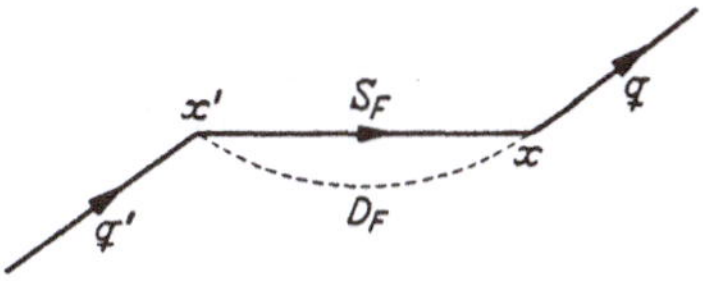

Fig. 2. The graph for the matrix element (22.2).

of a virtual photon, then its absorption, and finally the further propagation of the electron q. One should be cautioned against

1. A classification of the various terms in the S-matrix without the use of graphs has been given by E. R. Caianiello, Nuovo Cim. 10, 1634 (1954).

too literal an interpretation of the diagrams, however, since this can easily lead to false results.

In general a matrix element is not represented by a single term as in (22.2), but rather there is a sum of terms, because all non-vanishing normal products must be considered. We can obviously obtain all these terms if we draw all the diagrams which can arise by connecting n points with the required number of "external" and "internal" lines. In all practical cases, these diagrams can be drawn immediately.

The diagrams which correspond to the n-th approximation (22.1) to the S-matrix can be divided immediately into several groups. Each group contains exactly $n!$ diagrams which are distinguished from each other only by a different labeling of the variables x^i. All these terms are most simply taken into account by dropping the factor $n!$ in the denominator of (22.1) and actually evaluating only one of the diagrams in each group. A further simplification results if we neglect all "disconnected" graphs. A graph is called disconnected if it can be broken into two separate parts which are not connected with each other by either electron or photon lines.[1]

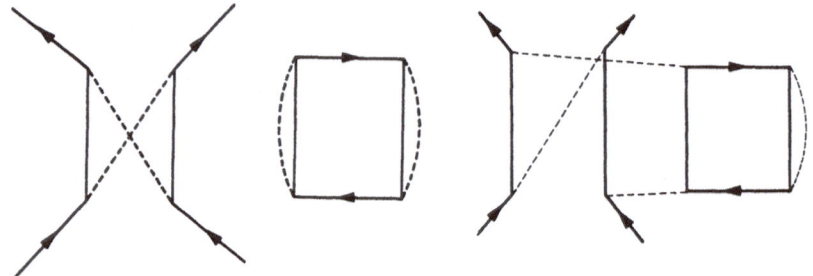

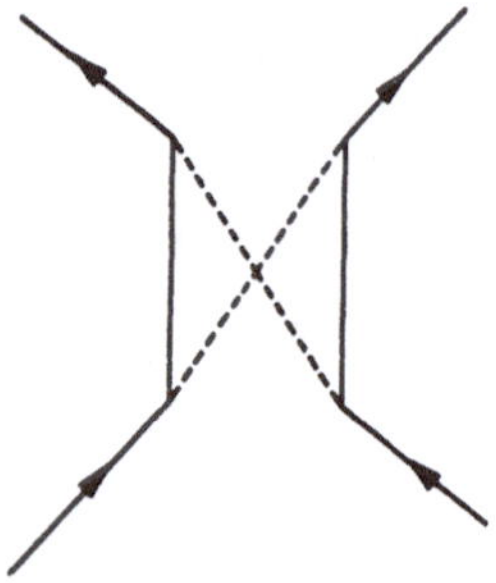

Fig. 3. A disconnected graph.

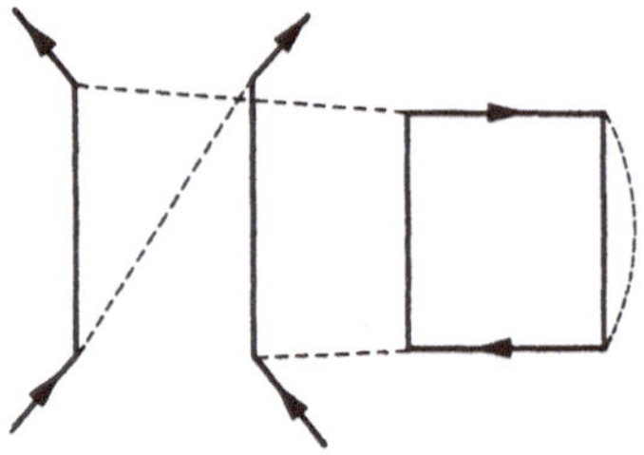

Fig. 4. A connected graph containing a closed loop.

Thus, for example, the graph shown in Fig. 3 is disconnected, while that of Fig. 4 is connected. The basis for the neglect of the disconnected graphs is that their inclusion serves only to multiply all matrix elements by the same numerical factor. This factor is clearly just the matrix element $\langle 0|S|0\rangle$, i.e., of the form $e^{i\delta}$, and therefore without physical significance.

The following rules are to be used for transforming a diagram into an analytical result:

1. Each internal photon line corresponds to a factor $\frac{1}{2}\,\delta_{\nu_i\nu_j}\,D_F(ij)$.

2. Each internal electron line from a point x^i to a point x^j corresponds to a factor $-\frac{1}{2}\,S_F(ji)$.

3. Each external line corresponds to a creation operator if the particle leaves the graph and to an annihilation operator if the

1. (Translator's note) The disconnected part is to have no external lines either! Some authors refer to these parts as "vacuum bubbles".

particle enters. Note that with our conventions about the arrows in the diagrams, a positron is to be considered as running "backwards". The photons are taken as entering if they are present in the initial state, otherwise they are leaving.

4. Each point x^i corresponds to a factor γ_{ν_i}.

5. The whole expression is multiplied by the factor $e^n(-1)^{l+n}$, where n is the number of points and l is the number of closed (electron) loops in the graph.

6. The whole term is multiplied by an overall factor $(-1)^P$, where P is determined by the permutation of the factors $\psi^{(0)}(x)$ in the normal product.

With the exception of the sign $(-1)^l$ in rule 5, these rules require no further discussion since they follow immediately from the considerations of Sec. 21. The expression "a closed loop" of electron lines is quite evident and requires no further elaboration. Thus, for example, the graph of Fig. 4 contains just one such loop. In order to prove the overall sign given in rule 5, we write all the factors $:\bar\psi(i)\,\gamma_{\nu_i}\,\psi(i):$ which enter the loop next to each other in the P-symbol. This does not affect the sign of the matrix element. We must then evaluate the following expression:

$$\dot{\bar\psi}(i_1)\,\gamma_{\nu_{i_1}}\ddot\psi(i_1)\,\ddot{\bar\psi}(i_2)\,\gamma_{\nu_{i_2}}\dddot\psi(i_2)\,\dddot{\bar\psi}(i_3)\,\gamma_{\nu_{i_3}}\cdots\gamma_{\nu_{i_n}}\dot\psi(i_n)\,. \tag{22.3}$$

The contractions can be carried out immediately and give the result $-\frac{1}{2}S_F(i_r\,i_{r+1})$, with the exception of the contraction $\dot{\bar\psi}(i_1)\,\dot\psi(i_n)$. The contraction $\dot{\bar\psi}(i_1)\,\dot\psi(i_n)$ gives the factor $+\frac{1}{2}S_F(i_n\,i_1)$ on account of (21.35). From this the rule for the overall sign in rule 5 follows immediately.

In conclusion, we shall discuss one further simplification which can be easily stated in this graphical representation. We can omit all graphs which contain a closed loop with an odd number of electron lines. As an example, consider the diagram of Fig. 5.

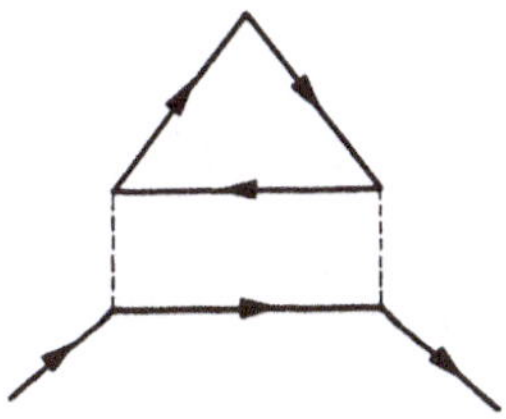

Fig. 5. A closed loop having three electron lines.

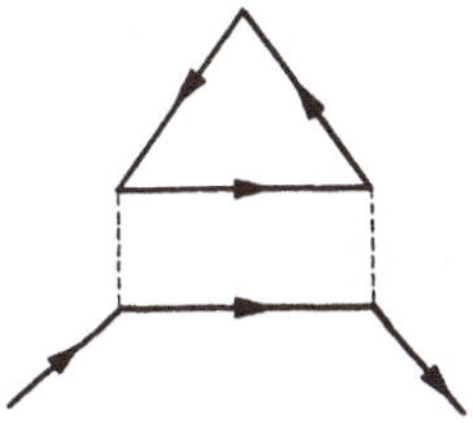

Fig. 6. The contribution of this graph exactly cancels that of the graph of Fig. 5.

The assertion is that it may be dropped. The basis for this is that for every diagram with a closed loop, there is another diagram which must be added, and in which the direction around the loop is just reversed. To the contribution of Fig. 5 we must therefore add the contribution of the graph in Fig. 6. For simplicity we shall designate the points which are vertices of the loop by $x^1, \ldots, x^n$. The contribution of the diagram with the loop contains a factor

$$\mathrm{Sp}\,[\gamma_{\nu_1} S_F(12)\gamma_{\nu_2} S_F(23)\ldots\gamma_{\nu_n} S_F(n1)]. \tag{22.4}$$

The contribution of the diagram with the reversed direction around the loop differs from the first expression only in the replacement of the spur by

$$\mathrm{Sp}\,[\gamma_{\nu_1} S_F(1n)\,\gamma_{\nu_n} S_F(n, n-1)\ldots\gamma_{\nu_2} S_F(21)]. \tag{22.5}$$

In the final result, the sum

$$\begin{aligned}&\mathrm{Sp}\,[\gamma_{\nu_1} S_F(12)\,\gamma_{\nu_2} S_F(23)\ldots\gamma_{\nu_n} S_F(n1)]+\\ &\quad+\mathrm{Sp}[\gamma_{\nu_1} S_F(1n)\gamma_{\nu_n} S_F(n,n-1)\ldots\gamma_{\nu_2} S_F(21)]\end{aligned} \tag{22.6}$$

will occur. In order to evaluate this expression, we need a few theorems about the traces of γ-matrices. In what follows, we shall often have to evaluate such traces, so we shall review their properties here.

From the fundamental property of the γ-matrices,

$$\gamma_\mu\gamma_\nu+\gamma_\nu\gamma_\mu=2\delta_{\mu\nu}, \tag{22.7}$$

we obtain the expression

$$\left.\begin{aligned}&\mathrm{Sp}\,[\gamma_{\nu_1}\gamma_{\nu_2}\gamma_{\nu_3}\ldots\gamma_{\nu_n}]=2\delta_{\nu_1\nu_2}\,\mathrm{Sp}\,[\gamma_{\nu_3}\ldots\gamma_{\nu_n}]-\mathrm{Sp}\,[\gamma_{\nu_2}\gamma_{\nu_1}\gamma_{\nu_3}\ldots\gamma_{\nu_n}]=\\ &=2\sum_{i=2}^{n}\delta_{\nu_1\nu_i}(-1)^i\,\mathrm{Sp}\,[\gamma_{\nu_2}\ldots\gamma_{\nu_{i-1}}\gamma_{\nu_{i+1}}\ldots\gamma_{\nu_n}]+(-1)^{n-1}\,\mathrm{Sp}\,[\gamma_{\nu_2}\ldots\gamma_{\nu_n}\gamma_{\nu_1}].\end{aligned}\right\} \tag{22.8}$$

Using the elementary property of a trace,

$$\mathrm{Sp}\,[AB]=\mathrm{Sp}\,[BA], \tag{22.9}$$

and (22.8), for n even, we have

$$\mathrm{Sp}\,[\gamma_{\nu_1}\ldots\gamma_{\nu_n}]=\sum_{i=2}^{n}\delta_{\nu_1\nu_i}(-1)^i\,\mathrm{Sp}\,[\gamma_{\nu_2}\ldots\gamma_{\nu_{i-1}}\gamma_{\nu_{i+1}}\ldots\gamma_{\nu_n}]. \tag{22.10}$$

Equation (22.10) allows us to calculate a trace with n γ-matrices if all traces with $n-2$ γ-matrices are known. In principle, this result can be used to calculate all traces of an even number of γ-matrices. For odd n we write

$$\mathrm{Sp}\,[\gamma_{\nu_1}\ldots\gamma_{\nu_n}]=\mathrm{Sp}\,[\gamma_{\nu_1}\ldots\gamma_{\nu_n}\gamma_5^2]=\mathrm{Sp}\,[\gamma_5\gamma_{\nu_1}\ldots\gamma_{\nu_n}\gamma_5], \tag{22.11}$$

where γ_5 is defined by

$$\gamma_5=\gamma_1\gamma_2\gamma_3\gamma_4, \tag{22.12}$$

and has the properties

$$\gamma_5^2=1, \tag{22.13}$$

$$\gamma_5\gamma_\mu+\gamma_\mu\gamma_5=0, \quad \mu=1,2,3,4. \tag{22.14}$$

From (22.14) and (22.11), it follows directly that

$$\left.\begin{aligned}\mathrm{Sp}[\gamma_{\nu_1}\cdots\gamma_{\nu_n}]=\mathrm{Sp}[\gamma_5\gamma_{\nu_1}\cdots\gamma_{\nu_n}\gamma_5]=(-1)^n\,\mathrm{Sp}[\gamma_5^2\gamma_{\nu_1}\cdots\gamma_{\nu_n}]=\\=(-1)^n\,\mathrm{Sp}[\gamma_{\nu_1}\cdots\gamma_{\nu_n}].\end{aligned}\right\} \tag{22.15}$$

Therefore, if n is odd, the spur must vanish:

$$\mathrm{Sp}[\gamma_{\nu_1}\cdots\gamma_{\nu_{2n+1}}]=0. \tag{22.16}$$

Another symmetry condition for traces is often useful:

$$\mathrm{Sp}[\gamma_{\nu_1}\cdots\gamma_{\nu_n}]=\mathrm{Sp}[\gamma_{\nu_n}\gamma_{\nu_{n-1}}\cdots\gamma_{\nu_2}\gamma_{\nu_1}]. \tag{22.17}$$

Equation (22.17) is non-trivial only if n is even. For $n=2$, it is identical with (22.9). Using the expansion (22.10), the proof for $n+2$ is readily obtained from the result for n.

The general expansion (22.10) gives the spur of n γ-matrices as a sum of $n(n-2)(n-4)\ldots4\cdot2$ terms. Because there are only four different γ-matrices, related by (22.7), it is to be expected that considerable simplifications are possible in the results for large n. Actually, it can be shown[1] that such simplifications occur for $n>12$. We shall not pursue the matter further because in our applications there will not be traces complicated enough to make these simplifications important.

From (22.17), (22.16), and from the symmetry properties of the functions $\Delta_F(x)$,

$$\Delta_F(x-x')=\Delta_F(x'-x), \quad \frac{\partial}{\partial x_\nu}\Delta_F(x-x')=-\frac{\partial}{\partial x'_\nu}\Delta_F(x'-x), \tag{22.18}$$

one can readily show

$$\left.\begin{aligned}&\mathrm{Sp}[\gamma_{\nu_1}S_F(1\,2)\gamma_{\nu_2}S_F(23)\ldots\gamma_{\nu_n}S_F(n\,1)]=\\&=(-1)^n\,\mathrm{Sp}[\gamma_{\nu_1}S_F(1\,n)\gamma_{\nu_n}S_F(n,n-1)\ldots\gamma_{\nu_2}S_F(2\,1)].\end{aligned}\right\} \tag{22.19}$$

If n is odd, the sum in (22.6) therefore vanishes. We have now proved that all diagrams with closed loops having an odd number of electron lines may be dropped. This result was originally proved by Furry,[2] using a completely different method.

23. The Physical Interpretation of the F-Functions

A characteristic feature of the above calculations is the appearance of the functions $D_F(x-x')$ and $S_F(x-x')$ in the matrix elements of the S-matrix. In the original integration of the equations for the field operators, we worked only with the retarded functions. The latter have the obvious interpretation that the value of a field operator at a point x can depend only on quantities at points x' for which the times x'_0 are smaller than the time x_0.

1. E. R. Caianiello and S. Fubini, Nuovo Cim. 9, 1218 (1952).
2. W. H. Furry, Phys. Rev. 51, 125 (1937).

This is a natural consequence of the causality of all phenomena, in complete correspondence with the classical treatment of a field theory by means of integrals over the retarded light cone. In our expression for the S-matrix this causality has apparently been lost. The integrals over the variables x extend over the entire space and the F-functions which are present differ from zero for both signs of the time difference. Despite this we can readily convince ourselves by means of a simple example that the F-functions have a "causal" interpretation under certain conditions. Of course it is not expected that the S-matrix will have as simple a structure as the field operators, since the time sequence of two integration points is not fixed from the beginning.

We choose the following term in the S-matrix[1] as an example for the analysis of the F-functions:

$$\left.\begin{aligned}\langle q, q'|\, S^{(2)}\,|q'', q'''\rangle = \frac{e^2}{2}\iint dx'\, dx'' \langle q|\,\overline{\psi}^{(0)}(x')\,\gamma_\lambda\,\psi^{(0)}(x')\,|q''\rangle \times \\ \times D_F(x'-x'')\langle q'|\,\overline{\psi}^{(0)}(x'')\,\gamma_\lambda\,\psi^{(0)}(x'')\,|q'''\rangle + \cdots\,.\end{aligned}\right\} \quad (23.1)$$

Clearly this integral is obtained in calculating the cross section that two electrons with energy-momentum vectors q'' and q''' collide and emerge with new energy-momentum vectors q and q'. It is a straightforward interpretation that the electron q'' emits a photon at the point x' and as a result has momentum q. At the point x'' the photon is absorbed by the other electron and consequently its momentum is changed from q''' to q'. The total probability of this transition contains the sum over all possible combinations of x' and x''. Serious objections can be made to this naive interpretation. The function $D_F(x)$ does not vanish for spatial separations; not even approximately. The emitted photons must therefore be propagated with speeds greater than the velocity of light. Following Fierz,[1] we shall show that a similar interpretation of the integral (23.1) is still possible if we consider not events at single points, but rather the probability that the photon is absorbed or emitted in a finite region of space-time R. In order to pursue this idea, we break up the whole four-dimensional space into regions R and rewrite the integral (23.1) as

$$\langle q, q'|\, S^{(2)}\,|q'', q'''\rangle = \sum_{R', R''} I_{R', R''} + \cdots\,, \quad (23.2)$$

$$\left.\begin{aligned}I_{R', R''} = \frac{e^2}{2}\int\limits_{R'} dx' \int\limits_{R''} dx'' \langle q|\,\overline{\psi}^{(0)}(x')\,\gamma_\lambda\,\psi^{(0)}(x')\,|q''\rangle \times \\ \times D_F(x'-x'')\langle q'|\,\overline{\psi}^{(0)}(x'')\,\gamma_\lambda\,\psi^{(0)}(x'')\,|q'''\rangle\,.\end{aligned}\right\} \quad (23.3)$$

For the following discussion it is not essential that the operator $\psi^{(0)}(x)$ be developed in terms of solutions of the Dirac equation with an external, time-independent field. The only property which

1. M. Fierz, Helv. Phys. Acta 23, 731 (1950).

we shall use is the well-defined energy of each state $|q\rangle$.

From the integral representations of the functions $D_F(x)$, $D_R(x)$, and $D_A(x)$,

$$D_F(x) = \frac{2}{i}\frac{1}{(2\pi)^4}\int dk\, e^{ikx}\left\{P\frac{1}{k^2} + i\pi\,\delta(k^2)\right\}, \tag{23.4}$$

$$D_R(x) = \frac{1}{(2\pi)^4}\int dk\, e^{ikx}\left\{P\frac{1}{k^2} + i\pi\,\varepsilon(k)\,\delta(k^2)\right\}, \tag{23.5}$$

$$D_A(x) = \frac{1}{(2\pi)^4}\int dk\, e^{ikx}\left\{P\frac{1}{k^2} - i\pi\,\varepsilon(k)\,\delta(k^2)\right\}, \tag{23.6}$$

we find the following relations upon taking the positive and negative frequency parts of these three functions:

$$D_F^{(+)}(x) = \frac{2}{i}\frac{1}{(2\pi)^4}\int_{k_0>0} dk\, e^{ikx}\left\{P\frac{1}{k^2} + i\pi\,\delta(k^2)\right\} = \frac{2}{i}D_R^{(+)}(x), \tag{23.7}$$

$$D_F^{(-)}(x) = \frac{2}{i}\frac{1}{(2\pi)^4}\int_{k_0<0} dk\, e^{ikx}\left\{P\frac{1}{k^2} + i\pi\,\delta(k^2)\right\} = \frac{2}{i}D_R^{(-)}(x). \tag{23.8}$$

We have, therefore,

$$\frac{i}{2}D_F(x) = D_R^{(+)}(x) + D_A^{(-)}(x). \tag{23.9}$$

Introducing the relation (23.9) into (23.3), we find

$$\left.\begin{aligned} I_{R'R''} = -i\,e^2\Big\{&\int_{R'} dx' \int_{R''} dx'' \langle q|\,\overline{\psi}^{(0)}(x')\,\gamma_\lambda\,\psi^{(0)}(x')\,|q''\rangle D_R^{(+)}(x'-x'')\times \\ &\times\langle q'|\,\overline{\psi}^{(0)}(x'')\,\gamma_\lambda\,\psi^{(0)}(x'')\,|q'''\rangle + \\ &+\int_{R'} dx' \int_{R''} dx'' \langle q|\,\overline{\psi}^{(0)}(x')\,\gamma_\lambda\,\psi^{(0)}(x')\,|q''\rangle D_A^{(-)}(x'-x'')\times \\ &\times\langle q'|\,\overline{\psi}^{(0)}(x'')\,\gamma_\lambda\,\psi^{(0)}(x'')\,|q'''\rangle\Big\}. \end{aligned}\right\} \tag{23.10}$$

The time dependence of the factor $\langle q|\,\overline{\psi}^{(0)}(x')\,\gamma_\lambda\,\psi^{(0)}(x')\,|q''\rangle$ is given by $e^{-ix_0'(q_0''-q_0)}$. A similar expression results for the other matrix element. In the time integrations in (23.10) the limits of integration are not $\pm\infty$, but T' and T'', the limits of the regions R' and R''. If we require that the two times T' and T'' are very large compared to the times $|q_0''-q_0|^{-1}$ and $|q_0'''-q_0'|^{-1}$, then the time integration is "almost" a delta function. According to (23.7), the first term of (23.10) gives a contribution only if

$$q_0 - q_0'' = q_0''' - q_0' > 0. \tag{23.11}$$

With these assumptions we have

$$\left.\begin{aligned} I_{R',R''} \approx -i\,e^2 &\int_{R'} dx' \int_{R''} dx'' \langle q|\,\overline{\psi}^{(0)}(x')\,\gamma_\lambda\,\psi^{(0)}(x')\,|q''\rangle D_R^{(+)}(x'-x'')\times \\ &\times\langle q'|\,\overline{\psi}^{(0)}(x'')\,\gamma_\lambda\,\psi^{(0)}(x'')\,|q'''\rangle \approx \\ \approx -i\,e^2 &\int_{R'} dx' \int_{R''} dx'' \langle q|\,\overline{\psi}^{(0)}(x')\,\gamma_\lambda\,\psi^{(0)}(x')\,|q''\rangle D_R(x'-x'')\times \\ &\times\langle q'|\,\overline{\psi}^{(0)}(x'')\,\gamma_\lambda\,\psi^{(0)}(x'')\,|q'''\rangle. \end{aligned}\right\} \tag{23.12}$$

The terms which are neglected in (23.12) are very small if

$$T(q_0 - q_0'') \gg 1, \quad \text{and} \quad T'(q_0 - q_0'') \gg 1. \tag{23.13}$$

Thus the integral (23.12) is non-zero if

$$x_0' > x_0''. \tag{23.14}$$

The condition (23.13) for the validity of (23.12) is in full accord with the usual uncertainty relation. In order for the energy to be defined with sufficient precision so that the differences (23.11) have a physical meaning, the system must be observed for a time which satisfies the inequality (23.13). The condition (23.11) then says that the energy of the electron in the region R' has been increased by the collision while the energy of the other electron has fallen by the same amount. This means that the electron in R'' must have emitted a photon with the energy $\omega = q_0 - q_0''$ and that this photon has been absorbed by the other electron in R'. Equation (23.14) therefore says that the emission of a photon must always take place prior to its absorption. From Eqs. (7.19) and (7.38) it follows that the photon is propagated with exactly the velocity of light.

In a similar way we can show that the second term in (23.10) gives the contribution

$$\left.\begin{aligned} -i e^2 \int_{R'} dx' \int_{R''} dx'' \langle q | \bar{\psi}^{(0)}(x') \gamma_\lambda \psi^{(0)}(x') | q'' \rangle D_A(x' - x'') \times \\ \times \langle q' | \bar{\psi}^{(0)}(x'') \gamma_\lambda \psi^{(0)}(x'') | q''' \rangle \end{aligned}\right\} \tag{23.15}$$

if the following conditions are satisfied:

$$q_0'' - q_0 = q_0' - q_0''' > 0, \tag{23.16}$$

$$T(q_0'' - q_0) \gg 1, \quad T'(q_0'' - q_0) \gg 1, \tag{23.17}$$

$$x_0'' > x_0'. \tag{23.18}$$

Here the photon is emitted by the electron in R' and absorbed by the other electron, and the inequality (23.18) again says that the emission must take place before the absorption. In this way the function $D_F(x)$ actually gives a "causal" description of the collision. It is often referred to in the literature as the "causal propagator of the photon", or simply as the "photon propagator". It is clear that the properties of the F-function discussed here do not enter only in this particular example, but that a similar discussion ought to be possible in other cases. We shall not carry the discussion further except to mention that these matters have been fully discussed by Stückelberg and his coworkers.[1]

1. See, for example, E. C. G. Stückelberg and D. Rivier, Helv. Phys. Acta 23, 215 (1950), where references to the older literature will be found.

CHAPTER V

ELEMENTARY APPLICATIONS

24. Scattering of Electrons and Pair Production by an External Field

In these next sections we shall give a few elementary applications of the theory developed so far. Since quite complete textbooks exist in this area[1] (see also footnotes 2, 3) and also because the details belong in Vol. XXXI of this handbook,[4] we shall not attempt to be complete. The following calculations are to be considered only as examples of results which are obtained rather easily with modern quantum electrodynamics, although they were usually discovered by other methods, long before the beginning of the modern theory. It seems to us that by the calculation of a few examples which have already been satisfactorily worked out, we demonstrate the great simplicity of the new methods compared to the old.

As the first example, we choose the scattering of an electron by an external field. The quantized electromagnetic field does not enter this very simple example, which can clearly be treated using the Dirac equation without the "second quantization". We work out this problem here because in the next chapter we are going to compute the radiative corrections to the interaction cross section, and then we shall want to compare with this simple calculation. Moreover, we can use this calculation to introduce a simple method for summing over the two directions of spin of the electron in the initial or final states.

Denoting the initial state of the electron by $|q\rangle$ and the final state by $|q'\rangle$, the quantity of interest is

$$\left.\begin{aligned}\langle q'|S|q\rangle = \sum_1^\infty \frac{(-e)^n}{n!}\int\cdots\int dx'\ldots dx^n \times\\ \times\langle q'|P\left(:\bar\psi^{(0)}(x')\,\gamma_{\nu_1}\psi^{(0)}(x'):A_{\nu_1}(x')\ldots:\bar\psi^{(0)}(x^n)\,\gamma_{\nu_n}\psi^{(0)}(x^n):A_{\nu_n}(x^n)\right)|q\rangle.\end{aligned}\right\}\quad(24.1)$$

Because there is an external field present, the "Schroedinger equation" is not (20.9), but rather

---

1. W. Heitler, Quantum Theory of Radiation, Third Ed., Oxford, 1954.

2. A. I. Akhiezer and V. B. Berestetski, Quantum Electrodynamics, translated from the second Russian edition, Interscience, 1965.

3. J. M. Jauch and F. Rohrlich, The Theory of Photons and Electrons, Addison-Wesley, Cambridge, Mass., 1955.

4. Handbuch der Physik, edited by S. Flügge, Springer-Verlag, Heidelberg.

$$i\frac{\partial U(x_0)}{\partial x_0} = -ie\int :\bar{\psi}^{(0)}(x)\gamma_\mu\psi^{(0)}(x): \left(A_\mu^{(0)}(x) + A_\mu^{\text{äuss}}(x)\right) d^3x\, U(x_0). \quad (24.2)$$

In (24.2) $A_\mu^{(0)}(x)$ is the quantized field and $A_\mu^{\text{äuss}}(x)$ is the external field.[1] The term $A_\mu(x)$ in (24.1) is therefore the sum of these two quantities. If we impose the further restriction that the external field need be considered only to first order (the Born approximation) and that the radiation field may be neglected, (24.1) is then simplified to

$$\langle q'|S|q\rangle = -e\int dx\,\langle q'|\bar{\psi}^{(0)}(x)|0\rangle\gamma_\mu\langle 0|\psi^{(0)}(x)|q\rangle A_\mu^{\text{äuss}}(x). \quad (24.3)$$

The matrix elements present in (24.3) are given by

$$\langle 0|\psi^{(0)}(x)|q\rangle = \frac{1}{\sqrt{V}}u^{(r)}(q)\,e^{iqx}, \quad (24.4)$$

where the functions $u^{(r)}(q)$ are given explicitly in Sec. 12. Since we are now studying the scattering of electrons (and not positrons), we must take the index r in (24.4) to be 1 or 2. These two possibilities correspond to the two possible orientations of electron spin as was discussed in Sec. 13. For a time-independent, external field

$$A_\mu^{\text{äuss}}(x) = A_\mu^{\text{äuss}}(\boldsymbol{x}), \quad (24.5)$$

we now obtain

$$\langle q'|S|q\rangle = -\frac{e}{V}\bar{u}^{(r')}(q')\gamma_\mu u^{(r)}(q)\,A_\mu^{\text{äuss}}(\boldsymbol{q}-\boldsymbol{q}')\cdot 2\pi\,\delta(q_0-q_0'), \quad (24.6)$$

where the notation

$$A_\mu^{\text{äuss}}(\boldsymbol{Q}) = \int d^3x\,A_\mu^{\text{äuss}}(\boldsymbol{x})\,e^{i\boldsymbol{Q}\boldsymbol{x}} \quad (24.7)$$

has been introduced. According to (20.22), the transition probability per unit time is then

$$\left.\begin{aligned} \frac{\partial w}{\partial t} &= \frac{e^2}{V^2}|\bar{u}^{(r')}(q')\gamma_\mu u^{(r)}(q)\,A_\mu^{\text{äuss}}(\boldsymbol{q}-\boldsymbol{q}')|^2\cdot 2\pi\,\delta(q_0-q_0') = \\ &= -2\pi\frac{e^2}{V^2}\bar{u}^{(r')}(q')\gamma_\mu u^{(r)}(q)\,\bar{u}^{(r)}(q)\gamma_\nu u^{(r')}(q')\,A_\mu^{\text{äuss}}(\boldsymbol{q}-\boldsymbol{q}')\times \\ &\quad\times A_\nu^{\text{äuss}}(\boldsymbol{q}'-\boldsymbol{q})\,\delta(q_0-q_0'). \end{aligned}\right\} \quad (24.8)$$

To obtain the last form in (24.8), we have made use of the reality properties of the potentials $A_\mu^{\text{äuss}}(x)$. Since $A_k^{\text{äuss}}(x)$ are real and $A_4^{\text{äuss}}(x)$ is pure imaginary, we have

$$\left.\begin{aligned} &(\bar{u}^{(r')}(q')\gamma_\mu u^{(r)}(q)\,A_\mu^{\text{äuss}}(\boldsymbol{q}-\boldsymbol{q}'))^* = u^{*(r)}(q)\gamma_k\gamma_4 u^{(r')}(q')\times \\ &\qquad\times A_k^{\text{äuss}}(\boldsymbol{q}'-\boldsymbol{q}) - u^{*(r)}(q)\,u^{(r')}(q')\,A_4^{\text{äuss}}(\boldsymbol{q}'-\boldsymbol{q}) = \\ &\qquad = -\bar{u}^{(r)}(q)\gamma_\mu u^{(r')}(q')\,A_\mu^{\text{äuss}}(\boldsymbol{q}'-\boldsymbol{q}). \end{aligned}\right\} \quad (24.9)$$

1. (Translator's Note) To avoid resetting the equations, we shall follow the German and use $A^{\text{äuss}}$ for the external field. This is to be distinguished from $A^{(\text{aus})}$, the out-field.

From (24.8) we obtain the cross section for the scattering of the state $|q\rangle$ with polarization r into the state $|q'\rangle$ with polarization r' after dividing by the number of incident particles per unit time and per unit surface area $\frac{|\boldsymbol{q}|}{q_0 V}$:

$$\left.\begin{aligned} d\sigma = -\int d q_0' \frac{2\pi e^2 q_0}{|\boldsymbol{q}|} \bar{u}^{(r')}(q')\gamma_\mu u^{(r)}(q)\,\bar{u}^{(r)}(q)\gamma_\nu u^{(r')}(q') \times \\ \times A_\mu^{\text{äuss}}(\boldsymbol{q}-\boldsymbol{q}')A_\nu^{\text{äuss}}(\boldsymbol{q}'-\boldsymbol{q})\,\delta(q_0-q_0')\frac{|\boldsymbol{q}'|\,q_0'}{(2\pi)^3}\,d\Omega. \end{aligned}\right\} \quad (24.10)$$

In (24.10) we have required only that the direction of propagation of the outgoing particle lie in a given solid angle and we have summed over all energies possible. The last factor in (24.10) is the number of states per energy interval divided by V [c.f. (1.4)]. Because of the delta function, the integral can be evaluated immediately and if we define $q_0 = q_0' = E$ and $|\boldsymbol{q}| = |\boldsymbol{q}'| = p$, then we find

$$\left.\begin{aligned} \frac{d\sigma}{d\Omega} = -\frac{e^2 E^2}{(2\pi)^2}\bar{u}^{(r')}(q')\gamma_\mu u^{(r)}(q)\,\bar{u}^{(r)}(q)\gamma_\nu u^{(r')}(q') \times \\ \times A_\mu^{\text{äuss}}(\boldsymbol{q}-\boldsymbol{q}')A_\nu^{\text{äuss}}(\boldsymbol{q}'-\boldsymbol{q}). \end{aligned}\right\} \quad (24.11)$$

If we wished, we could introduce the explicit form for the functions $u^{(r)}(q)$ and work out the cross section for polarized electrons. In most experiments, however, the incident electrons are unpolarized. That is, we must take the average over both directions of polarization for the incident electrons. In general, the polarization of the outgoing electron will not be measured either, and therefore the sum of the cross sections for the two polarizations must be calculated. This sum and average can be worked out directly in Eq. (24.11) without having to use the explicit forms of the functions $u^{(r)}(q)$, which depend on the particular representation employed. By means of Eq. (12.12), we have

$$\left.\begin{aligned} \frac{d\sigma}{d\Omega} &= -\frac{e^2 E^2}{(2\pi)^2}\frac{1}{2}\sum_{r=1}^{2}\sum_{r'=1}^{2}\bar{u}^{(r')}(q')\gamma_\mu u^{(r)}(q)\,\bar{u}^{(r)}(q)\gamma_\nu u^{(r')}(q') \times \\ &\quad \times A_\mu^{\text{äuss}}(\boldsymbol{q}-\boldsymbol{q}')A_\nu^{\text{äuss}}(\boldsymbol{q}'-\boldsymbol{q}) = \\ &= -\frac{e^2}{8(2\pi)^2}\,\mathrm{Sp}\,[\gamma_\mu(i\gamma q - m)\gamma_\nu(i\gamma q' - m)] \times \\ &\quad \times A_\mu^{\text{äuss}}(\boldsymbol{q}-\boldsymbol{q}')A_\nu^{\text{äuss}}(\boldsymbol{q}'-\boldsymbol{q}). \end{aligned}\right\} \quad (24.12)$$

The spur in (24.12) can be calculated immediately by the use of (22.10) and the relation

$$\mathrm{Sp}\,[\mathrm{I}] = 4\,. \qquad (24.13)$$

We obtain

$$\frac{d\sigma}{d\Omega} = -\frac{e^2}{8\pi^2}[\delta_{\mu\nu}(q q' + m^2) - q_\mu q_\nu' - q_\mu' q_\nu]\,A_\mu^{\text{äuss}}(\boldsymbol{q}-\boldsymbol{q}')\,A_\nu^{\text{äuss}}(\boldsymbol{q}'-\boldsymbol{q}). \qquad (24.14)$$

Equation (24.14) is the desired result.

The general result (24.14) can be specialized for different forms of the external field. Perhaps most important is the case of an electrostatic Coulomb field. Here we have

$$A_\mu^{\text{äuss}}(\boldsymbol{x}) = -\frac{Ze}{4\pi r} i\,\delta_{\mu 4}. \tag{24.15}$$

We obtain the form in p-space from (24.7):

$$A_\mu^{\text{äuss}}(\boldsymbol{Q}) = -i\,\delta_{\mu 4}\frac{Ze}{\boldsymbol{Q}^2}. \tag{24.16}$$

In this case $\boldsymbol{Q} = \boldsymbol{q} - \boldsymbol{q}'$ and letting Θ denote the scattering angle, we have

$$\boldsymbol{Q}^2 = \boldsymbol{q}^2 + \boldsymbol{q}'^2 - 2\boldsymbol{q}\,\boldsymbol{q}' = 2p^2(1-\cos\Theta) = 4p^2\sin^2\frac{\Theta}{2}. \tag{24.17}$$

After substituting in (24.14) and simplifying, there results

$$\frac{d\sigma}{d\Omega} = \left[\frac{Ze^2 m}{8\pi p^2 \sin^2\frac{\Theta}{2}}\right]^2\left(1 + \frac{p^2}{m^2}\cos^2\frac{\Theta}{2}\right). \tag{24.18}$$

This, therefore, is the relativistic generalization of the well known Rutherford scattering cross section.[1] In fact, if $p^2 \ll m^2$, then it follows from (24.18) that

$$\frac{d\sigma}{d\Omega} = \left[\frac{Ze^2}{16\pi E_k \sin^2\frac{\Theta}{2}}\right]^2, \tag{24.19}$$

where

$$E_k = \frac{p^2}{2m} \tag{24.20}$$

is the non-relativistic kinetic energy of the electron. Equation (24.19) is the familiar Rutherford cross section in the Heaviside units employed here.

A similar problem, whose solution is to be found from the previous formula, is the probability of pair production by a weak, external field. Again if the field can be treated in Born approximation, the S-matrix element of interest is given by equations similar to (24.3) and (24.6):

$$\left.\begin{aligned} \langle q, q'|S|0\rangle &= -e\int dx'\langle q|\overline{\psi}^{(0)}(x')|0\rangle\gamma_\mu\langle q'|\psi^{(0)}(x')|0\rangle A_\mu^{\text{äuss}}(x') = \\ &= -\frac{e}{V}\bar{u}^{(r)}(q)\,\gamma_\mu\, u^{(r')}(-q')\,A_\mu^{\text{äuss}}(-q-q') \end{aligned}\right\} \tag{24.21}$$

In (24.21) $|q'\rangle$ is a positron and $|q\rangle$ is an electron. The index r is therefore equal to 1 or 2 and the index r' is equal to 3 or 4. Furthermore, the external field has been assumed to be time dependent and we have introduced the notation

1. Equation (24.18) was first derived by N. F. Mott., Proc. Roy. Soc. Lond., A126, 259 (1930).

$$A_\mu^{\text{äuss}}(Q) = \int dx\, A_\mu^{\text{äuss}}(x)\, e^{iQx}. \tag{24.22}$$

In (24.7) the integration was only over the three spatial coordinates; here we integrate over the time coordinate as well. According to (19.7) the probability for the production of a pair $|q, q'\rangle$ is

$$\left.\begin{aligned} dw &= |\langle q, q' | S | 0\rangle|^2 \frac{V^2}{(2\pi)^6} d^3q\, d^3q' = -\frac{e^2}{(2\pi)^6} \bar{u}^{(r)}(q)\, \gamma_\mu\, u^{(r')}(-q') \times \\ &\times \bar{u}^{(r')}(-q')\, \gamma_\nu\, u^{(r)}(q)\, A_\mu^{\text{äuss}}(-q-q')\, A_\nu^{\text{äuss}}(q+q')\, d^3q\, d^3q'. \end{aligned}\right\} \tag{24.23}$$

The total probability that a pair is produced at all is therefore

$$\left.\begin{aligned} w &= \frac{e^2}{(2\pi)^6} \iint \frac{d^3q\, d^3q'}{4E_q E_{q'}} \mathrm{Sp}\left[(i\gamma q - m)\gamma_\mu(-i\gamma q' - m)\gamma_\nu\right] \times \\ &\quad \times A_\mu^{\text{äuss}}(-q-q')\, A_\nu^{\text{äuss}}(q+q') = \\ &= \frac{e^2}{(2\pi)^6} \int T_{\mu\nu}(Q)\, A_\mu^{\text{äuss}}(-Q)\, A_\nu^{\text{äuss}}(Q)\, dQ\,, \end{aligned}\right\} \tag{24.24}$$

where

$$\left.\begin{aligned} T_{\mu\nu}(Q) &= \iint \frac{d^3q\, d^3q'}{4E_q E_{q'}} \mathrm{Sp}\left[(i\gamma q - m)\gamma_\mu(-i\gamma q' - m)\gamma_\nu\right] \delta(Q - q - q') = \\ &= -\int dq'\, \delta(q'^2 + m^2)\, \delta((Q-q')^2 + m^2)\, \Theta(q')\, \Theta(Q-q') \times \\ &\quad \times \mathrm{Sp}\left[(i\gamma(Q-q') - m)\gamma_\mu(i\gamma q' + m)\gamma_\nu\right], \end{aligned}\right\} \tag{24.25}$$

$$\Theta(q) = \frac{1}{2}\left[1 + \varepsilon(q)\right]. \tag{24.25a}$$

In (24.24), Eq. (12.12) has been employed in the summation over r and Eq. (12.13) in the summation over r'. The tensor $T_{\mu\nu}$ obviously has the transformation properties specified by the indices μ and ν and depends only on the vector Q. It must therefore be of the following general form:

$$T_{\mu\nu} = A(Q^2)\, \delta_{\mu\nu} + B(Q^2)\, Q_\mu Q_\nu = T_{\nu\mu}. \tag{24.26}$$

Now from

$$\left.\begin{aligned} Q_\mu \mathrm{Sp}\left[(i\gamma(Q-q') - m)\gamma_\mu(i\gamma q' + m)\gamma_\nu\right] &= -4q'_\nu\left[(Q-q')^2 + m^2\right] - \\ &\quad - 4(Q_\nu - q'_\nu)\left[q'^2 + m^2\right] \end{aligned}\right\} \tag{24.27}$$

follows

$$T_{\mu\nu} Q_\mu = 0 = Q_\nu\left[A + BQ^2\right], \tag{24.28}$$

and hence

$$T_{\mu\nu} = B(Q^2)\left[Q_\mu Q_\nu - \delta_{\mu\nu} Q^2\right], \tag{24.29}$$

with

$$\left.\begin{aligned} B(Q^2) &= \frac{T_{\mu\mu}}{-3Q^2} = \frac{1}{3Q^2} \int dq'\, \delta(q'^2 + m^2)\, \delta((Q-q')^2 + m^2) \times \\ &\quad \times \Theta(q')\, \Theta(Q-q')\, \mathrm{Sp}\left[(i\gamma(Q-q') - m)\gamma_\mu(i\gamma q' + m)\gamma_\mu\right]. \end{aligned}\right\} \tag{24.30}$$

The spur in (24.30) can easily be worked out and the result is the following integral for $B(Q^2)$:

$$\left.\begin{aligned} B(Q^2) &= \frac{1}{3Q^2}\int dq'\,\delta(q'^2+m^2)\,\delta(Q^2-2Qq')\,\Theta(q')\,\Theta(Q-q')\,8[Qq'-m^2] = \\ &= \frac{4}{3}\int dq'\,\delta(q'^2+m^2)\,\delta(Q^2-2Qq')\,\Theta(q')\,\Theta(Q-q')\left[1-\frac{2m^2}{Q^2}\right]. \end{aligned}\right\} \quad (24.31)$$

We compute the invariant function $B(Q^2)$ in that coordinate system where the spatial components Q_k vanish. This assumes that Q is time-like, but from the delta function in (24.31) it follows immediately that the integral must vanish identically for space-like Q. In this way we find

$$\left.\begin{aligned} B(-Q_0^2) &= \frac{4}{3}\left[1+\frac{2m^2}{Q_0^2}\right]\int d^3q'\frac{1}{2\sqrt{\boldsymbol{q}'^2+m^2}}\left[\frac{|Q_0-\sqrt{\boldsymbol{q}'^2+m^2}|}{|Q_0-\sqrt{\boldsymbol{q}'^2+m^2}|}+1\right]\frac{1}{2}\times \\ &\quad\times\delta\left(2Q_0\sqrt{\boldsymbol{q}'^2+m^2}-Q_0^2\right) = \\ &= \frac{1}{3}\left[1+\frac{2m^2}{Q_0^2}\right]2\pi\sqrt{1-\frac{4m^2}{Q_0^2}}\;\Theta(Q_0^2-4m^2)\,\Theta(Q_0). \end{aligned}\right\} \quad (24.32)$$

Therefore in an arbitrary coordinate system we have

$$T_{\mu\nu}(Q) = \frac{2\pi}{3}\left[1-\frac{2m^2}{Q^2}\right]\sqrt{1+\frac{4m^2}{Q^2}}\,(Q_\mu Q_\nu-\delta_{\mu\nu}Q^2)\,\Theta(Q)\,\Theta\left(-\frac{Q^2}{4}-m^2\right). \quad (24.33)$$

By introducing the external current and using the symmetry in Q-space, we can write the total probability as

$$w = \frac{1}{(2\pi)^3}\int dQ\,\frac{j_\mu^{\text{äuss}}(Q)\,j_\mu^{\text{äuss}}(-Q)}{-2Q^2}\,\Pi^{(0)}(Q^2)\,, \quad (24.34)$$

$$\Pi^{(0)}(Q^2) = \frac{e^2}{12\pi^2}\left(1-\frac{2m^2}{Q^2}\right)\sqrt{1+\frac{4m^2}{Q^2}}\,\Theta\left(-\frac{Q^2}{4}-m^2\right), \quad (24.35)$$

$$j_\mu^{\text{äuss}}(Q) = (\delta_{\mu\nu}Q^2-Q_\mu Q_\nu)\,A_\mu^{\text{äuss}}(Q)\,, \quad (24.36)$$

$$Q_\mu j_\mu^{\text{äuss}}(Q) = 0. \quad (24.37)$$

In a later chapter we shall again find these equations in another connection. Here we remark only that the factor $\Theta\left(-\frac{Q^2}{4}-m^2\right)$ obviously expresses the condition that the energy given up by the external field must be greater than $2m$ in every coordinate system.

25. Scattering of Light by an Electron

As the next example, we shall work out the cross section for the Compton effect, i.e., for the scattering of a photon by an electron. The first non-vanishing approximation to the S-matrix is clearly

$$\left.\begin{aligned}\langle q', k'|S|k, q\rangle = -\frac{e^2}{2}\iint dx'\,dx''\langle q'|\bar{\psi}^{(0)}(x'')|0\rangle\gamma_{\nu_2}S_F(x''-x')\times\\ \times\gamma_{\nu_1}\langle 0|\psi^{(0)}(x')|q\rangle[\langle 0|A^{(0)}_{\nu_1}(x')|k\rangle\langle k'|A^{(0)}_{\nu_2}(x'')|0\rangle+\\ +\langle 0|A^{(0)}_{\nu_2}(x'')|k\rangle\langle k'|A^{(0)}_{\nu_1}(x')|0\rangle].\end{aligned}\right\}\quad(25.1)$$

In Eq. (25.1), $|k, q\rangle$ stands for the initial state and $|k', q'\rangle$ for the final state in the process. The above formula clearly corresponds to the two graphs of Fig. 7. After introducing the Fourier

Fig. 7. The graphs for Compton scattering.

representations of the various factors into Eq. (25.1), the two x-integrations can be carried out. The result is two delta functions, one of which expresses the conservation of energy-momentum for the process and the other of which allows the p-integration in the S_F-function to be done. We finally obtain

$$\left.\begin{aligned}\langle q', k'|S|k, q\rangle = \frac{i e^2}{2V^2}\frac{1}{\sqrt{\omega\omega'}}\bar{u}^{(r')}(q')\Big[\gamma e'\frac{i\gamma(q+k)-m}{(q+k)^2+m^2}\gamma e+\\ +\gamma e\frac{i\gamma(q-k')-m}{(q-k')^2+m^2}\gamma e'\Big]u^{(r)}(q)\,(2\pi)^4\,\delta(q+k-q'-k').\end{aligned}\right\}\quad(25.2)$$

As in Sec. 24, the two indices r and r' denote the polarization states of the electrons. The vectors e and e' are the polarization vectors of the two photons k and k'. In order not to complicate the calculation unnecessarily, we shall specify the initial state of the electron as being at rest; i.e., $\boldsymbol{q}=0$, $q_4 = im$. In addition, if the two photons are taken as "real", i.e., transverse, then we have

$$q\,e = q\,e' = k\,e = k'\,e' = 0. \quad (25.3)$$

Using (25.3) and the Dirac equation $(i\gamma q + m)\,u^{(r)}(q) = 0$, we can simplify (25.2) to

$$\left.\begin{aligned}\langle q', k'|S|k, q\rangle = -\frac{i e^2}{2V^2}\frac{1}{\sqrt{\omega\omega'}}\bar{u}^{(r')}(q')\Big[\gamma e'\gamma e\frac{i\gamma k}{2qk}+\gamma e\gamma e'\frac{i\gamma k'}{2qk'}\Big]\times\\ \times u^{(r)}(q)\,(2\pi)^4\,\delta(q+k-q'-k').\end{aligned}\right\}\quad(25.4)$$

The conditions

$$q^2 + m^2 = k^2 = 0$$

have also been used in the denominators. With the assumption that the polarization of the electron is random prior to the collision and that both polarization directions are observed[1] afterward, the transition probability per unit time for the process is

$$\frac{\delta w}{\delta t} = \frac{e^4}{4V^3}\frac{1}{\omega\omega'}\frac{1}{8EE'}\,\mathrm{Sp}\,[\cdots]\,(2\pi)^4\,\delta(q+k-q'-k')\,, \qquad (25.5)$$

according to (20.22). Here

$$\left.\begin{aligned}\mathrm{Sp}\,[\cdots] = \mathrm{Sp}\Big[\Big(\gamma e'\gamma e\frac{i\gamma k}{2qk} + \gamma e\gamma e'\frac{i\gamma k'}{2qk'}\Big)(i\gamma q - m)\,\times \\ \times\Big(\frac{i\gamma k'}{2qk'}\gamma e'\gamma e + \frac{i\gamma k}{2qk}\gamma e\gamma e'\Big)(i\gamma q' - m)\Big].\end{aligned}\right\} \qquad (25.6)$$

This gives the differential cross section for the outgoing photon:

$$\frac{d\sigma}{d\Omega_{k'}} = \frac{e^4}{128\pi^2}\frac{1}{m\omega}\int\frac{d^3q'}{E'}\int\omega' d\omega'\,\mathrm{Sp}\,[\cdots]\,\delta(\mathbf{k}-\mathbf{q}'-\mathbf{k}')\,\delta(m+\omega-E'-\omega'). \quad (25.7)$$

The indicated integrations over d^3q' and $d\omega'$ can be done by using the delta functions:

$$\frac{d\sigma}{d\Omega_{k'}} = \frac{e^4}{128\pi^2}\frac{\omega'^2}{m^2\omega^2}\,\mathrm{Sp}\,[\cdots]. \qquad (25.8)$$

The spur (25.6) should therefore be evaluated under the restriction

$$(m+\omega-\omega')^2 = m^2 + (\mathbf{k}-\mathbf{k}')^2, \qquad (25.9)$$

or

$$kk' = \mathbf{k}\mathbf{k}' - \omega\omega' = m(\omega'-\omega). \qquad (25.10)$$

From the conservation of energy and momentum, it follows that the energy ω' of the scattered photon is a single-valued function of the scattering angle. Therefore we can give (25.8) as a function of only ω and ω'.

To work out the spur (25.6), we write

$$\gamma e'\gamma e\frac{i\gamma k}{2qk} + \gamma e\gamma e'\frac{i\gamma k'}{2qk'} = \gamma e\gamma e' i\gamma a + 2ee'\frac{i\gamma k}{2qk}\,, \qquad (25.11)$$

with

$$a = \frac{k'}{2qk'} - \frac{k}{2qk}. \qquad (25.12)$$

The vector a has the property that

$$qa = 0. \qquad (25.13)$$

Now we break the spur into four parts

$$\mathrm{Sp}\,[\cdots] = \mathrm{Sp}\,[\mathrm{I}] + \mathrm{Sp}\,[\mathrm{II}] + \mathrm{Sp}\,[\mathrm{III}] + \mathrm{Sp}\,[\mathrm{IV}]\,, \qquad (25.14)$$

where

$$\mathrm{Sp}\,[\mathrm{I}] = \mathrm{Sp}\,[\gamma e\gamma e' i\gamma a\,(i\gamma q - m)\, i\gamma a\gamma e'\gamma e\,(i\gamma(q+k-k') - m)], \quad (25.15)$$

---

1. Explicit equations for the various possible polarizations of the electrons have been given by W. Franz, Ann. Phys. 33, 689 (1938). See also F. W. Lipps and H. A. Tolhoek, Physica, Haag 20, 85, 395 (1954).

$$\mathrm{Sp}\,[\mathrm{II}] = 4(e\,e')^2\,\mathrm{Sp}\left[\frac{i\gamma k}{2qk}(i\gamma q - m)\frac{i\gamma k}{2qk}(i\gamma(q+k-k') - m)\right], \quad (25.16)$$

$$\left.\begin{aligned} &\mathrm{Sp}\,[\mathrm{III}] = \mathrm{Sp}\,[\mathrm{IV}] = \\ &\quad = 2(e\,e')\,\mathrm{Sp}\left[\gamma e \gamma e' i\gamma a(i\gamma q - m)\frac{i\gamma k}{2qk}(i\gamma(q+k-k') - m)\right]. \end{aligned}\right\} \quad (25.17)$$

Using (25.13) and (25.3), (25.15) simplifies to

$$\left.\begin{aligned} \mathrm{Sp}\,[\mathrm{I}] &= -\mathrm{Sp}\,[(i\gamma q+m)\gamma e\gamma e' i\gamma a i\gamma a\gamma e'\gamma e(i\gamma(q+k-k')-m)] = \\ &= a^2\,\mathrm{Sp}\,[(i\gamma q+m)(i\gamma(q+k-k')-m)] = a^2\,\mathrm{Sp}\,[i\gamma q\, i\gamma(k-k')] = \\ &= -4a^2 q(k-k') = 2\frac{-k\,k'}{q\,k\cdot q\,k'}(q\,k' - q\,k) = 2\frac{(\omega-\omega')^2}{\omega\,\omega'}. \end{aligned}\right\} \quad (25.18)$$

Likewise, we obtain

$$\mathrm{Sp}\,[\mathrm{II}] = -4(e\,e')^2\,\mathrm{Sp}\left[\frac{i\gamma k}{2qk}i\gamma(q-k')\right] = 8(e\,e')^2\left(1 - \frac{k\,k'}{q\,k}\right), \quad (25.19)$$

$$\left.\begin{aligned} \mathrm{Sp}\,[\mathrm{III}] + \mathrm{Sp}\,[\mathrm{IV}] &= 4(e\,e')\,\mathrm{Sp}\left[(i\gamma q + m)\gamma e\gamma e' i\gamma a\frac{i\gamma k}{2qk}i\gamma k'\right] = \\ &= 4(e\,e')^2\,\mathrm{Sp}\left[i\gamma q\frac{i\gamma k'}{2qk'}\frac{i\gamma k}{2qk}i\gamma k'\right] = 8(e\,e')^2\frac{k\,k'}{q\,k}. \end{aligned}\right\} \quad (25.20)$$

Accordingly,

$$\mathrm{Sp}\,[\cdots] = 8(e\,e')^2 + 2\frac{(\omega-\omega')^2}{\omega\,\omega'}. \quad (25.21)$$

Using (25.21), Eq. (25.8) becomes

$$\frac{d\sigma}{d\Omega_{k'}} = \left(\frac{e^2}{4\pi}\right)^2 \frac{1}{4}\frac{1}{m^2}\left(\frac{\omega'}{\omega}\right)^2\left[\frac{(\omega-\omega')^2}{\omega\,\omega'} + 4(e\,e')^2\right]. \quad (25.22)$$

If we introduce the scattering angle Θ , we obtain the so-called "Klein-Nishina formula"[1] for the Compton effect:

$$\frac{d\sigma}{d\Omega_{k'}} = \left(\frac{e^2}{4\pi}\right)^2 \frac{1}{4[m+\omega(1-\cos\Theta)]^2}\left[\frac{\omega^2(1-\cos\Theta)^2}{m[m+\omega(1-\cos\Theta)]} + 4(e\,e')^2\right]. \quad (25.23)$$

In many applications, the directions of polarization of the incoming and outgoing photons are not known. Then we must sum and average over the directions of polarization of the photons in a manner similar to that used previously for electrons:

$$\tfrac{1}{2}\sum_{\lambda,\lambda'}(e\,e')^2 = \tfrac{1}{2}(1+\cos^2\Theta)\ , \quad (25.24)$$

$$\frac{d\sigma}{d\Omega_{k'}} = \left(\frac{e^2}{4\pi}\right)^2 \frac{1}{2}\frac{1}{[m+\omega(1-\cos\Theta)]^2}\left[\frac{\omega^2(1-\cos\Theta)^2}{m[m+\omega(1-\cos\Theta)]} + 1 + \cos^2\Theta\right]. \quad (25.25)$$

With a view toward later results, we shall look at the limiting case of (25.25) when $\omega/m \ll 1$. We find

1. O. Klein and Y. Nishina, Z. Physik 52, 853 (1929).

$$\frac{d\sigma}{d\Omega_{k'}} = \left(\frac{e^2}{4\pi m}\right)^2 \frac{1}{2} (1 + \cos^2\Theta). \tag{25.26}$$

The quantity $\frac{e^2}{4\pi m}$ is the so-called classical electron radius which we shall denote by r_0. Equation (25.26) agrees completely with the result that is obtained from classical electromagnetic theory. The quantum mechanical effects first show up when the energy of the incident photon is of the same order as the rest mass of the electron. Integrating (25.26) over the angles, we find

$$\sigma_{\text{tot}} = \frac{8\pi}{3} r_0^2 = \sigma_T, \tag{25.27}$$

where σ_T is the so-called Thomson cross section.

The complete expression (25.25) can also be integrated over solid angles. The calculation is elementary but tedious. As the result, we obtain

$$\sigma_{\text{tot}} = \pi r_0^2 \left[\log(1 + 2\lambda)\left(\frac{1}{\lambda} - 2\frac{1+\lambda}{\lambda^3}\right) + \frac{4(1+\lambda)^2}{\lambda^2(1+2\lambda)} - \frac{2(1+3\lambda)}{(1+2\lambda)^2}\right], \tag{25.28}$$

with

$$\lambda = \frac{\omega}{m}. \tag{25.29}$$

If λ becomes very large, the cross section in this approximation goes to zero at a rate given by

$$\sigma_{\text{tot}} \approx \pi r_0^2 \frac{1}{\lambda}\left[\log(1 + 2\lambda) + \frac{1}{2}\right]. \tag{25.30}$$

As we shall see later, the higher terms in the S-matrix have no effect on the result (25.27). This is to be expected because here we are dealing only with a classical effect. For very high energies, on the other hand, the higher-order corrections to (25.30) become very large and the result (25.30) actually loses its meaning. However, the energy at which these corrections are evident is so large that these limitations on the region of validity of (25.28) have no experimental interest at present. The Klein-Nishina formula has been fully confirmed by all experiments up to now. For the theory, we refer the reader to the book by Heitler, which has already been mentioned,[1] and to Vol. XXXIV of this handbook.[2]

26. Bremsstrahlung and Pair Production by Photons in an External Field

As the next, slightly more involved application, we consider the Born approximation to the cross section for bremsstrahlung. Here the process is that an electron, passing through an external electromagnetic field $A_\mu^{\text{äuss}}(x)$, emits a photon. As before, we shall take account of both the external field and the radiation

1. See, for example, footnote 1, p. 107.

2. Handbuch der Physik, edited by S. Flügge, Springer-Verlag, Heidelberg.

field only to first order.[1] With this approximation, we obtain the S-matrix element of interest from (25.1) by replacing the photon which is absorbed by the external field:

$$\left.\begin{aligned}\langle q', k|S|q\rangle = -\frac{e^2}{2}\int\int dx'\,dx''\langle q'|\bar{\psi}^{(0)}(x'')|0\rangle\gamma_{\nu_2}S_F(x''-x')\gamma_{\nu_1}\langle 0|\psi^{(0)}(x')|q\rangle\times\\ \times\left[\langle k|A^{(0)}_{\nu_2}(x'')|0\rangle A^{\text{äuss}}_{\nu_1}(x') + \langle k|A^{(0)}_{\nu_1}(x')|0\rangle A^{\text{äuss}}_{\nu_2}(x'')\right].\end{aligned}\right\}\quad(26.1)$$

As in Sec. 24, we assume that the external field is time-independent. Using the notation of (24.7), we obtain, in a manner analogous to the derivation of (25.2) from (25.1),

$$\left.\begin{aligned}\langle q', k|S|q\rangle = \frac{i e^2}{V^{\frac{3}{2}}}\frac{1}{\sqrt{2\omega}}\bar{u}^{(r')}(q')\left[\gamma_\nu\frac{i\gamma(q-k)-m}{-2qk}\gamma e + \gamma e\frac{i\gamma(q'+k)-m}{2q'k}\gamma_\nu\right]\times\\ \times\ u^{(r)}(q)\,A^{\text{äuss}}_\nu(\boldsymbol{q}-\boldsymbol{q}'-\boldsymbol{k})\,2\pi\,\delta(q_0-q'_0-k_0).\end{aligned}\right\}\quad(26.2)$$

As before, we sum over the two possible polarizations of the final electron and average over the polarizations of the initial electron. After a few transformations, we can write the cross section for the process in the following form:

$$d\sigma = \frac{e^4}{8(2\pi)^5}\frac{p'}{p}\frac{d\omega}{\omega}\,d\Omega_k\,d\Omega_{q'}\,T_{\mu\nu}A^{\text{äuss}}_\mu(\boldsymbol{q}-\boldsymbol{q}'-\boldsymbol{k})\,A^{\text{äuss}}_\nu(\boldsymbol{k}+\boldsymbol{q}'-\boldsymbol{q}).\quad(26.3)$$

In (26.3), p and p' stand for the absolute value of the momenta $\boldsymbol{q}$ and $\boldsymbol{q}'$, while ω is the frequency of the photon, $\omega=|\boldsymbol{k}|$. The solid angles of the final electron and photon are $d\Omega_{q'}$ and $d\Omega_k$ and $T_{\mu\nu}$ is used for the quantity

$$\left.\begin{aligned}T_{\mu\nu} = \frac{\omega^2}{2}\,\mathrm{Sp}\left[\left(\gamma_\mu\frac{i\gamma(q-k)-m}{-2qk}\gamma e + \gamma e\frac{i\gamma(q'+k)-m}{2q'k}\gamma_\mu\right)(i\gamma q - m)\times\right.\\ \left.\times\left(\gamma e\frac{i\gamma(q-k)-m}{-2qk}\gamma_\nu + \gamma_\nu\frac{i\gamma(q'+k)-m}{2q'k}\gamma e\right)(i\gamma q' - m)\right].\end{aligned}\right\}\quad(26.4)$$

The spur (26.4) and the cross section (26.3) must be evaluated subject to the condition

$$E \equiv \sqrt{p^2+m^2} = \sqrt{p'^2+m^2}+\omega \equiv E'+\omega.\quad(26.5)$$

Equation (26.5) obviously expresses energy conservation for the bremsstrahlung process. There is no momentum conservation because the external field can take up an arbitrary momentum. This is expressed in (26.2) by the fact that we have only a single delta function for the time component of the vector $q-q'-k$. The spatial component of this vector, which we shall call Q from now on, enters as the argument of the Fourier components of the ex-

1. For a more complete treatment of the external field, see H.A. Bethe and L. C. Maximon, Phys. Rev. 93, 768 (1954); H. Davies, H. A. Bethe and L. C. Maximon, Phys. Rev. 93, 788 (1954).

ternal field.

If we replace the polarization vector e by the vector k in one of the brackets of (26.4), the expression can be transformed as follows:

$$\left.\begin{aligned}&\gamma_\mu \frac{i\gamma(q-k)-m}{-2qk}\gamma k+\gamma k\frac{i\gamma(q'+k)-m}{2q'k}\gamma_\mu=\\&=-i\gamma_\mu+\gamma_\mu\gamma k\frac{i\gamma q+m}{2qk}+i\gamma_\mu-\frac{i\gamma q'+m}{2q'k}\gamma k\gamma_\mu=\\&=\gamma_\mu\gamma k\frac{i\gamma q+m}{2qk}-\frac{i\gamma q'+m}{2q'k}\gamma k\gamma_\mu.\end{aligned}\right\}\quad(26.6)$$

In (26.6) we have used the fact that $k^2=0$. The factor $i\gamma q+m$ clearly gives zero if it operates on the function $u^{(r)}(q)$ in (26.2) or on the factor $i\gamma q-m$ in (26.4). In a similar way, the factor $i\gamma q'+m$ gives zero everywhere and we see that the tensor $T_{\mu\nu}$ in (26.4) vanishes if one of the factors e is replaced by k. This is actually only a consequence of the gauge invariance of the theory and can be shown in x-space by the use of the continuity equation for the current operator $j_\mu^{(0)}(x)$.

If we are not interested in the polarization of the photon,[1] this will allow us to simplify the evaluation of (26.4) somewhat. In view of our previous remarks, we obtain the same value for the spur (26.4) if we take the polarization vector e as longitudinal or scalar. Therefore we have

$$\left.\begin{aligned}\sum_{\substack{\text{transv.}\\\text{Photons}}}T_{\mu\nu}=\frac{\omega^2}{2}\,\mathrm{Sp}\Big[&\Big(\gamma_\mu\frac{i\gamma(q-k)-m}{-2qk}\gamma_\lambda+\gamma_\lambda\frac{i\gamma(q'+k)-m}{2q'k}\gamma_\mu\Big)(i\gamma q-m)\times\\\times&\Big(\gamma_\lambda\frac{i\gamma(q-k)-m}{-2qk}\gamma_\nu+\gamma_\nu\frac{i\gamma(q'+k)-m}{2q'k}\gamma_\lambda\Big)(i\gamma q'-m)\Big].\end{aligned}\right\}(26.7)$$

The expression (26.7) can be further simplified by means of the identities

$$\gamma_\lambda\gamma_\mu\gamma_\lambda=-2\gamma_\mu\,, \qquad (26.8)$$

$$\gamma_\lambda\gamma_\mu\gamma_\nu\gamma_\lambda=4\delta_{\mu\nu}\,. \qquad (26.9)$$

The Eqs. (26.8) and (26.9) can be proved immediately from (12.2). In principle, further reduction of (26.7) does not involve any new problems. We therefore give only the result:

---

1. The cross section for arbitrary polarization of the photon has been worked out by M. M. May, Phys. Rev. 84, 265 (1951) and by R. L. Gluckstern, M. H. Hull and G. Breit, Phys. Rev. 90, 1030 (1953). For polarized electrons, the cross section has been given for a few specialized cases by K. W. McVoy, Phys. Rev. 106, 828 (1957) and K. W. McVoy and F. J. Dyson, Phys. Rev. 106, 1360 (1957). General results for polarized electrons and photons will be found in A. Claesson, Ark. Fysik 12, 569 (1957).

$$\sum_{\substack{\text{transv.}\\ \text{Photons}}} T_{\mu\nu} = \frac{\omega^2}{2}\left[\frac{A_{\mu\nu}}{(q\,k)^2} + \frac{B_{\mu\nu}}{(q'\,k)^2} + \frac{C_{\mu\nu}}{q\,k\cdot q'\,k}\right], \tag{26.10}$$

$$\begin{aligned} A_{\mu\nu} = {} & 4\,(m^2 - k\,q)\,[k_\mu q'_\nu + k_\nu q'_\mu - \delta_{\mu\nu}\,(k\,q' + m^2)] - \\ & - 4m^2\,[q_\mu q'_\nu + q_\nu q'_\mu - \delta_{\mu\nu}(q\,q' + 2m^2)]\,, \end{aligned} \tag{26.11}$$

$$\begin{aligned} B_{\mu\nu} = {} & -4\,(m^2 + k\,q')\,[k_\mu q_\nu + k_\nu q_\mu - \delta_{\mu\nu}\,(k\,q - m^2)] - \\ & - 4m^2\,[q_\mu q'_\nu + q_\nu q'_\mu - \delta_{\mu\nu}\,(q\,q' + 2m^2)]\,, \end{aligned} \tag{26.12}$$

$$\begin{aligned} C_{\mu\nu} = {} & -8m^2 k_\mu k_\nu + 4k\,q'(2q_\mu q_\nu + q_\mu q'_\nu + q_\nu q'_\mu - 2\delta_{\mu\nu}\,q\,q') - \\ & - 4k\,q(2q'_\mu q'_\nu + q'_\mu q_\nu + q'_\nu q_\mu - 2\delta_{\mu\nu}\,q\,q') + \\ & + 4q\,q'\,[k_\nu(q'_\mu - q_\mu) + k_\mu(q'_\nu - q_\nu) - 2q_\mu q'_\nu - 2q'_\mu q_\nu + \\ & + 2\delta_{\mu\nu}\,(q\,q' + m^2)]\,. \end{aligned} \tag{26.13}$$

The cross section must be independent of the choice of gauge of the external field. In p-space this means that the tensor $T_{\mu\nu}$ must satisfy the conditions

$$T_{\mu\nu}(q_\nu - q'_\nu - k_\nu) = T_{\mu\nu}(q_\mu - q'_\mu - k_\mu) = 0. \tag{26.14}$$

Equations (26.14) can be verified either by means of (26.4) and a transformation similar to (26.6) or by the use of the explicit forms (26.10) through (26.13). This last proof can serve as a check that no algebraic mistakes have been made in the rather tedious reduction of the spur (26.7). These considerations about the use of the gauge invariance of the theory could also have been used as a check in the previous sections: in the summation over polarizations of the photons in Compton scattering or in (24.14). In the Compton scattering it must be remembered that Eq. (25.3) is not satisfied for longitudinal and scalar photons; consequently the transformation which led from (25.2) to (25.4) cannot be done if the summation over transverse photons is carried out as in (26.7).

From (26.3) and (26.10) through (26.13), we can find the cross section for bremsstrahlung in an arbitrary external field if the field is weak enough so that the Born approximation is meaningful. In most applications the external field is the Coulomb field of the nucleus, and for this case the general equations specialize to

$$A_\mu^{\text{äuss}}(Q) = -i\,\delta_{\mu 4}\frac{Z e}{Q^2}\,, \tag{26.15}$$

$$d\sigma = \frac{Z^2}{(2\pi)^2}\left(\frac{e^2}{4\pi}\right)^3 \frac{p'}{p}\frac{d\omega}{\omega}\left[-\frac{T_{44}}{(Q^2)^2}\right] d\Omega_k\, d\Omega_{q'}\,. \tag{26.16}$$

The component T_{44} can be obtained from (26.10) through (26.13):

$$\left.\begin{aligned} -T_{44} = {} & \frac{2\omega}{-p\cos\Theta + E}\left[m^2 + \omega(p'\cos\Theta' + E')\right] - \frac{2m^2}{(p\cos\Theta - E)^2}\times \\ & \times\left[E'^2 - \omega p'\cos\Theta' + p p'\cos\vartheta + m^2\right] + \\ & + \frac{2\omega}{-p'\cos\Theta' + E'}\left[m^2 + \omega(p\cos\Theta + E)\right] - \frac{2m^2}{(p'\cos\Theta' - E')^2}\times \\ & \times\left[E^2 - \omega p\cos\Theta + p p'\cos\vartheta + m^2\right] + \\ & + \frac{4\omega E}{E' - p'\cos\Theta'}(E + E') - \frac{4\omega E'}{E - p\cos\Theta}(E + E') + \\ & + \frac{4(EE' - pp'\cos\vartheta)\left(2EE' + \omega^2 - \frac{Q^2}{2}\right) - 4m^2\omega^2}{(E - p\cos\Theta)(E' - p'\cos\Theta')}. \end{aligned}\right\} \quad (26.17)$$

In (26.17) the angles Θ, Θ', and ϑ are defined by

$$\boldsymbol{q}\boldsymbol{q}' = pp'\cos\vartheta\,, \quad (26.18)$$

$$\boldsymbol{q}\boldsymbol{k} = \omega p\cos\Theta\,, \quad (26.19)$$

$$\boldsymbol{q}'\boldsymbol{k} = \omega p'\cos\Theta'. \quad (26.20)$$

Rather than use ϑ, we introduce the angle Φ between the $(\boldsymbol{q}, \boldsymbol{k})$ and $(\boldsymbol{q}', \boldsymbol{k})$ planes:

$$\cos\vartheta = \cos\Theta\cos\Theta' + \sin\Theta\sin\Theta'\cos\Phi. \quad (26.21)$$

Then (26.17) can be written in the form given by Bethe and Heitler,[1]

$$\left.\begin{aligned} -T_{44} = {} & \frac{p^2\sin^2\Theta}{(E - p\cos\Theta)^2}(4E'^2 - Q^2) + \frac{p'^2\sin^2\Theta'}{(E' - p'\cos\Theta')^2}(4E^2 - Q^2) - \\ & - 2\frac{pp'\sin\Theta\sin\Theta'\cos\Phi}{(E - p\cos\Theta)(E' - p'\cos\Theta')}(2E^2 + 2E'^2 - Q^2) + \\ & + 2\omega^2\frac{p^2\sin^2\Theta + p'^2\sin^2\Theta'}{(E - p\cos\Theta)(E' - p'\cos\Theta')}. \end{aligned}\right\} \quad (26.22)$$

In (26.22) the quantity $Q^2 = (\boldsymbol{q} - \boldsymbol{q}' - \boldsymbol{k})^2$ has often been used, rather than $pp'\cos\vartheta$. For the comparison of (26.16) and (26.22) with experiments, and for integrations over the angles, etc., we refer the reader to Heitler's book.[1]

The cross section for another process can be obtained directly from the previous calculation: the production of an electron pair by a photon which goes through an external field. The corresponding element of the S-matrix is

$$\left.\begin{aligned} \langle q, q'|\, S\, |k\rangle = {} & -\frac{e^2}{2}\iint dx'\,dx''\langle q|\,\bar{\psi}^{(0)}(x'')\,|0\rangle\gamma_{\nu_2} S_F(x'' - x')\gamma_{\nu_1}\times \\ & \times\langle q'|\,\psi^{(0)}(x')\,|0\rangle\left[\langle 0|\,A^{(0)}_{\nu_2}(x'')\,|k\rangle A^{\text{äuss}}_{\nu_1}(x') + \langle 0|\,A^{(0)}_{\nu_1}(x')\,|k\rangle A^{\text{äuss}}_{\nu_2}(x'')\right]. \end{aligned}\right\} \quad (26.23)$$

1. H. A. Bethe and W. Heitler, Proc. Roy. Soc. Lond. A146, 83 (1934). See also Heitler, Quantum Theory of Radiation, Third Ed., Oxford, 1954.

In Eq. (26.23), $|k\rangle$ denotes the incident photon, $|q\rangle$ the electron produced, and $|q'\rangle$ the positron. Rather than (26.2), we find

$$\left.\begin{aligned}\langle q,q'|S|k\rangle = \frac{i e^2}{V^{\frac{3}{2}}}\frac{1}{\sqrt{2\omega}}\bar{u}^{(r)}(q)\left[\gamma_\mu\frac{i\gamma(q'-k)+m}{2q'k}\gamma e+\gamma e\frac{i\gamma(q-k)-m}{-2qk}\gamma_\mu\right]\times \\ \times u^{(r')}(-q')\,A_\mu^{\text{äuss}}(\boldsymbol{k}-\boldsymbol{q}-\boldsymbol{q}')\,2\pi\,\delta(\omega-E-E').\end{aligned}\right\}\quad(26.24)$$

As before, the index r in (26.24) is 1 or 2 and the index r' must now take the values 3 or 4. In working out the cross section it must be remembered that the density of final states is now different, that in the sum over spins in the final state, (12.13) rather than (12.12) must be used, and that the possible polarizations of the photon must be averaged over. Carrying out all this, we obtain

$$d\sigma = \frac{e^4}{8(2\pi)^5}\frac{p\,p'\,dE}{\omega^3}\,d\Omega_q\,d\Omega_{q'}\,T_{\mu\nu}^{(1)}\,A_\mu^{\text{äuss}}(\boldsymbol{k}-\boldsymbol{q}-\boldsymbol{q}')\,A_\nu^{\text{äuss}}(\boldsymbol{q}+\boldsymbol{q}'-\boldsymbol{k}),\quad(26.25)$$

$$\left.\begin{aligned}T_{\mu\nu}^{(1)} = \frac{\omega^2}{2}\,\mathrm{Sp}\left[\left(\gamma_\mu\frac{i\gamma(q'-k)+m}{2q'k}\gamma_\lambda+\gamma_\lambda\frac{i\gamma(q-k)-m}{-2qk}\gamma_\mu\right)(i\gamma q'+m)\times\right. \\ \left.\times\left(\gamma_\lambda\frac{i\gamma(q'-k)+m}{2q'k}\gamma_\nu+\gamma_\nu\frac{i\gamma(q-k)-m}{-2qk}\gamma_\lambda\right)(i\gamma q-m)\right].\end{aligned}\right\}\quad(26.26)$$

If we compare (26.26) with (26.7), we see that

$$T_{\mu\nu}^{(1)}(k,q,q') = -T_{\mu\nu}(-k,-q',q) = -T_{\mu\nu}(k,-q,q').\quad(26.27)$$

We can obtain the new spur from (26.10) through (26.13) if we replace q by $-q$ everywhere, or, what is the same thing, E by $-E$ and Θ by $\pi-\Theta$ and change the overall sign. In particular, for the special case of pair production by a light quantum in a Coulomb field, we obtain the cross section

$$d\sigma = \frac{Z^2}{(2\pi)^2}\left(\frac{e^2}{4\pi}\right)^3\frac{p\,p'\,dE}{\omega^3}\,d\Omega_q\,d\Omega_{q'}\left[-\frac{T_{44}^{(1)}}{(\boldsymbol{Q}^2)^2}\right],\quad(26.28)$$

with

$$\left.\begin{aligned}-T_{44}^{(1)} = -\left[\frac{p^2\sin^2\Theta}{(E-p\cos\Theta)^2}(4E'^2-\boldsymbol{Q}^2)+\frac{p'^2\sin^2\Theta'}{(E'-p'\cos\Theta')^2}(4E^2-\boldsymbol{Q}^2)-\right. \\ \left.-2\frac{p\,p'\sin\Theta\sin\Theta'\cos\Phi}{(E-p\cos\Theta)(E'-p'\cos\Theta')}\left[2(E^2+E'^2)-Q^2\right]-2\omega^2\frac{p^2\sin^2\Theta+p'^2\sin^2\Theta'}{(E-p\cos\Theta)(E'-p'\cos\Theta')}\right]\end{aligned}\right\}\quad(26.29)$$

and

$$\boldsymbol{Q} = \boldsymbol{k}-\boldsymbol{q}-\boldsymbol{q}'.\quad(26.30)$$

Equations (26.28) through (26.30) as well as (26.16) and (26.22) are well confirmed by experiment, at least in the energy region where the Born approximation is expected to be good.[1]

1. W. Heitler, Quantum Theory of Radiation, Third Ed., Oxford, 1954.

27. Scattering of Two Electrons from Each Other

We now consider the problem of two incident electrons with energy-momentum vectors p and q which collide with each other and emerge again with energy-momentum vectors p' and q'. For initial and final states we therefore have quantities of the type

$$|p, q\rangle = a^{*(r)}(p)\, a^{*(r')}(q)\, |0\rangle. \tag{27.1}$$

Since the operators $a^{*(r)}(p)$ and $a^{*(r')}(q)$ anticommute with each other, it is clear that

$$|p, q\rangle = -\,|q, p\rangle. \tag{27.2}$$

The appropriate S-matrix element is

$$\left.\begin{aligned}\langle q', p'|S|p, q\rangle = &-\frac{e^2}{2}\iint dx'\, dx''\, [\langle q'|\bar\psi^{(0)}(x')|0\rangle \gamma_{\nu_1}\langle 0|\psi^{(0)}(x')|q\rangle \times\\ &\times \langle p'|\bar\psi^{(0)}(x'')|0\rangle\gamma_{\nu_2}\langle 0|\psi^{(0)}(x'')|p\rangle - \langle q'|\bar\psi^{(0)}(x')|0\rangle\gamma_{\nu_1}\langle 0|\psi^{(0)}(x')|p\rangle\times\\ &\times\langle p'|\bar\psi^{(0)}(x'')|0\rangle\gamma_{\nu_2}\langle 0|\psi^{(0)}(x'')|q\rangle]\,\delta_{\nu_1\nu_2}\, D_F(x'-x'') =\\ &= \frac{i e^2}{V^2}\left[\frac{\bar u(q')\gamma_\lambda u(q)\,\bar u(p')\gamma_\lambda u(p)}{(p-p')^2} - \frac{\bar u(q')\gamma_\lambda u(p)\,\bar u(p')\gamma_\lambda u(q)}{(p-q')^2}\right](2\pi)^4\,\delta(p+q-p'-q').\end{aligned}\right\} \tag{27.3}$$

The two terms in (27.3) occur because either the operator $\psi^{(0)}(x')$ can annihilate the particle q and the operator $\psi^{(0)}(x'')$ can annihilate the particle p, or the roles of the two operators can be reversed. The corresponding two possibilities for the operators $\bar\psi^{(0)}(x')$ and $\bar\psi^{(0)}(x'')$ concern the particles p' and q' and obviously give two terms where only x' and x'' are interchanged. As usual, this has been taken into account by omitting the factor 2! in the denominator. The two terms in (27.3) have the opposite signs. This follows from (27.2) and the fact that the two particles p and q are absorbed in different order in the two terms.

After averaging and summing over the spin states,[1] the cross section follows directly from (27.3):

$$\left.\begin{aligned}\sigma = &\left(\frac{e^2}{4\pi}\right)^2 \frac{1}{v_{\text{rel}}}\iint \frac{d^3p'\, d^3q'}{16 E_p E_{p'} E_q E_{q'}}\times\\ &\times\left[\frac{A}{[(p-p')^2]^2} + \frac{B}{[(p-q')^2]^2} + \frac{C}{(p-p')^2(p-q')^2}\right]\delta(p+q-p'-q').\end{aligned}\right\} \tag{27.4}$$

Here we employ the abbreviations

$$A = \operatorname{Sp}[\gamma_\lambda(i\gamma p - m)\gamma_\nu(i\gamma p' - m)]\cdot\operatorname{Sp}[\gamma_\lambda(i\gamma q - m)\gamma_\nu(i\gamma q' - m)], \tag{27.5}$$

$$B = \operatorname{Sp}[\gamma_\lambda(i\gamma p - m)\gamma_\nu(i\gamma q' - m)]\cdot\operatorname{Sp}[\gamma_\lambda(i\gamma q - m)\gamma_\nu(i\gamma p' - m)], \tag{27.6}$$

$$C = -2\operatorname{Sp}[\gamma_\lambda(i\gamma p - m)\gamma_\nu(i\gamma q' - m)\gamma_\lambda(i\gamma q - m)\gamma_\nu(i\gamma p' - m)]. \tag{27.7}$$

The quantity v_{rel} in (27.4) is the "relative velocity" of the two

1. For longitudinal polarization of the particles the cross section has been calculated by A. M. Bincer, Phys. Rev. 107, 1434. (1957).

incident particles. That is, it is a quantity which, in a coordinate system where one particle is at rest, reduces to the velocity of the other particle. For our purposes, it is not appropriate to define this quantity as the velocity with which an observer in the rest frame of one particle sees the other particle approach. Rather, we shall define it as a quantity such that $v_{\text{rel}} E_p E_q$ is an invariant. By this, we make the scattering cross section itself an invariant. In the rest system of the particle p we have

$$v_{\text{rel}} E_p E_q = m E_q \frac{|\boldsymbol{q}|}{E_q} = m \sqrt{E_q^2 - m^2} = \sqrt{m^2 E_q^2 - m^4}. \tag{27.8}$$

We therefore define

$$v_{\text{rel}} = \frac{1}{E_p E_q} \sqrt{(p\,q)^2 - m^4}. \tag{27.9}$$

In a coordinate system where the two particles are moving in the same direction with velocities v_1 and v_2, we have

$$v_{\text{rel}} = |v_1 - v_2|. \tag{27.10}$$

The expression (27.10) can reach the value 2 under certain conditions, that is, double the speed of light.

Because of the conservation of energy and momentum in (27.4), we can, for example, express the two scattering angles and the energy $E_{q'}$ as functions of $E_{p'}$ and the quantities describing the incident particles. In order to do this in an invariant fashion, we introduce the three parameters λ_1, λ_2, and γ by

$$\lambda_1 = \frac{(p-p')^2}{2m^2} = -\left(1 + \frac{p\,p'}{m^2}\right), \tag{27.11}$$

$$\lambda_2 = \frac{(p-q')^2}{2m^2} = -\left(1 + \frac{p\,q'}{m^2}\right), \tag{27.12}$$

$$\gamma = -\frac{p\,q}{m^2} = 1 + \lambda_1 + \lambda_2 . \tag{27.13}$$

The quantity γ is therefore given by the incident particles and (27.13) expresses λ_2 as a function of λ_1. We now define an invariant differential cross section by

$$\frac{d\sigma}{d\lambda_1} = \frac{r_0^2}{\sqrt{\gamma^2 - 1}} \frac{1}{\lambda_1^2 \lambda_2^2} \frac{1}{64 m^4} [A \lambda_2^2 + B \lambda_1^2 + C \lambda_1 \lambda_2] \cdot I. \tag{27.14}$$

The invariant integral I is given by

$$\left.\begin{aligned} I &= \iint \frac{d^3p'\,d^3q'}{E_{p'} E_{q'}} \delta(p + q - p' - q')\, \delta\left(\lambda_1 - \frac{(p-p')^2}{2m^2}\right) = \\ &= 4 \iint dp'\,dq'\,\delta(p'^2 + m^2)\,\delta(q'^2 + m^2)\,\Theta(p')\,\Theta(q')\,\delta(p+q-p'-q')\,\delta\left(\lambda_1 - \frac{(p-p')^2}{2m^2}\right). \end{aligned}\right\} \tag{27.15}$$

As in (24.32), the integrations present in (27.15) may be carried out easily in the coordinate frame in which $\boldsymbol{p} = 0$. The result is

$$I = 4\int dp'\,\delta(p'^2+m^2)\,\Theta(p')\,\delta((p+q-p')^2+m^2)\,\Theta(p+q-p')\times \\ \times\,\delta\left(\lambda_1 - \frac{(p-p')^2}{2m^2}\right) = \frac{2\pi m}{|\mathbf{q}|}\,\Theta(\lambda_1)\,\Theta\left(\frac{E_q}{m} - 1 - \lambda_1\right). \quad (27.16)$$

Consequently, in an arbitrary coordinate system we have

$$I = \frac{2\pi}{\sqrt{\gamma^2 - 1}}\,\Theta(\lambda_1)\,\Theta(\lambda_2). \quad (27.17)$$

The three quantities A, B, and C in (27.5) through (27.7) can easily be expressed as functions of the three invariants λ_1, λ_2, and γ by methods given previously. After some algebra and using the following equations (which follow from the conservation laws):

$$\left.\begin{aligned} p\,q &= p'\,q', \\ p\,q' &= p'\,q, \\ p\,p' &= q\,q', \end{aligned}\right\} \quad (27.18)$$

the result may be written as

$$\frac{1}{64m^4}[A\,\lambda_2^2 + B\,\lambda_1^2 + C\,\lambda_1\,\lambda_2] = \gamma^2(\lambda_1^2+\lambda_2^2) - \lambda_1\,\lambda_2(2\gamma - 1 - \lambda_1\,\lambda_2). \quad (27.19)$$

For the differential cross section (27.14) we therefore obtain

$$\frac{d\sigma}{d\lambda_1} = \frac{2\pi r_0^2}{\gamma^2 - 1}\,\frac{1}{\lambda_1^2\,\lambda_2^2}\left[\gamma^2(\lambda_1^2+\lambda_2^2) - \lambda_1\,\lambda_2(2\gamma - 1 - \lambda_1\,\lambda_2)\right]. \quad (27.20)$$

In (27.20) the two λ_i are positive and related by Eq. (27.13). The length r_0 has been defined by (25.26). The expression (27.20) was first derived by Møller[1] and therefore the problem discussed here is usually referred to as "Møller scattering".

Equation (27.20) is thoroughly confirmed by comparison with experiment. We shall not go into further details; however, such a discussion may be found in various textbooks.[2] We shall make a comparison with the formula (24.18) for the scattering in an external field because it illustrates some matters of principle. For this, we limit ourselves to the non-relativistic case. Here, in the center-of-momentum frame,

$$E_q = E_p = E_{q'} = E_{p'} = m + \frac{p^2}{2m}, \quad (27.21)$$

$$\mathbf{p} + \mathbf{q} = \mathbf{p}' + \mathbf{q}' = 0. \quad (27.22)$$

We introduce the scattering angle Θ by

1. C. Møller, Ann. Phys. 14, 531 (1932).

2. For example, N. F. Mott and H. S. W. Massey, Theory of Atomic Collisions, Second Ed., Oxford, 1949, p. 369. See also A. Ashkin, L. A. Page and W. M. Woodward, Phys. Rev. 94, 357 (1954).

$$\mathbf{p}\,\mathbf{p}' = p^2 \cos\Theta\,, \tag{27.23}$$

so that

$$\lambda_1 = \frac{2p^2}{m^2}\sin^2\frac{\Theta}{2}\,, \tag{27.24}$$

$$\lambda_2 = \frac{2p^2}{m^2}\cos^2\frac{\Theta}{2}\,. \tag{27.25}$$

In this approximation the cross section becomes

$$d\sigma = \frac{\pi}{8} r_0^2 \frac{m^4}{p^4}\left[\frac{1}{\sin^4\frac{\Theta}{2}} + \frac{1}{\cos^4\frac{\Theta}{2}} - \frac{1}{\cos^2\frac{\Theta}{2}\cdot\sin^2\frac{\Theta}{2}}\right]\sin\Theta\, d\Theta\,. \tag{27.26}$$

With $Z = 1$, and using the present notation, we can write (24.19) as

$$d\sigma = \frac{\pi}{8} r_0^2 \frac{m^4}{p^4 \sin^4\frac{\Theta}{2}} \sin\Theta\, d\Theta\,. \tag{27.27}$$

In (27.27) the polar angle Φ has been integrated out, i.e.,

$$d\Omega = 2\pi \sin\Theta\, d\Theta\,. \tag{27.28}$$

One might wonder why the two expressions (27.26) and (27.27) do not agree. Formally, the two "extra" terms on the right side of (27.26) arise because of the second term on the right side of (27.3). If only the first term in (27.3) had been considered, the definition of the scattering angle (27.23) would have given exactly (27.27). We obtained two terms in (27.3) because the two states $|p, q\rangle$ and $|q, p\rangle$ cannot be distinguished from each other. In other words, it is impossible to decide, for example, whether the electron p' was originally the electron p or the electron q. If we arbitrarily determine the scattering angle by (27.23), i.e., we consider p and p' as the "same" electron, then there is a certain probability that the two electrons in the final state have been "exchanged". We can then denote the "extra" terms in (27.26) as the exchange terms. Obviously, it is equally valid to denote the first or the second term in (27.3) as the exchange term, provided that the definition of the scattering angle is in terms of the appropriate electrons. The non-relativistic form (27.26), including the exchange effects, was first derived by Mott.[1]

28. Natural Line Width[2]

As the final example, we shall consider the emission of light by an electron in an external field which is strong enough so that the Born approximation cannot be used. A typical case is a time-

1. N. F. Mott, Proc. Roy. Soc. Lond. A126, 259 (1930).

2. See also the article written from the experimental point of view by R. G. Breene, Jr. in Vol. XXVII of this handbook (Handbuch der Physik, edited by S. Flügge, Springer-Verlag, Heidelberg).

independent external field with one or more bound states. For this we shall employ the method used in Sec. 16 for treating the external field, i.e., we assume that the solutions of the eigenvalue equation

$$\left[\alpha_k\left(-i\frac{\partial}{\partial x_k} - e A_k(x)\right) + m\gamma_4 + e A_0(x)\right] u_n(x) = E_n u_n(x) \qquad (28.1)$$

are known. In general the functions $u_n(x)$ range over both bound states and scattering states for the electron. We take $u_n(x)$ as the in-fields, i.e., the incoming ψ-field is developed in terms of the eigenstates $u_n(x)$ which are given by (16.8) with the operators (16.11) and (16.12):

$$\psi^{(0)}(x) = \sum_{E_n>0} [u_n(x)\, e^{-iE_n x_0} a^{(n)} + u_n'(x)\, e^{iE_n x_0} b^{*(n)}], \qquad (28.2)$$

$$\{a^{*(n)}, a^{(n')}\} = \{b^{*(n)}, b^{(n')}\} = \delta_{nn'}. \qquad (28.3)$$

As before, we treat the electromagnetic radiation field as a perturbation which causes transitions between the states $u_n(x)$. Using the procedure given in previous sections, we obtain the first non-vanishing S-matrix element for a transition from $u_n(x)$ to $u_{n'}(x)$ with the emission of a light quantum k:

$$\langle n', k|S|n\rangle = -\frac{e}{\sqrt{2V\omega}}\int d^3x\, \bar{u}_{n'}(x)\,\gamma_\nu\, u_n(x)\, e^{-ikx}\, e_\nu^{(\lambda)}\, 2\pi\,\delta(E_{n'}+\omega-E_n). \quad (28.4)$$

The transition probability per unit time, with the photon emitted into solid angle $d\Omega_k$, is therefore

$$\frac{\delta w}{\delta t} = \frac{e^2\omega}{8\pi^2}\left|\int d^3x\, \bar{u}_{n'}(x)\,\gamma\, e^{(\lambda)}\, u_n(x)\, e^{-ikx}\right|^2 d\Omega_k. \qquad (28.5)$$

If the initial and final states of the electron in (28.5) are both bound states, the whole expression is independent of V. That is, the transition probability per unit time cannot be made arbitrarily small if only V is sufficiently large. Consequently, the formalism developed in Sec. 20 cannot be correct, at least in the higher orders, since the transition probability is not time-independent.

In order to proceed further, we have to modify slightly the integration procedure for our differential equations, so that it is not required that the lifetime of the initial state be arbitrarily large. We shall restrict ourselves to an approximation which was originally developed by Weisskopf and Wigner.[1] A more general theory of these processes, in which higher orders can in principle be considered, has been developed by Heitler and coworkers.[2] Because the initial state is now "short-lived" we cannot prescribe

1. V. F. Weisskopf and E. Wigner, Z. Physik 63, 54 (1930).

2. E. Arnous and W. Heitler, Proc. Roy. Soc. Lond. A220, 290 (1953). This paper also has references to the older literature. See also F. Low, Phys. Rev. 88, 53 (1952).

the initial conditions of this problem for $x_0 = -\infty$ for, if we did, the initial state would no longer be occupied at finite times. Following Weisskopf and Wigner, we shall require that our system be in the state $|n\rangle$ at time $x_0 = 0$ and we shall use an "interaction picture" to treat the problem. This picture is defined by the following equations:

$$|x_0\rangle = U(x_0)\,|n\rangle\,, \tag{28.6}$$

$$i\,\frac{\partial U(x_0)}{\partial x_0} = H_1\big(\psi^{(0)}(x),\,A_\mu^{(0)}(x)\big)\,U(x_0)\,, \tag{28.7}$$

$$U(0) = 1. \tag{28.8}$$

Here $|x_0\rangle$ stands for the state vector at the time x_0 and H_1 is the interaction energy as a function of the operators $\psi^{(0)}$ of (28.2) as well as $A_\mu^{(0)}(x)$ of, for example, (5.28). The essential idea in this approach is actually only the boundary condition (28.8). In principle, the system of equations (28.6) through (28.8) is exact. However, following Weisskopf and Wigner, we shall approximate these exact equations by considering only the states $|n\rangle$ and $|n', k\rangle$ in the matrix multiplication of (28.7). That is, we consider only those states which are either the electron in the initial state and no photon or the electron in the final state and one photon. The physical significance of this approximation is not easy to grasp. Certainly it is clear that the probability that several photons should be present in the final state must be small, since the coupling with the electromagnetic field is weak. It does not follow directly from this that such states can also be neglected in the matrix multiplication of (28.7). At any rate, the suitability of this approximation is probably best decided by comparing the results obtained with those of experiment.

We define

$$\langle n\,|\,U(x_0)\,|\,n\rangle = a(x_0)\,, \tag{28.9}$$

$$\langle n', k\,|\,U(x_0)\,|\,n\rangle = b_k(x_0), \tag{28.10}$$

so that the equations to be solved in this approximation are

$$i\,\frac{\partial a(x_0)}{\partial x_0} = \sum_k c_k^*\, e^{-i\Delta\omega x_0}\, b_k(x_0)\,, \tag{28.11}$$

$$i\,\frac{\partial b_k(x_0)}{\partial x_0} = c_k\, e^{i\Delta\omega x_0}\, a(x_0)\,. \tag{28.12}$$

Here we have used the following notation:

$$\Delta\omega = E_{n'} + \omega - E_n\,, \tag{28.13}$$

$$\langle n', k\,|\,H_1\,|\,n\rangle = c_k\, e^{i\Delta\omega x_0}\,, \tag{28.14}$$

$$c_k = -\,i\,e\int \bar{u}_{n'}(\boldsymbol{x})\,\gamma\, e^{(\lambda)}\, u_n(\boldsymbol{x})\,\frac{e^{-i\boldsymbol{k}\boldsymbol{x}}}{\sqrt{2V\omega}}\,d^3x. \tag{28.15}$$

The boundary conditions for (28.11) and (28.12) are obtained from (28.8):

$$a(0) = 1\,, \tag{28.16}$$

$$b_k(0) = 0. \tag{28.17}$$

To solve these differential equations we can, for example, first eliminate the b_k:

$$b_k(x_0) = -i\,c_k \int_0^{x_0} dx_0'\, e^{i\Delta\omega x_0'}\, a(x_0')\,, \tag{28.18}$$

$$\frac{\partial a(x_0)}{\partial x_0} = -\sum_k |c_k|^2 \int_0^{x_0} dx_0'\, e^{-i\Delta\omega(x_0-x_0')}\, a(x_0'). \tag{28.19}$$

The equation for $a(x_0)$ can then be solved, for example, by means of the Laplace transformation. We set

$$a(E) = \int_0^\infty e^{-Ex_0}\, a(x_0)\, dx_0 \tag{28.20}$$

and transform (28.19) into an equation for $a(E)$. For the left side we obtain

$$\int_0^\infty e^{-Ex_0} \frac{\partial a(x_0)}{\partial x_0}\, dx_0 = [e^{-Ex_0}\, a(x_0)]_0^\infty + E\, a(E) = E\, a(E) - 1. \tag{28.21}$$

The integral on the right side can be transformed in a similar way:

$$\left.\begin{aligned} &\int_0^\infty e^{-Ex_0}\, dx_0 \int_0^{x_0} dx_0'\, e^{-i\Delta\omega(x_0-x_0')}\, a(x_0') = \\ &\qquad = \int_0^\infty dx_0'\, a(x_0') \int_{x_0'}^\infty dx_0\, e^{-(E+i\Delta\omega)x_0 + i\Delta\omega x_0'} = \frac{a(E)}{E+i\Delta\omega}. \end{aligned}\right\} \tag{28.22}$$

Substitution of (28.21) and (28.22) into (28.19) leads to the result for $a(E)$,

$$a(E) = \frac{1}{E + \sum_k \frac{|c_k|^2}{E+i\Delta\omega}}. \tag{28.23}$$

By means of the well known formula for inverting the Laplace transformation, we have

$$a(x_0) = \frac{1}{2\pi i} \int_{\varepsilon-i\infty}^{\varepsilon+i\infty} \frac{dE\, e^{Ex_0}}{E + \sum_k \frac{|c_k|^2}{E+i\Delta\omega}}. \tag{28.24}$$

In Eq. (28.24) the integration is along a path immediately to the right of the imaginary axis in the complex E-plane. If we introduce a real variable of integration z by $E = \varepsilon + iz$, then the integral (28.24) can be written in the following way:

$$\left.\begin{aligned} a(x_0) &= \frac{1}{2\pi i}\int_{-\infty}^{+\infty}\frac{dz\, e^{izx_0}e^{\varepsilon x_0}}{z - i\varepsilon - \sum_k \frac{|c_k|^2}{z+\Delta\omega - i\varepsilon}} = \\ &= \frac{1}{2\pi i}\int_{-\infty}^{+\infty}\frac{dz\, e^{izx_0}}{z - P\sum_k \frac{|c_k|^2}{z+\Delta\omega} - i\pi\sum_k |c_k|^2\delta(z-\Delta\omega)} . \end{aligned}\right\} \tag{28.25}$$

The last form on the right side of (28.25) follows if we let ε go to zero and use the relation

$$\lim_{\varepsilon\to 0}\frac{1}{a - i\varepsilon} = P\frac{1}{a} + i\pi\,\delta(a). \tag{28.26}$$

The evaluation of the integral (28.25) for arbitrary times is quite complicated and not very interesting physically. Because the electromagnetic coupling is "suddenly switched on" at $x_0=0$, it is to be expected that there will be "transients" for small times which will depend strongly on the choice of boundary conditions. These are therefore uninteresting. Consequently, we shall estimate the integral only for times so large that $x_0 E_n \gg 1$, and $x_0 E_{n'} \gg 1$. With the additional restriction that

$$\sum_k |c_k|^2\,\delta(\Delta\omega) \ll E_n - E_{n'}\,, \tag{28.27}$$

we can write

$$a(x_0) \approx \frac{1}{2\pi i}\int_{-\infty}^{+\infty}\frac{dz\, e^{izx_0}}{z - P\sum_k \frac{|c_k|^2}{\Delta\omega} - i\pi\sum_k |c_k|^2\delta(\Delta\omega)} = e^{-i\Delta E\, x_0 - \frac{\gamma}{2}x_0}, \tag{28.28}$$

$$\gamma = 2\pi\sum_k |c_k|^2\,\delta(\Delta\omega), \tag{28.29}$$

$$\Delta E = -P\sum_k \frac{|c_k|^2}{\Delta\omega}. \tag{28.30}$$

The probability that at time x_0 the system is still in the initial state is therefore

$$|a(x_0)|^2 = e^{-\gamma x_0}. \tag{28.31}$$

The lifetime of the state is therefore $1/\gamma$, where γ is just the total transition probability per unit time, calculated from (28.5).

For times such that the result (28.28) is correct, we obtain from (28.18),

$$b_k(x_0) = \frac{-c_k}{\Delta\omega - \Delta E + i\frac{\gamma}{2}}\left[e^{\left(i\Delta\omega - i\Delta E - \frac{\gamma}{2}\right)x_0} - 1\right]. \tag{28.32}$$

For small times ($\gamma x_0 \ll 1$) we obtain from this the probability that a photon has been emitted into solid angle $d\Omega_k$,

$$\left.\begin{aligned} d\Omega_k \int_0^\infty \frac{V\omega^2}{(2\pi)^3} |b_k(x_0)|^2\, d\omega = \\ = d\Omega_k \int_0^\infty \frac{|c_k|^2 \omega^2 V}{(2\pi)^3} \frac{\sin^2\left(\frac{\Delta'\omega\, x_0}{2}\right)}{\left(\frac{\Delta'\omega}{2}\right)^2 + \frac{\gamma^2}{16}} d\omega \approx 2\pi\, x_0 \frac{|c_k|^2 \omega^2 V}{(2\pi)^3}\Bigg|_{\Delta'\omega=0} d\Omega_k . \end{aligned}\right\} \quad (28.33)$$

Here we have used

$$\Delta'\omega = \Delta\omega - \Delta E. \quad (28.34)$$

Apart from a shift ΔE of the emitted frequency, this is just the result (28.5) with a transition probability proportional to the time. According to (28.31), during this time the probability that the system remains in the original state is practically 1. With these restrictions, we can use the simpler formalism which we developed earlier. However, for large times ($\gamma x_0 \gg 1$) we obtain

$$|b_k(x_0)|^2 = \frac{|c_k|^2}{(\Delta'\omega)^2 + \frac{1}{4}\gamma^2}. \quad (28.35)$$

The frequency of the emitted photon does not have a sharp value, but the emitted spectral line has a width of order γ. In addition, the maximum of the line is shifted by an amount ΔE .

The level shift is not of interest at the moment. We shall return to this problem in greater detail in Sec. 37 when we have further developed the formal techniques of the theory. Here we note that the result (28.35) for the line breadth can be understood at least qualitatively in terms of the uncertainty principle. If the state $|n\rangle$ has a finite lifetime γ^{-1}, then the state can be physically realized only during a time which is shorter than γ^{-1}. Now, since it is impossible in principle to define the energy of the state with greater accuracy than γ, the emitted spectral line must also have this uncertainty. According to this, it would be expected that if <u>both</u> the levels $|n\rangle$ and $|n'\rangle$ have finite lifetimes, the width of the emitted line ought to be the <u>sum</u> of two quantities γ. In the preceding calculation it has been implicitly assumed that the final state $|n'\rangle$ is the ground state of the system and that in the state $|n\rangle$ only transitions to the ground state of the system are allowed. If these restrictions are not made, then the other transitions have to be included in Eqs. (28.11) and (28.12). The calculation can be carried out following the method used above and the result is what would be expected from the uncertainty principle. We shall not explore the matter further here, but refer the reader to the work of Weisskopf and Wigner for the details.[1]

The fact that the result (28.35) can be understood by means of

1. V. F. Weisskopf and E. Wigner, Z. Physik <u>63</u>, 54 (1930).

the uncertainty relation at least makes it plausible that the approximate equations (28.11) and (28.12) have a physical meaning. Certainly this argument should not be overemphasized, since a similar result must come out of every approximation using a unitary matrix $U(x_0)$. In the usual perturbation calculations a power series is assumed for U and then it is terminated at the n-th term. The matrix

$$U = 1 + U^{(1)} + \cdots + U^{(n)} \tag{28.36}$$

is not exactly unitary and such a matrix is not suitable for treating the problem of the line width.

Accurate measurements of the natural line widths can be made only with difficulty since several effects such as doppler shift, collisions between the excited atoms, etc. also broaden the spectral lines. At the usual densities and temperatures these effects are very much larger than the natural line widths.[1] At least the previous measurements of the natural line widths[2] are not in contradiction to the above equations, although they cannot be regarded as definitely confirming them. More detailed discussion of this will be found in the article of R. G. Breene, Jr. in Vol. XXVII of this handbook.

---

1. See, for example, H. Margenau and W. W. Watson, Rev. Mod. Phys. 8, 22 (1936).

2. See W. Heitler, Quantum Theory of Radiation, Third Ed., Oxford, 1954, p. 188.

CHAPTER VI

RADIATIVE CORRECTIONS IN THE LOWEST ORDER*

29. Vacuum Polarization in an External Field. Charge Renormalization

Again we consider a system containing an external field which is weak enough so that the Born approximation can be used. In this section we shall not only be concerned with the S-matrix, but also with the field operators in the Heisenberg picture. We write the equations of motion for them as

$$\left(\gamma \frac{\partial}{\partial x} + m\right) \psi(x) = i e \gamma \left(A(x) + A^{\text{äuss}}(x)\right) \psi(x)\,, \tag{29.1}$$

$$\Box A_\mu(x) = -\frac{ie}{2}\,[\overline{\psi}(x), \gamma_\mu \psi(x)] \equiv -j_\mu(x)\,. \tag{29.2}$$

With the assumption that we can expand in powers of the quantized field $A_\mu(x)$ as well as the external field $A_\mu^{\text{äuss}}(x)$, we obtain the lowest order approximation to the ψ-operator:

$$\psi(x) = \psi^{(0)}(x) - i e \int S_R(x-x')\, \gamma\, [A^{(0)}(x') + A^{\text{äuss}}(x')]\, \psi^{(0)}(x')\, dx' + \cdots\,. \tag{29.3}$$

For the right side of (29.2), i.e., for the current operator, the corresponding expansion is

$$\left.\begin{aligned} j_\mu(x) = \frac{ie}{2}[\overline{\psi}^{(0)}(x), \gamma_\mu \psi^{(0)}(x)] + \frac{e^2}{2}\int dx' \,([\overline{\psi}^{(0)}(x), \gamma_\mu\, S_R(x-x')\, \gamma_\nu \psi^{(0)}(x')] + \\ + [\overline{\psi}^{(0)}(x')\, \gamma_\nu\, S_A(x'-x)\, \gamma_\mu, \psi^{(0)}(x)])\left(A_\nu^{(0)}(x') + A_\nu^{\text{äuss}}(x')\right) + \cdots\,. \end{aligned}\right\} \tag{29.4}$$

From (29.4) we see that the vacuum expectation value of the current operator does not vanish identically if an external field is present. With the aid of (15.29) we obtain in this approximation

---

* (Translator's note) The detailed comparison of the predictions of quantum electrodynamics with experiment is an active area of continuing research. Although this chapter provides an excellent introduction, a discussion of the higher-order calculations and the more recent experiments must be sought in the current literature. At the present time, a very useful starting point is the article "The Present Status of Quantum Electrodynamics", by Stanley J. Brodsky and Sidney D. Drell in Annual Review of Nuclear Science, edited by Emilio Segrè, J. Robb Grover, and H. Pierre Noyes, Vol. 20, Annual Reviews, Inc., Palo Alto, Calif., 1970.

$$\langle 0 | j_\mu(x) | 0 \rangle = \int dx' \, K_{\mu\nu}(x - x') \, A_\nu^{\text{äuss}}(x') \,, \tag{29.5}$$

$$\left. \begin{aligned} K_{\mu\nu}(x - x') = \frac{e^2}{2} \big(&\mathrm{Sp}\, [\gamma_\mu S_R(x - x') \gamma_\nu S^{(1)}(x' - x)] + \\ &+ \mathrm{Sp}\, [\gamma_\mu S^{(1)}(x - x') \gamma_\nu S_A(x' - x)] \big) . \end{aligned} \right\} \tag{29.6}$$

In order to study the structure of the integral (29.6) more closely, we introduce the Fourier representation

$$K_{\mu\nu}(x - x') = \frac{1}{(2\pi)^4} \int dp \, e^{ip(x-x')} K_{\mu\nu}(p) . \tag{29.7}$$

After using (15.20), (15.21), and (15.30), we obtain

$$\left. \begin{aligned} K_{\mu\nu}(p) = \frac{e^2}{16\pi^3} \iint & dp' \, dp'' \, \delta(p - p' + p'') \, \mathrm{Sp}\, [\gamma_\mu (i\gamma p' - m) \gamma_\nu (i\gamma p'' - m)] \times \\ \times \Big\{ & \delta(p'^2 + m^2) \left[P \frac{1}{p''^2 + m^2} - i\pi \varepsilon(p'') \, \delta(p''^2 + m^2) \right] + \\ + & \delta(p''^2 + m^2) \left[P \frac{1}{p'^2 + m^2} + i\pi \varepsilon(p') \, \delta(p'^2 + m^2) \right] \Big\} . \end{aligned} \right\} \tag{29.8}$$

The expression (29.8) differs from elements of the S-matrix obtained in the previous chapter in various respects, one of them being that it contains two four-dimensional p-integrations but only one "conservation law", $\delta(p - p' + p'')$. For the first time we are summing over an infinite number of intermediate states, and therefore we must deal with convergence difficulties. Before working out the details of (29.8), we shall study a few general properties of the kernel $K_{\mu\nu}(p)$.

From the conservation of charge,

$$\frac{\partial j_\mu(x)}{\partial x_\mu} = 0 \,, \tag{29.9}$$

we have

$$\frac{\partial}{\partial x_\mu} K_{\mu\nu}(x - x') = 0 , \tag{29.10}$$

or

$$p_\mu K_{\mu\nu}(p) = 0 . \tag{29.11}$$

According to (29.8), $K_{\mu\nu}(p)$ is a tensor which depends only upon the vector p. Consequently $K_{\mu\nu}(p)$ must be symmetric and of the form

$$K_{\mu\nu}(p) = G(p) \, p_\mu p_\nu + H(p) \, \delta_{\mu\nu} . \tag{29.12}$$

From (29.11) we obtain

$$p_\mu K_{\mu\nu}(p) = p_\nu [G(p) \, p^2 + H(p)] = 0 , \tag{29.13}$$

or

$$K_{\mu\nu}(p) = G(p) \, [p_\mu p_\nu - \delta_{\mu\nu} p^2] . \tag{29.14}$$

Finally from (29.14) it follows that

$$\frac{\partial}{\partial x'_\nu} K_{\mu\nu}(x-x') = 0. \tag{29.15}$$

Obviously this expresses the independence of the expression (29.5) from the gauge of the external field.

Equation (29.11) can be verified formally by means of (29.8). If we multiply the latter equation by $i p_\mu$, we find

$$\left.\begin{aligned} i p_\mu K_{\mu\nu}(p) = \frac{e^2}{16\pi^3} \iint dp'\, dp''\, \delta(p-p'+p'') \times \\ \times [\mathrm{Sp}\,[(i\gamma p''-m)(i\gamma p'+m)(i\gamma p'-m)\gamma_\nu] - \\ - \mathrm{Sp}\,[(i\gamma p''-m)(i\gamma p''+m)(i\gamma p'-m)\gamma_\nu] \times \{\cdots\} = \\ = \frac{i e^2}{4\pi^3} \iint dp'\, dp''\, \delta(p-p'+p'') [p'_\nu(p''^2+m^2) - p''_\nu(p'^2+m^2)] \times \{\cdots\}. \end{aligned}\right\} \tag{29.16}$$

By the use of $x\delta(x)=0$, we can simplify (29.16) to

$$i p_\mu K_{\mu\nu}(p) = \frac{i e^2}{4\pi^3} \iint dp'\, dp''\, \delta(p-p'+p'') [p'_\nu\, \delta(p'^2+m^2) - p''_\nu\, \delta(p''^2+m^2)]. \tag{29.17}$$

Each of the two terms in (29.17) is zero because integrals such as

$$I = \int dp'\, p'_\nu\, \delta(p'^2+m^2) \tag{29.18}$$

must vanish by considerations of symmetry. Here it must be noted that the integral (29.18) is actually divergent. Consequently, by a change of the origin of coordinates, an integral which does not vanish can readily be obtained from an integral such as (29.18). Despite this, we must require on physical grounds that (29.11) hold and therefore (29.18) may be taken as a definition of the value of the integral. If we observe that a calculation similar to (29.16) and (29.17) can also be done in x-space, we shall obtain this same convention. Thus we have

$$\frac{\partial}{\partial x_\mu} K_{\mu\nu}(x-x') = -e^2\, \delta(x-x')\, \mathrm{Sp}\,[\gamma_\nu S^{(1)}(0)] = 2 i e\, \delta(x-x') \langle 0 | j_\nu^{(0)}(x) | 0 \rangle. \tag{29.19}$$

As one can readily convince himself, the vacuum expectation value of the operator $j_\nu^{(0)}(x)$ contains exactly the integral I of (29.18). It must vanish, as the preceding discussion shows in detail.

By the use of (29.14) we can now write (29.5) in the following form:

$$\left.\begin{aligned} \langle 0 | j_\mu(x) | 0 \rangle = \frac{1}{(2\pi)^4} \iint dp\, dx'\, e^{ip(x-x')} G(p) \left[\Box A_\mu^{\text{äuss}}(x') - \frac{\partial^2 A_\nu^{\text{äuss}}(x')}{\partial x'_\mu \partial x'_\nu}\right] = \\ = -\frac{1}{(2\pi)^4} \iint dp\, dx'\, e^{ip(x-x')} G(p)\, j_\mu^{\text{äuss}}(x'), \end{aligned}\right\} \tag{29.20}$$

$$\left.\begin{aligned} G(p) = -\frac{1}{3p^2} K_{\mu\mu}(p) = \\ = -\frac{e^2}{48\pi^3 p^2} \iint dp'\, dp''\, \delta(p-p'+p'')\, \mathrm{Sp}\,[\gamma_\mu(i\gamma p'-m)\gamma_\mu(i\gamma p''-m)] \times \\ \times \left\{\delta(p'^2+m^2)\left[P\frac{1}{p''^2+m^2} - i\pi\,\varepsilon(p'')\,\delta(p''^2+m^2)\right] + \right. \\ \left. + \delta(p''^2+m^2)\left[P\frac{1}{p'^2+m^2} + i\pi\,\varepsilon(p')\,\delta(p'^2+m^2)\right]\right\}. \end{aligned}\right\} \tag{29.21}$$

Here the experimentally observable quantity is the sum of the original external current and the induced current (29.20). That is,

$$j_\mu^{\text{äuss}}(x) + \langle 0 | j_\mu(x) | 0 \rangle = \frac{1}{(2\pi)^4} \iint dp\, dx' \, [1 - G(p)]\, j_\mu^{\text{äuss}}(x')\, e^{ip(x-x')}. \quad (29.22)$$

From (29.22) or from (29.20), we see that if the function G were a constant independent of p, the effect of the interaction of the external field and the electron field would be only a multiplication of all currents by a constant. This would mean only a change in the unit of charge of the external current and therefore would be unmeasurable in principle. In fact, G is clearly not a constant; however, as this argument shows, we can add an arbitrary constant to G and this addition means only a change in the unit of charge. Observationally, the quantity of major interest is thus the change in G when p varies. For an unequivocal comparison of (29.22) with experiment, it is necessary to determine the arbitrary constant in (29.22) according to a certain convention. In principle, we can do this by specifying the value of $1-G$ for some arbitrary value of the "frequency" p. In particular, we shall require, for external fields which vary extremely slowly in time and space, that the external current and the observable current be identical. This may also be expressed by saying that the function G in (29.22) should vanish for $p=0$. Certainly the definition (29.21) for G does not contain an arbitrary constant and it is not to be expected that $G(0)=0$. We therefore define the observable currents as

$$j^{\text{beob}}(x) = \frac{1}{(2\pi)^4} \iint dp\, dx' \, e^{ip(x-x')} \, [1 - G(p) + G(0)]\, j_\mu^{\text{äuss}}(x'). \quad (29.23)$$

Again we stress that the introduction of the constant $G(0)$ in (29.23) signifies only an unequivocal determination of the unit of charge.[1] This procedure is known as "charge renormalization" in the literature.

In working out the function G explicitly, we note that according to (29.21) the imaginary part of G can be calculated easily because of the delta function occurring in it. We get

$$\left.\begin{aligned}
\operatorname{Im} G(p) &= -\frac{e^2}{6\pi^2 p^2} \int dp' \, [p'(p'-p) + 2m^2] \times \\
&\quad \times \delta(p'^2 + m^2)\, \delta\big((p-p')^2 + m^2\big)\, [\varepsilon(p') + \varepsilon(p-p')] = \\
&= -\frac{e^2}{12\pi^2 p^2} \left(m^2 - \frac{p^2}{2}\right) \int \frac{d^3 p'}{\sqrt{\mathbf{p}'^2 + m^2}} \times \\
&\quad \times \left\{ \delta\Big(p^2 - 2\mathbf{p}\mathbf{p}' + 2p_0 \sqrt{\mathbf{p}'^2 + m^2}\Big) \left[1 + \frac{p_0 - \sqrt{\mathbf{p}'^2 + m^2}}{|p_0 - \sqrt{\mathbf{p}'^2 + m^2}|} \right] + \right. \\
&\quad \left. + \delta\Big(p^2 - 2\mathbf{p}\mathbf{p}' - 2p_0 \sqrt{\mathbf{p}'^2 + m^2}\Big) \left[-1 + \frac{p_0 + \sqrt{\mathbf{p}'^2 + m^2}}{|p_0 + \sqrt{\mathbf{p}'^2 + m^2}|} \right] \right\}.
\end{aligned}\right\} \quad (29.24)$$

---

1. Certainly one could have changed the unit of charge by multiplying by a constant rather than by adding a term. However, in the approximation considered here, both of these procedures are completely equivalent.

The integral (29.24) is invariant and clearly vanishes if $-p^2$ is smaller than $4m^2$. Just as in the last chapter, we shall also evaluate this integral in a special coordinate frame: the frame in which the components of $\boldsymbol{p}$ vanish. In this system, (29.24) can be written in the following way:

$$\left.\begin{aligned} \operatorname{Im} G(p) &= \frac{e^2}{3\pi}\left(1+\frac{2m^2}{p_0^2}\right)\int\limits_0^{\sqrt{p_0^2-m^2}} \frac{x^2\,dx}{\sqrt{x^2+m^2}}\,\delta\!\left(2\,|p_0|\sqrt{x^2+m^2}-p_0^2\right)\frac{p_0}{|p_0|} = \\ &= \frac{e^2}{6\pi}\left(1+\frac{2m^2}{p_0^2}\right)\frac{1}{p_0}\sqrt{\frac{p_0^2}{4}-m^2}\,\Theta\!\left(\frac{|p_0|}{2}-m\right) = \\ &= \frac{e^2}{12\pi}\left(1+\frac{2m^2}{p_0^2}\right)\sqrt{1-\frac{4m^2}{p_0^2}}\,\frac{p_0}{|p_0|}\,\Theta\!\left(\frac{p_0^2}{4}-m^2\right), \end{aligned}\right\} \quad (29.25)$$

$$\Theta(x) = \frac{1}{2}\left[1+\varepsilon(x)\right]. \quad (29.25\text{a})$$

Thus in an arbitrary system of coordinates we have

$$\operatorname{Im} G(p) = \varepsilon(p)\,\frac{e^2}{12\pi}\left(1-\frac{2m^2}{p^2}\right)\sqrt{1+\frac{4m^2}{p^2}}\,\Theta\!\left(-\frac{p^2}{4}-m^2\right). \quad (29.26)$$

The imaginary part of G therefore contributes nothing to the term $G(0)$.

The direct evaluation of the real part of $G(p)$ from the definition (29.21) is quite tedious. We spare ourselves a considerable effort if we note that the function

$$G(x-x') = \frac{1}{(2\pi)^4}\int G(p)\,e^{ip(x-x')}\,dp \quad (29.27)$$

must vanish for $x_0' > x_0$ according to (29.6). This is also necessary if the induced current is to depend on the value of the external current only within the retarded light cone. If this were not the case, the theory would clearly be "acausal". From this property of the function (29.27), it follows that the integral

$$G(\boldsymbol{p}, p_0+i\eta) = \int dx\, G(x)\, e^{-i(\boldsymbol{p}\boldsymbol{x}-(p_0+i\eta)x_0)} \quad (29.28)$$

exists if η is greater than zero. The function $G(\boldsymbol{p}, z)$ in (29.28) is an analytic function of z with no singularities in the upper half plane. By the well known theorems of analysis[1] for such a function, the real and imaginary values, on the real axis, satisfy the relation

$$\operatorname{Re} G(\boldsymbol{p}, p_0) = \frac{1}{\pi}\,P\int\limits_{-\infty}^{+\infty}\frac{\operatorname{Im} G(\boldsymbol{p}, x)\,dx}{x-p_0}. \quad (29.29)$$

From the imaginary part (29.26), we can therefore compute the real part by a simple integration. Using the notation $\Pi^{(0)}(p^2)$ introduced

1. See, for example, B. A. Hurwitz and R. Courant, Funktionentheorie, Third Ed., Berlin, 1929, p. 335.

in (24.35), we find

$$\operatorname{Im} G(p) = \pi\,\varepsilon(p)\,\Pi^{(0)}(p^2), \tag{29.30}$$

$$\left.\begin{aligned}\operatorname{Re} G(p) &= P\int_{-\infty}^{+\infty} \frac{x}{|x|}\,\frac{\Pi^{(0)}(\mathbf{p}^2 - x^2)}{x - p_0}\,dx = P\int_0^\infty \Pi^{(0)}(\mathbf{p}^2 - x^2)\left[\frac{1}{x-p_0} + \frac{1}{x+p_0}\right]dx = \\ &= P\int_0^\infty \frac{\Pi^{(0)}(\mathbf{p}^2 - x^2)}{x^2 - p_0^2}\,d(x^2) = P\int_{4m^2}^\infty \frac{\Pi^{(0)}(-a)\,da}{a + p^2} = \overline{\Pi}^{(0)}(p^2)\,,\end{aligned}\right\} \tag{29.31}$$

$$\overline{\Pi}^{(0)}(p^2) = P\int_0^\infty \frac{\Pi^{(0)}(-a)\,da}{a + p^2}. \tag{29.32}$$

Hence we have

$$G(p) - G(0) = \overline{\Pi}^{(0)}(p^2) - \overline{\Pi}^{(0)}(0) + i\pi\,\varepsilon(p)\,\Pi^{(0)}(p^2). \tag{29.33}$$

From (24.35) we see directly that the function $\Pi^{(0)}(-a)$ has the value $\frac{e^2}{12\pi^2}$ for very large values of a. Consequently the integral (29.32) does not converge and the functions $\overline{\Pi}^{(0)}(p^2)$ and $G(p)$ actually do not exist. This is clearly a result of the summation over an infinite number of states in (29.21). Until now, we have not paid attention to the convergence of the sum. One might therefore suspect that the theory is not capable of treating the problem of the induced current. If we ignore all this for the moment and just write down a formal expression for the difference $\overline{\Pi}^{(0)}(p^2) - \overline{\Pi}^{(0)}(0)$, we get

$$\overline{\Pi}^{(0)}(p^2) - \overline{\Pi}^{(0)}(0) = P\int_0^\infty \Pi^{(0)}(-a)\,da\left[\frac{1}{a+p^2} - \frac{1}{a}\right] = -p^2\,P\int_0^\infty \frac{\Pi^{(0)}(-a)\,da}{a(a+p^2)}. \tag{29.34}$$

The integral in (29.34) is convergent and can even be done by elementary methods. The result is

$$\overline{\Pi}^{(0)}(p^2) - \overline{\Pi}^{(0)}(0) = \frac{e^2}{12\pi^2}\left[\frac{5}{3} - \frac{4m^2}{p^2} - \left(1 - \frac{2m^2}{p^2}\right)\sqrt{1 + \frac{4m^2}{p^2}}\,\log\frac{1 + \sqrt{1 + \frac{4m^2}{p^2}}}{\left|1 - \sqrt{1 + \frac{4m^2}{p^2}}\right|}\right]. \tag{29.35}$$

Equation (29.35) assumes that $1 + \frac{4m^2}{p^2} > 0$. Otherwise the logarithm must be replaced by an arctan function. For small values of $\left|\frac{p^2}{m^2}\right|$ the result is

$$\overline{\Pi}^{(0)}(p^2) - \overline{\Pi}^{(0)}(0) = -\frac{p^2}{m^2}\,\frac{e^2}{60\pi^2} + \cdots. \tag{29.36}$$

In this way the theory gives well-defined expressions for the experimentally observable effects although the "charge renormalization" $\overline{\Pi}^{(0)}(0)$ is infinite. First, the renormalization of the charge is necessary in principle if the expressions of the theory are to be compared with experimental results. In addition, the infinite term in the expectation value of the current is thereby eliminated. This is the great practical significance of the renormalization and this is the reason that a satisfactory formulation of quantum electrodynamics was not possible until a clear formulation of the renormalization principle had been given. Despite this, the result (29.36) was derived long ago by Uehling,[1] although the method was not completely satisfactory from a theoretical standpoint. The first "modern" deduction of (29.35) was given by Schwinger.[2]

According to this calculation, the vacuum behaves like a polarizable medium with a "dielectric constant"

$$\varepsilon(p^2) = 1 - \overline{\Pi}^{(0)}(p^2) + \overline{\Pi}^{(0)}(0) - i\pi\,\varepsilon(p)\,\Pi^{(0)}(p^2). \qquad (29.37)$$

The expression (29.37) has an imaginary part. In the usual way, this must signify absorption of energy from the external field. According to (18.22) and (18.24), the total energy given up by the external field is

$$\left.\begin{aligned} \delta E &= -\int \frac{\partial A_\mu^{\text{äuss}}(x)}{\partial x_0} \langle 0|\, j_\mu(x)\,|0\rangle\, dx = \frac{-i}{(2\pi)^4} \int dp\, p_0\, j_\mu(p)\, A_\mu^{\text{äuss}}(-p) = \\ &= \frac{i}{(2\pi)^4} \int dp\, \frac{j_\mu^{\text{äuss}}(p)\, j_\mu^{\text{äuss}}(-p)}{-p^2}\, p_0 \,\times \\ &\quad \times \left[1 - \overline{\Pi}^{(0)}(p^2) + \overline{\Pi}^{(0)}(0) - i\pi\,\varepsilon(p)\,\Pi^{(0)}(p^2)\right] = \\ &= \frac{1}{(2\pi)^3} \int dp\, |p_0|\, \frac{j_\mu^{\text{äuss}}(p)\, j_\mu^{\text{äuss}}(-p)}{-2p^2}\, \Pi^{(0)}(p^2). \end{aligned}\right\} \qquad (29.38)$$

Comparison with (24.34) shows that this energy agrees with the energy of all pairs produced by an external field.

A direct experimental verification of the equations derived here has not yet been made. In a later section we shall see that the term (29.36) contributes to the level shift in the hydrogen atom an amount which is about a hundred times larger than the experimental uncertainty. In this way we can regard (29.36) as having been indirectly confirmed. In addition there are several effects,[3] for example, the energy levels in a so-called "μ-mesic atom" (an atom where the electron has been replaced by a μ-meson) and in the proton-proton scattering, where a small improvement in the present experimental accuracy should show the effect of the vac-

---

1. E. A. Uehling, Phys. Rev. 48, 55 (1935). See also W. Heisenberg, Z. Physik 90, 209 (1934); W. Heisenberg and H. Euler, Z. Physik 98, 714 (1936); V. F. Weisskopf, Dan. Mat. Fys. Medd. 14, No. 6 (1936).

2. J. Schwinger, Phys. Rev. 75, 651 (1949).

3. L. L. Foldy and E. Eriksen, Phys. Rev. 95, 1048 (1954), 98, 775 (1955); E. Eriksen, L. L. Foldy and W. Rarita, Phys. Rev. 103, 781 (1956).

uum polarization. In certain measurements[1] on the x-rays from μ-mesic atoms, it has even proved necessary to make a small correction for the vacuum polarization. This is done in order to obtain agreement between the mass of the μ-meson determined from these measurements and from other independent measurements.

30. Regularization and the Self-Energy of the Photon

The use of general considerations of symmetry such as gauge invariance and causality [in Eq. (29.27)] was essential to our calculation of the effects of vacuum polarization in Sec. 29. In the present section we are going to show that these considerations saved us some work in the calculation as well as giving a precise meaning to the otherwise ambiguous, divergent integrals in (29.8). In order to show more clearly the ambiguity of the expressions (29.8), we attempt a direct evaluation of these integrals. We are here concerned only with the real part, since, by the previous calculation, the imaginary part is unique and convergent. We have

$$\left.\begin{aligned} \operatorname{Re} K_{\mu\nu}(p) \equiv R_{\mu\nu}(p) = -\frac{e^2}{4\pi^3}\int dp' \times \\ \times [p'_\mu(p'_\nu - p_\nu) + p'_\nu(p'_\mu - p_\mu) - \delta_{\mu\nu}(p'^2 - p\,p' + m^2)] \times \\ \times P\left\{\frac{\delta(p'^2+m^2)}{(p-p')^2+m^2} + \frac{\delta((p-p')^2+m^2)}{p'^2+m^2}\right\}. \end{aligned}\right\} \quad (30.1)$$

To compute the last terms in brackets, we need the integral representations

$$\delta(a) = \frac{1}{2\pi}\int_{-\infty}^{+\infty} dw\, e^{iwa}, \quad (30.2)$$

$$P\frac{1}{b} = \frac{1}{2i}\int_{-\infty}^{+\infty} dw\,\frac{w}{|w|}\,e^{iwb}, \quad (30.3)$$

according to which

$$P\left\{\frac{\delta(a)}{b} + \frac{\delta(b)}{a}\right\} = \frac{1}{4\pi i}\iint dw_1\, dw_2\left[\frac{w_1}{|w_1|} + \frac{w_2}{|w_2|}\right] e^{iw_1 a + iw_2 b}. \quad (30.4)$$

With the change of variables

$$w_1 = w\cdot\alpha, \quad (30.5a)$$

$$w_2 = w(1-\alpha), \quad (30.5b)$$

we can transform (30.4) as follows:

$$\left.\begin{aligned} P\left\{\frac{\delta(a)}{b} + \frac{\delta(b)}{a}\right\} = \frac{1}{4\pi i}\iint dw\, d\alpha\, |w|\frac{w}{|w|}\left[\frac{\alpha}{|\alpha|} + \frac{1-\alpha}{|1-\alpha|}\right] e^{iw[\alpha a + (1-\alpha)b]} = \\ = \frac{1}{2\pi i}\int_{-\infty}^{+\infty} w\, dw \int_0^1 d\alpha\, e^{iw[\alpha a + (1-\alpha)b]}. \end{aligned}\right\} \quad (30.6)$$

1. S. Koslov, V. Fitch and J. Rainwater, Phys. Rev. 95, 291 (1954). See also the article by S. Flügge in Vol. XLII of this handbook (Handbuch der Physik, edited by S. Flügge, Springer-Verlag, Heidelberg).

Using (30.6) and making the change of variable $q = p' - \alpha p$ in (30.1), we obtain

$$\left.\begin{aligned} R_{\mu\nu}(p) = \frac{i e^2}{8\pi^4} \int_{-\infty}^{+\infty} w\, dw \int_0^1 d\alpha \int dq\, [2 q_\mu q_\nu - (1-2\alpha)(p_\mu q_\nu + q_\mu p_\nu) - \\ - 2\alpha(1-\alpha) p_\mu p_\nu - \delta_{\mu\nu}(q^2 - p q (1-2\alpha) - \\ - p^2 \alpha(1-\alpha) + m^2)]\, e^{i w [q^2 + p^2 \alpha(1-\alpha) + m^2]}\,. \end{aligned}\right\} \quad (30.7)$$

On the basis of symmetry, we can drop the term linear in q inside the square brackets and replace $q_\mu q_\nu$ by $\frac{1}{4}\delta_{\mu\nu} q^2$:

$$\left.\begin{aligned} R_{\mu\nu}(p) = -\frac{i e^2}{8\pi^4} \int_{-\infty}^{+\infty} w\, dw \int_0^1 d\alpha \int dq \Big[2\alpha(1-\alpha)(p_\mu p_\nu - p^2 \delta_{\mu\nu}) + \\ + \delta_{\mu\nu}\left(\frac{q^2}{2} + p^2\alpha(1-\alpha) + m^2\right)\Big]\, e^{i w [q^2 + p^2\alpha(1-\alpha) + m^2]}\,. \end{aligned}\right\} \quad (30.8)$$

It is useful to split (30.8) into two parts:

$$R_{\mu\nu}(p) = (p_\mu p_\nu - \delta_{\mu\nu} p^2)\, F_1(p^2) + \delta_{\mu\nu} F_2(p^2)\,, \quad (30.9)$$

$$F_1(p^2) = -\frac{i e^2}{4\pi^4} \int_0^1 d\alpha \int_{-\infty}^{+\infty} w\, dw\, \alpha(1-\alpha) \int dq\, e^{i w [q^2 + p^2\alpha(1-\alpha) + m^2]}, \quad (30.10)$$

$$\left.\begin{aligned} F_2(p^2) = -\frac{i e^2}{16\pi^4} \int_0^1 d\alpha \int_{-\infty}^{+\infty} w\, dw \int dq\, [q^2 + 2p^2\alpha(1-\alpha) + \\ + 2m^2]\, e^{i w [q^2 + p^2\alpha(1-\alpha) + m^2]}\,. \end{aligned}\right\} \quad (30.11)$$

According to the considerations of Sec. 29, F_2 should vanish identically and F_1 should contain the result (29.35). First we evaluate F_1. Since we can write [c.f. Eq. (16.39)]

$$\int dq\, e^{i w q^2} = \frac{i\pi^2}{w|w|}\,, \quad (30.12)$$

we therefore have

$$F_1(p^2) = \frac{e^2}{4\pi^2} \int_0^1 \alpha(1-\alpha)\, d\alpha \int_{-\infty}^{+\infty} \frac{dw}{|w|}\, e^{i w [p^2\alpha(1-\alpha) + m^2]}\,. \quad (30.13)$$

The w-integration in (30.13) diverges for $w = 0$. This corresponds to the divergence of the integral (29.32) for $a = \infty$. Evaluating this according to the charge renormalization program, we find

$$\left.\begin{aligned} F_1(p^2) - F_1(0) &= \frac{e^2}{4\pi^2} \int_0^1 \alpha(1-\alpha)\, d\alpha \int_{-\infty}^{+\infty} \frac{dw}{|w|} [e^{i w p^2 \alpha(1-\alpha)} - 1]\, e^{i w m^2} = \\ &= \frac{e^2}{4\pi^2} \int_0^1 \alpha(1-\alpha)\, d\alpha \int_0^1 d\beta \int_{-\infty}^{+\infty} i w p^2 \alpha(1-\alpha)\, e^{i w [p^2 \beta\alpha(1-\alpha) + m^2]} \frac{dw}{|w|} = \\ &= -\frac{e^2}{2\pi^2} \int_0^1 \alpha(1-\alpha)\, d\alpha \log\left|1 + \frac{p^2}{m^2}\alpha(1-\alpha)\right|. \end{aligned}\right\} \quad (30.14)$$

The α-integral in (30.14) can be done by elementary methods and the result agrees with (29.35). As expected, we have

$$F_1(p^2) - F_1(0) = \overline{\Pi}^{(0)}(p^2) - \overline{\Pi}^{(0)}(0). \tag{30.15}$$

For calculating F_2, we shall need the integral

$$\int q^2\, dq\, e^{iwq^2} = -\frac{2\pi^2}{w^2|w|}, \tag{30.16}$$

which can be proved from (30.12). After performing the q-integration, we can write (30.11) as

$$\left.\begin{aligned} F_2(p^2) &= \frac{e^2}{8\pi^2}\int_0^1 d\alpha \int_{-\infty}^{+\infty} dw \frac{w}{|w|}\left[\frac{p^2\alpha(1-\alpha)+m^2}{w} + \frac{i}{w^2}\right] e^{iw[p^2\alpha(1-\alpha)+m^2]} = \\ &= \frac{-ie^2}{8\pi^2}\int_0^1 d\alpha \int_{-\infty}^{+\infty} dw \frac{w}{|w|}\frac{\partial}{\partial w}\left[\frac{1}{w} e^{iw[p^2\alpha(1-\alpha)+m^2]}\right] = \\ &= \frac{ie^2}{4\pi^2}\int_0^1 d\alpha \left(\frac{e^{iw[p^2\alpha(1-\alpha)+m^2]}}{w}\right)_{w=0}. \end{aligned}\right\} \tag{30.17}$$

The function $F_2(p^2)$ does not vanish; indeed, it is infinite! This is in flat contradiction to our considerations of Sec. 29, which led to (29.14). It looks as if our theory were not gauge invariant, although this would seem to be "impossible" on physical grounds. In Eqs. (29.17) and (29.18) we even "proved" explicitly the gauge invariance of the theory. As was observed after (29.18), what was essential to the proof was that certain integrals which actually are strongly divergent, were defined to be zero by symmetry. Also, it is immediately evident for physical reasons that the gauge invariant result of Sec. 29 must be the correct one. The symmetry properties of the integral (29.18) have obviously been lost in our explicit calculation above, where we interchanged the orders of integration several times, displaced the origin of the coordinate system, etc. In order to obtain trustworthy results from our calculations, we should cut off the integrals (29.18) in a symmetric and invariant fashion. Then we should carry through the whole calculation with a finite cutoff, and take the limit of the cutoff going to infinity only in the final answer. Such a calculation has been done by Pauli and Villars.[1] In their method, the integral (29.18) alone is not considered. Rather, several hypothetical particles with various masses m_i are introduced. We multiply the charges of these particles with "weight factors" C_i so that in place of (29.18), we have

$$I = \int dp' \sum_i C_i\, \delta(p'^2 + m_i^2)\, p'_\nu. \tag{30.18}$$

If the quantities m_i and C_i satisfy the relations

1. W. Pauli and F. Villars, Revs. Mod. Phys. 21, 434 (1949).

$$\sum_i C_i = 0\,, \tag{30.19}$$

$$\sum_i C_i m_i^2 = 0\,. \tag{30.20}$$

then (30.18) is convergent and actually vanishes. The new masses serve as cutoffs, and in the final result all masses, except the original one, must tend to infinity. If $m_1 = m$ is the original mass, then we must also have $C_1 = 1$ in this limit. By this process--which is called regularization--a convergent integral will clearly not have its value changed. A divergent integral will be evaluated[1] exactly according to the prescription (29.18).

In order that (30.19) and (30.20) be satisfied, some of the C_i have to be negative. If we try to regard the auxiliary particles as real particles of spin 1/2, then we have to give them imaginary charges--something which is hard to ignore and which will probably have unphysical consequences. It can be shown[2] that relations similar to (30.19) and (30.20) can be satisfied by real charges and masses if some of the particles have spin zero. However, in this procedure we obtain a relation between the masses of the elementary particles which does not seem to exist in fact. We shall not pursue the matter further, and regard the regularization only as a mathematical aid for cutting off our integrals in an invariant manner.

After regularization, rather than (30.17), we have

$$\left.\begin{aligned} F_2^{\text{Reg}}(p^2) &= \frac{i e^2}{4\pi^2}\int_0^1 d\alpha \left(\frac{1}{w}\sum_i C_i\, e^{iw[p^2\alpha(1-\alpha)+m_i^2]}\right)_{w=0} = \\ &= \frac{i e^2}{4\pi^2}\int_0^1 d\alpha\left[\frac{1}{w}\sum_i C_i + i\sum_i C_i[p^2\alpha(1-\alpha)+m_i^2] + O(w)\right]_{w=0}. \end{aligned}\right\} \tag{30.21}$$

Using only (30.19) gives

$$F_2^{\text{Reg}}(p^2) = -\frac{e^2}{4\pi^2}\sum_i C_i m_i^2. \tag{30.22}$$

From (30.20) we conclude that

$$F_2^{\text{Reg}}(p^2) = 0. \tag{30.23}$$

Thus the results of this section agree completely with our previous results.

The use of only (30.19) for evaluating the vacuum polarization seems a trifle artificial and must obviously lead to a false result.

1. Other methods which give the same result are due to J. Schwinger, Phys. Rev. 82, 664 (1951); D. C. Peaslee, Phys. Rev. 81, 107 (1951); G. Källén, Ark. Fysik 5, 130 (1952); and S. N. Gupta, Proc. Phys. Soc. Lond. A 66, 129 (1953).

2. R. Jost and J. Rayski, Helv. Phys. Acta 22, 457 (1949); see also J. Rayski, Acta Phys. Polon. 9, 129 (1948).

If one is not particularly careful, it is quite possible to carry out the calculation without regularization, so as to obtain[1,2] the result

$$F_2(p^2) = -\frac{e^2}{4\pi^2} m^2 . \tag{30.24}$$

For the moment, if we were to treat[3] a light quantum as we have treated the external field, we would conclude that the photon had a self-energy

$$\delta E = \sqrt{\omega^2 + \frac{e^2 m^2}{4\pi^2}} - \omega \approx \frac{e^2}{8\pi^2} \frac{m^2}{\omega} . \tag{30.25}$$

This is also the result of Wentzel.

According to the preceding arguments, such results are to be regarded as more or less accidental and one should expect different results from the various methods of calculation. In the old calculations, which were not formally covariant, the self-energy of the photon was often given[4] as infinite. By the requirement that all the results of the theory be gauge invariant, the self-energy of the photon has been unequivocally determined to be zero.

In conclusion, we can say that the vacuum polarization gives an infinite charge renormalization ("self-charge") which, in principle, is not observable. The physical consequences of the theory are all gauge invariant and finite if the calculations are done with sufficient care. The preceding calculation verified this for the lowest order term in a theory with spin 1/2 particles, for a series expansion in both the radiation field and the external field. The charge renormalization has been calculated to the next higher order in the radiation field by Jost and Luttinger[5] and the observable terms have been calculated by Baranger, Dyson and Salpeter[6] and by Källén and Sabry.[7] As was already mentioned, the calculation has been done by Kroll and Wichmann[8] without expanding

1. G. Wentzel, Phys. Rev. 74, 1070 (1948).

2. One obtains this result from (30.17), for example, if the integral

$$\int dw \frac{w}{|w|} \frac{\partial}{\partial w} \left(\frac{1}{w} \cos \left[w \left(p^2 \alpha (1-\alpha) + m^2 \right) \right] \right)$$

is set equal to zero on the basis of symmetry. If the external field is a photon ($p^2 = 0$), then (30.24) follows from (30.17).

3. See Eq. (29.4), where the external field and the radiation field enter symmetrically.

4. See, for example, W. Heitler, Quantum Theory of Radiation, Second Ed., Oxford, 1944, p. 194.

5. R. Jost and J. M. Luttinger, Helv. Phys. Acta 23, 201 (1950).

6. M. Baranger, F. J. Dyson and E. E. Salpeter, Phys. Rev. 88, 680 (1952).

7. G. Källén and A. Sabry, Dan. Mat. Fys. Medd. 29, No. 17 (1955).

8. E. H. Wichmann and N. M. Kroll, Phys. Rev. 96, 232 (1954); 101, 843 (1956).

in powers of the external field for the case that this field is a Coulomb field. The higher powers of the external field[1] in a general series expansion in this field, as well as the regularization[2] in the order e^4 have been treated. The results of all these calculations are essentially the same as those of the lowest order. For particles of spin different from 1/2, the vacuum polarization has been evaluated by Umezawa and Kawabe and by Feldman.[3] Here it has been shown that for particles of spin ≥ 1 the observable terms contain divergent integrals. For these problems, therefore, the charge renormalization is not able to give usable expressions for the physically important quantities.

31. The Lowest Order Radiative Corrections to the Current Operator

Now we shall consider a problem which is formally more complicated: the next higher approximation to the current operator. For simplicity, we omit the external field and so we have just the operator equations which were studied in Sec. 17 and 18. We reproduce here the expression (18.12) for the current operator to order e^3:

$$\left.\begin{aligned} j_\mu^{(2)}(x) = \frac{i}{8}\int\int dx'dx''[\bar{\psi}^{(0)}(x), \gamma_\mu S_R(x-x')\gamma_\nu\{\psi^{(0)}(x'), [\bar{\psi}^{(0)}(x''), \gamma_\nu\psi^{(0)}(x'')]\}]D_R(x'-x'') - \\ -\frac{i}{4}\int\int dx'dx''[\bar{\psi}^{(0)}(x), \gamma_\mu S_R(x-x')\gamma_{\nu_1} S_R(x'-x'')\gamma_{\nu_2}\psi^{(0)}(x'')]\{A_{\nu_1}^{(0)}(x'), A_{\nu_2}^{(0)}(x'')\} - \\ -\frac{i}{4}\int\int dx'dx''[\bar{\psi}^{(0)}(x')\gamma_{\nu_1} S_A(x'-x), \gamma_\mu S_R(x-x'')\gamma_{\nu_2}\psi^{(0)}(x'')]\{A_{\nu_1}^{(0)}(x'), A_{\nu_2}^{(0)}(x'')\} + \\ +\frac{i}{8}\int\int dx'dx''[\{[\bar{\psi}^{(0)}(x''), \gamma_\nu\psi^{(0)}(x'')], \psi^{(0)}(x')\}\gamma_\nu S_A(x'-x), \gamma_\mu\psi^{(0)}(x)]D_R(x'-x'') - \\ -\frac{i}{4}\int\int dx'dx''[\bar{\psi}^{(0)}(x'')\gamma_{\nu_2} S_A(x''-x')\gamma_{\nu_1} S_A(x'-x), \gamma_\mu\psi^{(0)}(x)]\{A_{\nu_1}^{(0)}(x'), A_{\nu_2}^{(0)}(x'')\}. \end{aligned}\right\} \quad (31.1)$$

The operator (31.1) has non-vanishing matrix elements from the vacuum to the following states: states with an electron-positron pair and no photons, states with one pair and two photons, and states with two pairs. We are not concerned with these two last classes of matrix elements here, because all the p-integrations can be done just by using the conservation of energy and momentum. Therefore there will not be any "inner" p-integrations left over, i.e., there are no summations over an infinite number of intermediate states. Things are quite different if there is only one pair present in the final state. Then there is one p-integration remaining and so we can study this lowest, non-vanishing correction to the matrix element $\langle q,q'|j_\mu(x)|0\rangle$ or $\langle 0|j_\mu(x)|q,q'\rangle$.

1. G. Källén, Helv. Phys. Acta 22, 637 (1949).
2. E. Karlson, Ark. Fysik 7, 221 (1954).
3. H. Umezawa and R. Kawabe, Prog. Theor. Phys. 4, 443 (1949), and earlier articles. D. Feldman, Phys. Rev. 76, 1369 (1949). See also J. McConnell, Phys. Rev. 81, 275 (1951).

Similar considerations show that the quantities $\langle q|j_\mu(x)|q'\rangle$ are the only matrix elements which contain inner p-integrations if neither of the states is the vacuum. In this last expression, both states $|q\rangle$ and $|q'\rangle$ are either one-electron states or both are one-positron states. Using the methods of Sec. 18, we have

$$\left.\begin{aligned} e^3\langle q|j_\mu^{(2)}(x)|q'\rangle = \frac{ie}{2}\int dx'\langle q|[\overline{\psi}^{(0)}(x), \gamma_\mu S_R(x-x')\,\Phi(x')]|q'\rangle + \\ + \frac{ie}{2}\int dx'\langle q|[\overline{\Phi}(x')\,S_A(x'-x), \gamma_\mu\psi^{(0)}(x)]|q'\rangle + \\ + \frac{ie}{2}\iint dx'\,dx''\langle q|[\overline{\psi}^{(0)}(x'), K_\mu(x'-x, x-x'')\,\psi^{(0)}(x'')]|q'\rangle + \\ + \frac{ie}{2}\iint dx'\,dx''K_{\mu\nu}(x-x')\,D_R(x'-x'')\langle q|[\overline{\psi}^{(0)}(x''), \gamma_\nu\psi^{(0)}(x'')]|q'\rangle \end{aligned}\right\} \quad (31.2)$$

and

$$\left.\begin{aligned} e^3\langle 0|j_\mu^{(2)}(x)|q,q'\rangle = \frac{ie}{2}\int dx'\langle 0|[\overline{\psi}^{(0)}(x), \gamma_\mu S_R(x-x')\,\Phi(x')]|q,q'\rangle + \\ + \frac{ie}{2}\int dx'\langle 0|[\overline{\Phi}(x')\,S_A(x'-x), \gamma_\mu\psi^{(0)}(x)]|q,q'\rangle + \\ + \frac{ie}{2}\iint dx'\,dx''\langle 0|[\overline{\psi}^{(0)}(x'), K_\mu(x'-x, x-x'')\,\psi^{(0)}(x'')]|q,q'\rangle + \\ + \frac{ie}{2}\iint dx'dx''K_{\mu\nu}(x-x')\,D_R(x'-x'')\langle 0|[\overline{\psi}^{(0)}(x''), \gamma_\nu\psi^{(0)}(x'')]|q,q'\rangle . \end{aligned}\right\} \quad (31.2a)$$

In (31.2) and (31.2a), $K_\mu(x'-x, x-x'')$ is a c-number in the Hilbert space of the particles but a matrix in the "spin space" of the γ-matrices:

$$\left.\begin{aligned} K_\mu(x'-x, x-x'') = -\frac{e^2}{2}\gamma_\lambda[S^{(1)}(x'-x)\,\gamma_\mu S_R(x-x'')\,D_R(x''-x') + \\ + S_A(x'-x)\,\gamma_\mu S^{(1)}(x-x'')\,D_R(x'-x'') + \\ + S_A(x'-x)\,\gamma_\mu S_R(x-x'')\,D^{(1)}(x'-x'')]\,\gamma_\lambda\ . \end{aligned}\right\} \quad (31.3)$$

Here $\Phi(x)$ is the operator

$$\Phi(x) = -\frac{e^2}{2}\int\gamma_\lambda[S^{(1)}(x-x')\,D_A(x'-x) + S_R(x-x')\,D^{(1)}(x'-x)]\,\gamma_\lambda\psi^{(0)}(x')\,dx'. \quad (31.4)$$

The function $K_{\mu\nu}(x-x')$ is just the function which has been defined in Eq. (29.6) and studied in detail in Sec. 29 and 30. Because this quantity,

$$\frac{ie}{2}\int dx''\,D_R(x'-x'')\,[\overline{\psi}^{(0)}(x''), \gamma_\mu\psi^{(0)}(x'')]\,, \quad (31.5)$$

is just the potential produced by the current $\frac{ie}{2}[\overline{\psi}^{(0)}(x''), \gamma_\mu\psi^{(0)}(x'')]$ we can interpret the last terms in (31.2) and (31.2a) as the polarization of the vacuum by the electron current. We shall now work out (31.4). It is convenient to go over to p-space:

$$\Phi(x) = \int dx' F(x-x')\,\psi^{(0)}(x'), \tag{31.6}$$

$$\left.\begin{aligned} F(x-x') &= -\frac{e^2}{2}\gamma_\lambda\left[S^{(1)}(x-x')\,D_A(x'-x) + S_R(x-x')\,D^{(1)}(x'-x)\right]\gamma_\lambda \\ &= \frac{1}{(2\pi)^4}\int dq\, e^{iq(x-x')}\,F(q), \end{aligned}\right\} \tag{31.7}$$

$$\left.\begin{aligned} F(q) = &-\frac{e^2}{2(2\pi)^3}\iint dp\,dk\,\delta(q-p+k)\,\gamma_\lambda(i\gamma p - m)\,\gamma_\lambda \times \\ &\times\left[\delta(p^2+m^2)\left[P\frac{1}{k^2} - i\pi\,\varepsilon(k)\,\delta(k^2)\right] + \right. \\ &\left. + \delta(k^2)\left[P\frac{1}{p^2+m^2} + i\pi\,\varepsilon(p)\,\delta(p^2+m^2)\right]\right]. \end{aligned}\right\} \tag{31.8}$$

The function $F(x-x')$ in (31.7) is again a retarded function and consequently there should be a relation between real and imaginary parts of $F(q)$ similar to (29.29). As before, we shall first calculate the imaginary part. From considerations of invariance, we can write

$$\operatorname{Im} F(q) = F_1(q) + (i\gamma q + m)\,F_2(q), \tag{31.9}$$

with

$$4i\,q_\mu F_2(q) = \operatorname{Sp}\left[\gamma_\mu \operatorname{Im} F(q)\right], \tag{31.10}$$

$$4\left(F_1(q) + m\,F_2(q)\right) = \operatorname{Sp}\left[\operatorname{Im} F(q)\right]. \tag{31.11}$$

From (31.8) we have

$$\left.\begin{aligned} 4i\,q_\mu F_2(q) = \frac{e^2}{16\pi^2}\iint dp\,dk\,\delta(q-p+k)\,\operatorname{Sp}\left[\gamma_\mu\gamma_\lambda(i\gamma p - m)\,\gamma_\lambda\right] \times \\ \times\,\delta(p^2+m^2)\,\delta(k^2)\left[\varepsilon(k) - \varepsilon(p)\right] = \\ = \frac{-i e^2}{2\pi^2}\int dk\,(q_\mu + k_\mu)\,\delta\left((q+k)^2+m^2\right)\delta(k^2)\left[\varepsilon(k) - \varepsilon(q+k)\right], \end{aligned}\right\} \tag{31.12}$$

$$4\left(F_1(q) + m\,F_2(q)\right) = \frac{-m e^2}{\pi^2}\int dk\,\delta\left((q+k)^2+m^2\right)\delta(k^2)\left[\varepsilon(k) - \varepsilon(q+k)\right]. \tag{31.13}$$

If we multiply (31.12) by $-\frac{1}{4}i\,q_\mu$ and use the properties of the delta function, after a calculation similar to (29.24) and (29.25), we find

$$\left.\begin{aligned} F_2(q) &= \frac{-e^2}{16\pi^2}\left(1 - \frac{m^2}{q^2}\right)\int dk\,\delta(k^2)\,\delta(q^2+m^2+2qk)\left[\varepsilon(k) - \varepsilon(q+k)\right] = \\ &= \pi\,\varepsilon(q)\,\Sigma_2^{(0)}(q^2), \end{aligned}\right\} \tag{31.14}$$

$$\Sigma_2^{(0)}(q^2) = \frac{e^2}{16\pi^2}\left[1 - \left(\frac{m^2}{q^2}\right)^2\right]\Theta(-q^2-m^2), \tag{31.15}$$

$$F_1(q) = \pi\,\varepsilon(q)\,\Sigma_1^{(0)}(q^2), \tag{31.16}$$

$$\Sigma_1^{(0)}(q^2) = \frac{m\,e^2}{16\pi^2}\left(3+\frac{m^2}{q^2}\right)\left(1+\frac{m^2}{q^2}\right)\Theta(-q^2-m^2). \tag{31.17}$$

As before, we introduce the Hilbert transforms of these functions,

$$\overline{\Sigma}_i^{(0)}(q^2) = P\int_0^\infty \frac{\Sigma_i^{(0)}(-a)}{a+p^2}\,da\,. \tag{31.18}$$

Thus we finally obtain for the function $F(q)$,

$$F(q) = \Sigma_1^{(0)}(q^2) + i\pi\,\varepsilon(q)\Sigma_1^{(0)}(q^2) + (i\gamma q+m)[\overline{\Sigma}_2^{(0)}(q^2) + i\pi\,\varepsilon(q)\Sigma_2^{(0)}(q^2)]\,. \tag{31.19}$$

For large values of $-p^2$, the functions $\Sigma_i^{(0)}(p^2)$ tend to finite limits, just as the function $\Pi^{(0)}(p^2)$ did. The integrals in (31.18) therefore diverge at their upper limits, as did (29.32). In Sec. 32 we shall discuss further the interpretation of these divergences.

The evaluation of (31.3) in full generality is possible and can be done by methods similar to those already used. In order not to complicate the formal operations excessively, we shall work out only the special case which is of interest: that $K_\mu(x'-x,\,x-x'')$ appears with the operator $\psi^{(0)}(x)$ as in (31.2) and (31.2a). We write

$$K_\mu(x'-x,\,x-x'') = \frac{1}{(2\pi)^8}\iint dq\,dq'\,e^{iq(x'-x)+iq'(x-x'')}\,K_\mu(q,q'), \tag{31.20}$$

and consequently we are concerned with the function $K_\mu(q,q')$ only when $q^2=q'^2=-m^2$ and where a factor $i\gamma q+m$ on the left or a factor $i\gamma q'+m$ on the right can be set equal to zero. From (31.3) we obtain quite generally

$$\left.\begin{aligned}
K_\mu(q,q') = &-\frac{e^2}{2}\frac{1}{(2\pi)^3}\int dk\Big\{P\frac{\delta(k^2)}{[(q-k)^2+m^2][(q'-k)^2+m^2]}+\\
&+P\frac{\delta((q-k)^2+m^2)}{k^2[(q'-k)^2+m^2]}+P\frac{\delta((q'-k)^2+m^2)}{k^2[(q-k)^2+m^2]}+\\
&+i\pi P\frac{1}{k^2}\delta((q-k)^2+m^2)\,\delta((q'-k)^2-m^2)[1-\varepsilon(q-k)\,\varepsilon(q'-k)]\,\varepsilon(q'-q)+\\
&+i\pi P\frac{1}{(q-k)^2+m^2}\delta(k^2)\,\delta((q'-k)^2+m^2)\,[1-\varepsilon(k)\,\varepsilon(k-q')]\,\varepsilon(q')-\\
&-i\pi P\frac{1}{(q'-k)^2+m^2}\delta(k^2)\,\delta((q-k)^2+m^2)\,[1-\varepsilon(k)\,\varepsilon(k-q)]\,\varepsilon(q)+\\
&+\pi^2\delta(k^2)\,\delta((q'-k)^2+m^2)\,\delta((q-k)^2+m^2)\,[\varepsilon(k)\,\varepsilon(q-k)+\varepsilon(k)\times\\
&\times\varepsilon(q'-k)+\varepsilon(q-k)\,\varepsilon(q'-k)]\Big\}\,\gamma_\lambda(i\gamma(q-k)-m)\times\\
&\times\gamma_\mu(i\gamma(q'-k)-m)\,\gamma_\lambda.
\end{aligned}\right\} \tag{31.21}$$

Now because of $q^2+m^2=0$, the expressions k^2 and $(q-k)^2+m^2$

cannot vanish simultaneously.[1] We can therefore drop many terms in (31.21) and write

$$K_\mu(q,q') = K_\mu^{(1)}(q,q') + i\pi\,\varepsilon(q'-q)\,K_\mu^{(2)}(q,q')\,, \qquad (31.22)$$

$$\left.\begin{aligned} K_\mu^{(1)}(q,q') = -\frac{e^2}{2}\frac{1}{(2\pi)^3}\int dk \Big\{ P\frac{\delta(k^2)}{[(q-k)^2+m^2][(q'-k)^2+m^2]} + P\frac{\delta((q-k)^2+m^2)}{k^2[(q'-k)^2+m^2]} + \\ + P\frac{\delta((q'-k)^2+m^2)}{k^2[(q-k)^2+m^2]}\Big\}\gamma_\lambda[i\gamma(q-k)-m]\gamma_\mu[i\gamma(q'-k)-m]\gamma_\lambda\,, \end{aligned}\right\} \quad (31.23)$$

$$\left.\begin{aligned} K_\mu^{(2)}(q,q') = -\frac{e^2}{2}\frac{1}{(2\pi)^3}\int dk\,\delta((q-k)^2+m^2)\,\delta((q'-k)^2+m^2)\,P\frac{1}{k^2}\times \\ \times[1-\varepsilon(q-k)\,\varepsilon(q'-k)]\gamma_\lambda[i\gamma(q-k)-m]\gamma_\mu[i\gamma(q'-k)-m]\gamma_\lambda. \end{aligned}\right\} \quad (31.24)$$

As before, we start by evaluating the term with the function $\varepsilon(q'-q)$, i.e., (31.24). The factor with the γ-matrices can be simplified considerably by means of

$$\left.\begin{aligned} i\gamma q\gamma_\mu i\gamma q' = (i\gamma q+m)\gamma_\mu(i\gamma q'+m) - m\gamma_\mu(i\gamma q'+m) - \\ -(i\gamma q+m)\gamma_\mu m + m^2\gamma_\mu \sim m^2\gamma_\mu\,, \end{aligned}\right\} \quad (31.25)$$

$$i\gamma q'\gamma_\mu i\gamma q \sim -2im(q_\mu+q'_\mu) - \gamma_\mu((q'-q)^2+3m^2), \qquad (31.26)$$

etc. In a few terms the k-integration can be done directly, by elementary methods. By methods used several times in previous sections, we obtain

$$\left.\begin{aligned} P\int\frac{k_\mu k_\nu}{k^2}dk\,\delta((q-k)^2+m^2)\,\delta((q'-k)^2+m^2)\,[1-\varepsilon(q-k)\,\varepsilon(q'-k)] = \\ = \Big\{\frac{1}{2}\Big(1+\frac{2m^2}{Q^2}\Big)\Big(q_\mu q_\nu + q'_\mu q'_\nu - \frac{1}{2}\delta_{\mu\nu}Q^2\Big) - \frac{m^2}{Q^2}\Big(q'_\mu q_\nu + q_\mu q'_\nu + \frac{1}{2}\delta_{\mu\nu}Q^2\Big)\Big\}\times \\ \times\frac{\pi}{-Q^2}\frac{\Theta(-Q^2-4m^2)}{\sqrt{1+\frac{4m^2}{Q^2}}}\,, \end{aligned}\right\} \quad (31.27)$$

$$Q = q'-q\,, \qquad (31.27\text{a})$$

$$\left.\begin{aligned} P\int\frac{k_\mu dk}{k^2}\,\delta((q-k)^2+m^2)\,\delta((q'-k)^2+m^2)\times \\ \times[1-\varepsilon(q-k)\,\varepsilon(q'-k)] = (q_\mu+q'_\mu)\frac{\pi}{-Q^2}\frac{\Theta(-Q^2-4m^2)}{\sqrt{1+\frac{4m^2}{Q^2}}}. \end{aligned}\right\} \quad (31.28)$$

The angular integration in

$$P\int\frac{dk}{k^2}\,\delta((q-k)^2+m^2)\,\delta((q'-k)^2+m^2)\,[1-\varepsilon(q-k)\,\varepsilon(q'-k)] \quad (31.29)$$

diverges. This is because the denominator vanishes at the limits

1. Here we are ignoring an "infrared term" in which all components of k_μ vanish. As we shall see below, we have to deal with infrared terms by introducing a small photon mass μ. We have omitted it here.

of integration and so the principal value does not exist. We shall make this integral finite by formally introducing a small photon mass μ. Equation (31.29) is therefore replaced by

$$\left.\begin{aligned} P\int \frac{dk}{k^2+\mu^2}\,\delta\left((q-k)^2+m^2\right)\delta\left((q'-k)^2+m^2\right)\left[1-\varepsilon(q-k)\,\varepsilon(q'-k)\right]= \\ =\frac{\pi}{-Q^2}\frac{\Theta(-Q^2-4m^2)}{\sqrt{1+\frac{4m^2}{Q^2}}}\log\left[1-\frac{Q^2+4m^2}{\mu^2}\right]. \end{aligned}\right\} \quad (31.30)$$

Of course the cutoff μ must drop out of the final result.

With the aid of (31.27), (31.28), (31.30), and transformations like (31.25) and (31.26), we can evaluate (31.24) completely. The result is

$$K_\mu^{(2)}(q,q')=\gamma_\mu R^{(0)}(Q^2)+i\frac{q_\mu+q'_\mu}{2m}S^{(0)}(Q^2), \quad (31.31)$$

$$\left.\begin{aligned} R^{(0)}(Q^2)=-\frac{e^2}{8\pi^2}\left\{-\left(1+\frac{2m^2}{Q^2}\right)\log\left[1-\frac{Q^2+4m^2}{\mu^2}\right]+\frac{3}{2}\left[1+\frac{4m^2}{Q^2}\right]\right\} \\ \times\frac{\Theta(-Q^2-4m^2)}{\sqrt{1+\frac{4m^2}{Q^2}}}, \end{aligned}\right\} \quad (31.32)$$

$$S^{(0)}(Q^2)=-\frac{e^2}{4\pi^2}\frac{m^2}{Q^2}\frac{\Theta(-Q^2-4m^2)}{\sqrt{1+\frac{4m^2}{Q^2}}}. \quad (31.33)$$

The quantity (31.23) must have the same general form:

$$K_\mu^{(1)}(q,q')=\gamma_\mu\overline{R}^{(0)}(Q^2)+i\frac{q_\mu+q'_\mu}{2m}\overline{S}^{(0)}(Q^2). \quad (31.34)$$

From (31.3), we see that the function $K_\mu(x'-x,\,x-x'')$ vanishes if $x'_0>x_0$ or if $x''_0>x_0$. Since the functions $K_\mu^{(i)}(q,q')$ depend essentially only on Q, it is a natural assumption that these "causal" properties arise because of the following relations between $R^{(0)}$, $S^{(0)}$ and $\overline{R}^{(0)}$, $\overline{S}^{(0)}$:

$$\overline{R}^{(0)}(Q^2)=P\int_0^\infty\frac{R^{(0)}(-a)}{a+Q^2}\,da, \quad (31.35)$$

$$\overline{S}^{(0)}(Q^2)=P\int_0^\infty\frac{S^{(0)}(-a)}{a+Q^2}\,da. \quad (31.36)$$

This guess can be verified from (31.21) through (31.24) by an explicit calculation.[1] The functions $S^{(0)}(Q^2)$ go to zero like $\frac{e^2}{4\pi^2}\frac{m^2}{-Q^2}$ for large values of $-Q^2$, so that the integral (31.36) is convergent. On the other hand, the integral (31.35) is divergent because the function $R^{(0)}(Q^2)$ goes like $\frac{e^2}{8\pi^2}\log\frac{-Q^2}{\mu^2}$ for large values of $-Q^2$.

32. Mass Renormalization

In Sec. 31 we have seen that the matrix elements (31.2) and (31.2a) contain several divergent integrals. Our next task is to study these terms and, if possible, to interpret them physically and to remove them. We note first that the operator $\Phi(x)$ in (31.4) was also present in Eq. (18.21) and therefore that it can be taken as the first radiative correction to the matrix element $\langle 0|\,\psi(x)\,|q\rangle$. In fact,

$$\langle 0|\,\psi(x)\,|q\rangle = \langle 0|\,\psi^{(0)}(x)\,|q\rangle + \int S_R(x-x')\,\langle 0|\,\Phi(x')\,|q\rangle\,dx' + \cdots \tag{32.1}$$

or

$$\left(\gamma\frac{\partial}{\partial x}+m\right)\langle 0|\,\psi(x)\,|q\rangle = -\langle 0|\,\Phi(x)\,|q\rangle. \tag{32.2}$$

According to (31.15) and (31.17), the quantities $\Sigma_i^{(0)}(-m^2)$ vanish. With these and (31.6), (31.7), and (31.19), we obtain from (32.2)

$$\left.\begin{aligned}\left(\gamma\frac{\partial}{\partial x}+m\right)\langle 0|\,\psi(x)\,|q\rangle &= -\left[\bar{\Sigma}_1^{(0)}(-m^2)+(i\gamma q+m)\,\bar{\Sigma}_2^{(0)}(-m^2)\right]\langle 0|\,\psi^{(0)}(x)\,|q\rangle = \\ &= -\bar{\Sigma}_1^{(0)}(-m^2)\,\langle 0|\,\psi^{(0)}(x)\,|q\rangle.\end{aligned}\right\} \tag{32.3}$$

To the present order of approximation, we can write

$$\left[\gamma\frac{\partial}{\partial x}+m+\bar{\Sigma}_1^{(0)}(-m^2)\right]\langle 0|\,\psi(x)\,|q\rangle = 0, \tag{32.4}$$

rather than (32.3). From this, it follows that the infinite quantity $\bar{\Sigma}_1^{(0)}(-m^2)$ can be regarded as the mass difference between the

---

1. From (31.21) through (31.24) we see that

$$K_\mu(q,q') = K_\mu^{(1)}(q,q') + i\pi\,\varepsilon(q'-q)\,K_\mu^{(2)}(q,q') = \lim_{\varepsilon,\varepsilon'\to 0} K_\mu^{(1)}(\mathbf{q},q_0-i\varepsilon;\,\mathbf{q}',q_0'+i\varepsilon').$$

By a transformation similar to (30.6), we can write $K_\mu^{(1)}(q,q')$ as

$$\begin{aligned}K_\mu^{(1)}(q,q') = -\frac{e^2}{16\pi^2}\int_0^1\!\!\int_0^1\!\!\int_0^1 d\alpha\,d\beta\,d\gamma\,\delta(1-\alpha-\beta-\gamma)\times\\ \times\left\{\frac{1}{A}\gamma_\lambda[i\gamma(q(1-\alpha)-q'\beta)-m]\,\gamma_\mu\,[i\gamma(q'(1-\beta)-q\alpha)-m]\,\gamma_\lambda -\right.\\ \left. -2\gamma_\mu\left[\log\frac{A}{m^2}-\frac{1}{2}\int_{-\infty}^{+\infty}\frac{dw}{|w|}e^{iwm^2}\right]\right\},\end{aligned}$$

$$A = q^2\alpha\gamma + q'^2\beta\gamma + Q^2\alpha\beta + m^2(1-\gamma) = Q^2\alpha\beta + m^2(1-\gamma)^2.$$

It follows from this representation that for $q^2 = q'^2 = -m^2$, $K_\mu^{(1)}(q,q')$ is an analytic function of Q^2 which is regular for all complex values of Q^2 except the positive, real axis. From this, (31.35) and (31.36) follow immediately.

free particles at $x_0 = -\infty$ and the "physical" particles which are coupled to the electromagnetic field. In fact, the free particles at $-\infty$ are only a mathematical fiction (however, one which is essential for the whole interpretation of the formalism) and their mass is not an observable quantity. Indeed, we can choose these masses arbitrarily, as long as the mass of the physical particles, i.e., the quantity $m + \Sigma_1^{(0)}(-m^2)$, has the correct experimental value. It is now expedient to take the mass of the free particles equal to the experimental mass of the electron. We can do this by transforming the differential equation for the operator $\psi(x)$ as follows:

$$\left(\gamma \frac{\partial}{\partial x} + m_{\text{exp}}\right) \psi(x) = i e \gamma A(x) \psi(x) + \delta m \psi(x) . \tag{32.5}$$

That is, we add a term $\delta m \psi(x)$ to both sides so that just the experimental mass appears on the left. From now on, we shall denote this (observed) mass simply by m. It is always this mass which is to be used in the S-functions and other quantities that appear in the theory. The quantity δm is then to be defined as such a function of e that the mass of the state $|q\rangle$ remains unchanged during the adiabatic switching. Clearly $\delta m \to 0$ for $e \to 0$ and the previous calculation shows that the first term in a series development of δm in powers of e is given by

$$\delta m = \Sigma_1^{(0)}(-m^2) + \cdots . \tag{32.6}$$

Actually Eq. (32.5) means no change in the original differential equation, since the same term has been added to both sides, but it does change the adiabatic switching. The change is that only the self-mass on the right side is to be switched off for $x_0 \to -\infty$, while the experimental mass on the left is formally independent of e. This procedure is known[1] as "mass renormalization".

If now m is to be the experimental mass, then we can conclude from the differential equation (32.4) or from

$$\left(\gamma \frac{\partial}{\partial x} + m\right) \langle 0 | \psi(x) | q \rangle = 0 \tag{32.7}$$

that

$$\langle 0 | \psi(x) | q \rangle = N \langle 0 | \psi^{(0)}(x) | q \rangle . \tag{32.8}$$

---

1. Even in the classical electron theory of Lorentz, a similar procedure was used in order to treat the effect of the self-field of an electron on its motion. The self-energy of a relativistic electron was first studied in quantum electrodynamics by V. F. Weisskopf, Z. Physik 89, 27 (1934), 90, 817 (1934). The principle of mass renormalization was first enunciated by, among others, A. Kramers, Report of the Solvay Conference, 1948. It was further developed by H. A. Bethe, Phys. Rev. 72, 339 (1947); Z. Koba and S. Tomonaga, Progr. Theor. Phys. 3, 290 (1948); T. Tati and S. Tomonaga, Progr. Theor. Phys. 3, 391 (1948); and J. Schwinger, Phys. Rev. 75, 651 (1949).

In principle the constant N can be calculated from (32.1), but some care is necessary, since an expansion such as

$$\int S_R(x-x')\,\Sigma_2^{(0)}(-m^2)\left(\gamma\frac{\partial}{\partial x'}+m\right)\langle 0|\,\psi^{(0)}(x')\,|q\rangle\,dx'\,, \tag{32.9}$$

for example, is actually indeterminate. Because of the differential equation for $\psi^{(0)}(x)$, one might suspect that the expression vanishes. However, by a partial integration the derivatives may be moved over to the singular functions, so that the integral then appears as proportional to $\psi^{(0)}(x)$. This alternative form differs only by a "surface term" for $x_0=-\infty$. Consequently, we expect that this ambiguity is removed if the boundary condition for $x_0\to-\infty$ (i.e., the adiabatic switching) is treated with sufficient care.[1] In place of (31.6), (31.7), we therefore write

$$\left.\begin{aligned}\langle 0|\,\Phi(x,\alpha)\,|q\rangle &= \frac{e^2}{2}\int\limits_{-\infty}^{x} dx'\, e^{\alpha(x_0+x_0')}\,\gamma_\lambda\times\\ &[S(x-x')\,D^{(1)}(x'-x)-S^{(1)}(x-x')\,D(x'-x)]\,\gamma_\lambda\langle 0|\,\psi^{(0)}(x)\,|q\rangle\,.\end{aligned}\right\}\tag{32.10}$$

We have the factor $e^{\alpha(x_0+x_0')}$ here because one factor e is associated with the point x and the other with x' according to (31.1) through (31.4). Using the definitions (31.15) and (31.17), we can write (32.10) as

$$\left.\begin{aligned}\langle 0|\,\Phi(x,\alpha)\,|q\rangle &= \frac{i}{(2\pi)^3}\int dp\int\limits_{-\infty}^{x} dx'\, e^{\alpha(x_0+x_0')+ip(x-x')}\times\\ &\times[\Sigma_1^{(0)}(p^2)+(i\gamma p+m)\,\Sigma_2^{(0)}(p^2)]\,\varepsilon(p)\,u(q)\,e^{iqx'}=\\ &= i\,e^{2\alpha x_0+iqx}\int\limits_{-\infty}^{+\infty} dp_0\,\frac{\varepsilon(p)}{\alpha+i(p_0-q_0)}\times\\ &\times[\Sigma_1^{(0)}(\mathbf{q}^2-p_0^2)+(i\gamma_k q_k-\gamma_4 p_0+m)\,\Sigma_2^{(0)}(\mathbf{q}^2-p_0^2)]\,u(q)\,.\end{aligned}\right\}\tag{32.11}$$

We note that there is a factor $e^{2\alpha x_0}$ present in (32.11). The p_0-integral can be transformed as follows:

$$\left.\begin{aligned}\int\limits_{-\infty}^{+\infty}\frac{dp_0\,\varepsilon(p)}{p_0-q_0-i\alpha}\,\Sigma_i^{(0)}(\mathbf{q}^2-p_0^2) &= \int\limits_0^{\infty} dp_0\,\Sigma_i^{(0)}(\mathbf{q}^2-p_0^2)\left[\frac{1}{p_0-q_0-i\alpha}+\frac{1}{p_0+q_0+i\alpha}\right]=\\ &=\int\limits_0^{\infty}\frac{2p_0\,dp_0\,\Sigma_i^{(0)}(\mathbf{q}^2-p_0^2)}{p_0^2-(q_0+i\alpha)^2}=\int\limits_0^{\infty}\frac{da\,\Sigma_i^{(0)}(-a)}{a+\mathbf{q}^2-(q_0+i\alpha)^2}\,,\end{aligned}\right\}\tag{32.12}$$

$$\int\limits_{-\infty}^{+\infty}\frac{p_0\,dp_0\,\varepsilon(p)}{p_0-q_0-i\alpha}\,\Sigma_2^{(0)}(\mathbf{q}^2-p_0^2)=(q_0+i\alpha)\int\limits_0^{\infty}\frac{da\,\Sigma_2^{(0)}(-a)}{a+\mathbf{q}^2-(q_0+i\alpha)^2}\,, \tag{32.13}$$

1. See, for example, F. J. Dyson, Phys. Rev. 83, 608 (1951); G. Lüders, Z. Naturforsch. 7a, 206 (1952).

$$\left.\begin{aligned}\langle 0|\Phi(x,\alpha)|q\rangle = e^{2\alpha x_0}\int_0^\infty \frac{da}{a+\boldsymbol{q}^2-(q_0+i\alpha)^2}\ \times \\ \times\left[\Sigma_1^{(0)}(-a)-i\alpha\gamma_4\Sigma_2^{(0)}(-a)\right]\langle 0|\psi^{(0)}(x)|q\rangle.\end{aligned}\right\}\quad (32.14)$$

For very small values of α, (32.14) becomes

$$\langle 0|\Phi(x)|q\rangle = \overline{\Sigma}_1^{(0)}(-m^2)\langle 0|\psi^{(0)}(x)|q\rangle, \quad (32.15)$$

as is required by (32.3). We now subtract the self-mass according to (32.5), so that we have

$$\left.\begin{aligned}&\langle 0|\Phi(x,\alpha)|q\rangle - e^{2\alpha x_0}\overline{\Sigma}_1^{(0)}(-m^2)\langle 0|\psi^{(0)}(x)|q\rangle = \\ &= e^{2\alpha x_0}\int_0^\infty da\left\{\Sigma_1^{(0)}(-a)\frac{2iq_0\alpha-\alpha^2}{(a-m^2)[a+\boldsymbol{q}^2-(q_0+i\alpha)^2]} - i\gamma_4\alpha\frac{\Sigma_2^{(0)}(-a)}{a+\boldsymbol{q}^2-(q_0+i\alpha)^2}\right\}\times \\ &\qquad\times\langle 0|\psi^{(0)}(x)|q\rangle.\end{aligned}\right\}(32.16)$$

The overall factor $e^{2\alpha x_0}$ enters because we have to introduce the self-mass by means of a time-dependent charge. Instead of (32.1) we now have

$$\left.\begin{aligned}\langle 0|\psi(x)|q\rangle = \langle 0|\psi^{(0)}(x)|q\rangle + \frac{i}{(2\pi)^3}\int dp\int_{-\infty}^{x} dx'\, e^{ip(x-x')}\ \times \\ \times\ \delta(p^2+m^2)\,\varepsilon(p)\,(i\gamma p - m)\,e^{2\alpha x_0'}\int_0^\infty da\{\ldots\}\,u(q)\,e^{iqx'}.\end{aligned}\right\}\quad (32.17)$$

The integrations over x and p in (32.17) can be done immediately and give

$$\left.\begin{aligned}&\langle 0|\psi(x)|q\rangle = \langle 0|\psi^{(0)}(x)|q\rangle + \frac{1}{2q_0}\,e^{2\alpha x_0}\left[\frac{1}{2\alpha}(i\gamma q - m) - \frac{1}{2(\alpha - iq_0)}\times\right. \\ &\times\left.(i\gamma_k q_k + \gamma_4 q_0 - m)\right]\int_0^\infty da\left\{\Sigma_1^{(0)}(-a)\frac{-2q_0\alpha - i\alpha^2}{(a-m^2)[a+\boldsymbol{q}^2-(q_0+i\alpha)^2]}+\right. \\ &\left. + \alpha\gamma_4\frac{\Sigma_1^{(0)}(-a)}{a+\boldsymbol{q}^2-(q_0+i\alpha)^2}\right\}\langle 0|\psi^{(0)}(x)|q\rangle.\end{aligned}\right\}(32.18)$$

Here we have succeeded in making the calculation without encountering undetermined expressions like (32.9). In the limit $\alpha\to 0$, we obtain from (32.18),

$$\left.\begin{aligned}&\langle 0|\psi(x)|q\rangle = \langle 0|\psi^{(0)}(x)|q\rangle - \frac{1}{2}(i\gamma q - m)\int_0^\infty da\,\times \\ &\qquad\times\left\{\frac{\Sigma_1^{(0)}(-a)}{(a-m^2)^2} - \frac{\gamma_4}{2q_0}\frac{\Sigma_2^{(0)}(-a)}{a-m^2}\right\}\langle 0|\psi^{(0)}(x)|q\rangle = \\ &= \left[1 - \frac{1}{2}\int_0^\infty da\left[\frac{\Sigma_2^{(0)}(-a)}{a-m^2} - 2m\frac{\Sigma_1^{(0)}(-a)}{(a-m^2)^2}\right]\right]\langle 0|\psi^{(0)}(x)|q\rangle.\end{aligned}\right\}(32.19)$$

In this approximation, the constant N in (32.8) is therefore given by

$$N = 1 - \tfrac{1}{2}\left(\overline{\Sigma}_2^{(0)}(-m^2) + 2m\,\overline{\Sigma}_1^{(0)\prime}(-m^2)\right), \quad (32.20)$$

where $\overline{\Sigma}_1^{(0)\prime}(p^2)$ is the derivative of $\overline{\Sigma}_1^{(0)}(p^2)$ with respect to p^2. As is evident from (31.15) and (31.17), the integral for this derivative converges for $a\to\infty$, although the integral for $\overline{\Sigma}_2^{(0)}(-m^2)$ diverges. However, in the former integral another difficulty arises because the function $\Sigma_1^{(0)}(-a)$ only vanishes linearly for $a=m^2$. This leads to the divergence of the integral

$$\int\limits_0^\infty \frac{\Sigma_1^{(0)}(-a)\,da}{(a-m^2)^2} = \int\limits_{m^2}^\infty \frac{\Sigma_1^{(0)}(-a)}{(a-m^2)^2}\,da \tag{32.21}$$

at the lower limit. This divergence is of the same character as the divergence discussed in (31.29) and (31.30). We can circumvent it most readily by introducing formally a small but finite photon mass as a cutoff. That is, we replace the function $\delta(k^2)$ in (32.12) and (32.13) by $\delta(k^2+\mu^2)$. With this the function $\Sigma_1^{(0)}(p^2)$ becomes

$$\Sigma_1^{(0)}(p^2) = \frac{m\,e^2}{16\pi^2}\left(3-\frac{m^2-\mu^2}{-p^2}\right)\sqrt{\left(1-\frac{m^2-\mu^2}{-p^2}\right)^2-\frac{4\mu^2}{-p^2}}\;\Theta\left(-p^2-(m+\mu)^2\right). \tag{32.22}$$

The function (32.22) already vanishes for $-p^2<(m+\mu)^2$ and the integral (32.21) is also convergent at the lower limit.

We now return to the matrix element of the current operator (31.2). If we renormalize the mass in the equations of motion according to (32.5), we do not obtain $\Phi(x)$ in (31.2). Rather, for the first two terms, we obtain forms like (32.16) which have been folded with a function $S_R(x-x')$. By a calculation similar to (32.16) through (32.19) and after carrying out the <u>same</u> charge renormalization as for an external field, we finally get

$$\left.\begin{aligned}\langle q|\,j_\mu^{(2)}(x)\,|q'\rangle = {}&\langle q|\,j_\mu^{(0)}(x)\,|q'\rangle\,[-\overline{\Pi}^{(0)}(Q^2)+\overline{\Pi}^{(0)}(0)-i\pi\,\varepsilon(Q)\,\Pi^{(0)}(Q^2)-\\ &-\overline{\Sigma}_2^{(0)}(-m^2)-2m\,\overline{\Sigma}_1^{(0)\prime}(-m^2)+\overline{R}^{(0)}(Q^2)+i\pi\,\varepsilon(Q)\,R^{(0)}(Q^2)]-\\ &-\frac{e}{2m}\,(q_\mu+q'_\mu)\langle q|:\overline{\psi}^{(0)}(x)\,\psi^{(0)}(x):|q'\rangle\,[\overline{S}^{(0)}(Q^2)+i\pi\,\varepsilon(Q)\,S^{(0)}(Q^2)]\,,\end{aligned}\right\} \tag{32.23}$$

$$Q=q'-q. \tag{32.24}$$

For the matrix element (31.2a) we have a similar expression, except that q is replaced by $-q$ at the appropriate places.

33. The Magnetic Moment of the Electron

Considered as functions of q and q', the two terms in Eq. (32.23) show quite different behavior. One term is proportional to $j_\mu^{(0)}(x)$ and therefore has the same character as the "dielectric constant" of the vacuum in (29.37). In fact, part of (32.23) is the same as (29.37) and can be viewed as a vacuum polarization by the electron current, as was already noted in connection with (31.5). The other parts of (32.23) have to be understood as the difference of the interactions of an external current and the radiation field, as opposed to the interaction of an electron and the radiation field. For example, an electron emitting a virtual photon picks up recoil momentum which is included in (32.23). Of course, this effect would be absent for an external field. Moreover, (32.23) contains a term of completely different character; the term proportional to

$$-\frac{e}{2m}(q_\mu+q'_\mu)\langle q\,|:\bar\psi^{(0)}(x)\,\psi^{(0)}(x):|\,q'\rangle. \qquad (33.1)$$

Using the Dirac equation for the free electron, we can write (33.1) as follows:[1]

$$\left.\begin{aligned} &-\frac{e}{2m}(q_\mu+q'_\mu)\langle q\,|:\bar\psi^{(0)}(x)\,\psi^{(0)}(x):|\,q'\rangle = \\ &\quad=\frac{ie}{2m}Q_\nu\langle q\,|:\bar\psi^{(0)}(x)\,\sigma_{\mu\nu}\psi^{(0)}(x):|\,q'\rangle-\langle q\,|j_\mu^{(0)}(x)|\,q'\rangle\,, \end{aligned}\right\} \qquad (33.2)$$

$$\sigma_{\mu\nu}=\frac{i}{2}(\gamma_\nu\gamma_\mu-\gamma_\mu\gamma_\nu). \qquad (33.3)$$

We see that the electron gets an anomalous magnetic moment because of its coupling with the electromagnetic field:

$$\frac{e}{2m}\langle q\,|:\bar\psi^{(0)}(x)\,\sigma_{\mu\nu}\psi^{(0)}(x):|\,q'\rangle\left[\bar S^{(0)}(Q^2)+i\pi\,\varepsilon(Q)\,S^{(0)}(Q^2)\right]. \qquad (33.4)$$

For a one-electron state the expectation value of (33.4) is

$$\frac{e}{2m}\langle q\,|:\bar\psi^{(0)}(x)\,\sigma_{\mu\nu}\psi^{(0)}(x):|\,q\rangle\,\bar S^{(0)}(0). \qquad (33.5)$$

Since the zero-th order current operator corresponds to a magnetic moment given by

$$\langle q\,|m_{\mu\nu}^{(0)}(x)|\,q\rangle=\frac{e}{2m}\langle q\,|:\bar\psi^{(0)}(x)\,\sigma_{\mu\nu}\psi^{(0)}(x):|\,q\rangle\,, \qquad (33.6)$$

in this approximation the total magnetic moment of the electron becomes

$$\left[1+\bar S^{(0)}(0)\right]\langle q\,|m_{\mu\nu}^{(0)}(x)|\,q\rangle. \qquad (33.7)$$

As has been noted above, the integral $\bar S^{(0)}(0)$ is convergent, and from (31.33) we have

$$\bar S^{(0)}(0)=\int\limits_{4m^2}^{\infty}\frac{da}{a}\,\frac{e^2}{4\pi^2}\,\frac{m^2}{a}\,\frac{1}{\sqrt{1-\dfrac{4m^2}{a}}}=\frac{e^2}{8\pi^2}=\frac{\alpha}{2\pi}\,, \qquad (33.8)$$

where[2]

$$\alpha=\frac{e^2}{4\pi}=\frac{1}{137.0384}. \qquad (33.8a)$$

Thus the electron has a g-factor which is not exactly 2, as it would be in the Dirac theory. In this approximation its value is[3]

$$g=2\left(1+\frac{\alpha}{2\pi}\right)=2\cdot 1.0011614\,. \qquad (33.9)$$

---

1. W. Gordon, Z. Physik 50, 630 (1928).
2. C. f. Sec. 36 and 37.
3. The anomalous magnetic moment of the electron was first calculated by J. Schwinger, Phys. Rev. 73, 416 (1948).

It is remarkable that the finite magnetic moment of the electron can be singled out even in (31.31) and (31.34) by considerations of invariance alone. It is not absolutely necessary to carry out mass and charge renormalizations if only the magnetic moment is to be calculated.[1]

Experimental investigations of the magnetic moment of the electron were performed in 1947 with sufficient precision so that there were doubts about the validity of the Dirac theory.[2] This was due partly to the measurements of the hyperfine structure of the hydrogen atom and of the deuterium atom and partly to exact measurements of the Zeeman effect in a few complex atoms (Na, Ga, In). G. Breit had shown rather early[3] that the equality of the present deviations of the hyperfine structure in hydrogen and deuterium could be interpreted in a natural way as an anomalous magnetic moment of the electron. The later theoretical value (33.9) deduced by Schwinger agreed very well with the original measurements. Recent more exact measurements[4] of the magnetic moment of the electron do not differ very much from (33.9). Franken and Liebes[5] find the value

$$g = 2 \cdot (1.001\,167 \pm 0.000\,005) . \tag{33.10}$$

Equation (33.10) agrees completely with (33.9) to within the limits of error. Since Eq. (33.9) is only the first term in a series in α, one may ask how large the next term in this series is and whether it will disturb the agreement between theory and experiment. The term proportional to α^2 has been calculated by Karplus and Kroll,[6] Sommerfield,[7] and Petermann.[5] By quite involved calculations which we shall not go into here, they obtain the result

$$\left.\begin{aligned} g &= 2\left[1 + \frac{\alpha}{2\pi} + \frac{\alpha^2}{\pi^2}\left(\frac{197}{144} + \frac{\pi^2}{12} + \frac{3}{4}\zeta(3) - \frac{1}{2}\pi^2 \log 2\right)\right] = \\ &= 2\left[1 + 0.5\frac{\alpha}{\pi} - 0.328\frac{\alpha^2}{\pi^2}\right] = 2 \cdot 1.0011596 , \end{aligned}\right\} \tag{33.11}$$

1. See also J. Luttinger, Phys. Rev. 74, 893 (1948).

2. J. E. Nafe, E. B. Nelson and I. I. Rabi, Phys. Rev. 71, 914 (1947); D. E. Nagel, R. S. Julian and J. R. Zacharias, Phys. Rev. 72, 971 (1947); P. Kusch and H. M. Foley, Phys. Rev. 72, 1256 (1947).

3. G. Breit, Phys. Rev. 72, 984 (1947).

4. S. H. Koenig, A. G. Prodell and P. Kusch, Phys. Rev. 88, 191 (1952); R. Beringer and M. A. Heald, Phys. Rev. 95, 1474 (1954); P. Franken and S. Liebes, Phys. Rev. 104, 1197 (1957). A review of the older measurements is given by F. Bloch, Physica, Haag 19, 821 (1953). More complete references are to be found here. See also E. R. Cohen and J. W. M. Du Mond in Vol. XXXV of this handbook (Handbuch der Physik, edited by S. Flügge, Springer-Verlag, Heidelberg).

5. A. Petermann, Helv. Phys. Acta 30, 407 (1957).

6. R. Karplus and N. M. Kroll, Phys. Rev. 77, 536 (1950). See also A. Petermann, Nuclear Phys. 3, 689 (1957).

7. C. M. Sommerfield, Phys. Rev. 107, 328 (1957).

with

$$\zeta(3) = \sum_{1}^{\infty} \frac{1}{n^3} = 1.20206. \tag{33.11a}$$

The numerical agreement of (33.10) and (33.11) is excellent. The experimental limits of error are so large that the term proportional to α^2 in (33.11) can scarcely be viewed as confirmed. On the other hand, there is no discrepancy between theory and experiment: the quantum electrodynamic description of the magnetic moment of the electron is certainly correct in its essential points. This is a great success of the theory and gives one justification for the mathematically faulty formalism in which infinite expressions are treated as if they were finite.

34. The Charge Renormalization of the Electron State

We now return to the complete current operator (32.23) and write it as

$$\left.\begin{aligned}\langle q | j_\mu^{(2)}(x) | q'\rangle = \langle q | j_\mu^{(0)}(x) | q'\rangle \{ & -\overline{\Pi}^{(0)}(Q^2) + \overline{\Pi}^{(0)}(0) - i\pi\varepsilon(Q)\Pi^{(0)}(Q^2) + \\ & + \overline{R}^{(0)}(Q^2) - \overline{R}^{(0)}(0) + i\pi\varepsilon(Q) R^{(0)}(Q^2) - \overline{S}^{(0)}(Q^2) + \\ & + \overline{S}^{(0)}(0) - i\pi\varepsilon(Q) S^{(0)}(Q^2) + \overline{R}^{(0)}(0) - \overline{S}^{(0)}(0) - \\ & - \overline{\Sigma}_2^{(0)}(-m^2) - 2m\overline{\Sigma}_1^{(0)\prime}(-m^2)\} + \\ & + iQ_\nu \langle q | m_{\mu\nu}^{(0)}(x) | q'\rangle [\overline{S}^{(0)}(Q^2) + i\pi\varepsilon(Q) S^{(0)}(Q^2)].\end{aligned}\right\} \tag{34.1}$$

We put $\mu = 4$ in (34.1) and integrate over three-dimensional space, so that we obtain the renormalized charge of the electron:

$$\left.\begin{aligned}\langle q | Q | q'\rangle = -i\int d^3x \langle q | j_4(x) | q'\rangle = -i\int d^3x \langle q | j_4^{(0)}(x) + j_4^{(2)}(x) + \cdots | q'\rangle = \\ = e\langle q | q'\rangle [1 + \overline{R}^{(0)}(0) - \overline{S}^{(0)}(0) - \\ - \overline{\Sigma}_2^{(0)}(-m^2) - 2m\overline{\Sigma}_1^{(0)\prime}(-m^2) + \cdots].\end{aligned}\right\} \tag{34.2}$$

In order to obtain (34.2) we have used the same charge renormalization for the electron as for the external field. However, from (34.2) we see that the charge of an electron state does not seem to be equal to e, but that an additional charge renormalization

$$\overline{R}^{(0)}(0) - \overline{S}^{(0)}(0) - \overline{\Sigma}_2^{(0)}(-m^2) - 2m\overline{\Sigma}_1^{(0)\prime}(-m^2) \tag{34.3}$$

enters. This expression is not present for an external field. It contains two infinite integrals, namely $\overline{R}^{(0)}(0)$ and $\overline{\Sigma}_2^{(0)}(-m^2)$, and these are present with opposite signs. All other expressions in (34.1) are finite. We have two possible choices at this point. The first is to say that an additional charge renormalization for the electron, if it exists, must be unobservable and therefore that the corresponding terms in (34.1) can be dropped. This choice was made by the earlier workers in the subject. A second choice is possible, however. We can cut off the two infinite terms in (34.3) by a suitable limiting process and then consider a possible cancellation of both. The assertion that an expression like (34.3) (which actually contains infinite quantities) has a finite value or vanishes is clearly nonsense; one can give it a well defined meaning only by means of a limiting process. To do this, we cannot simply introduce an upper limit in the integrals (31.35) and (31.18) and then set these two cutoffs equal to each other. This is not possible because the two variables of integration a have nothing to do with each other and cannot be directly compared with each other. Instead of this, we must return to the original integrals (31.8) and (31.21), cut off both of these integrals in the same way and then work out the expression (34.3) once more. We shall take the regularization method developed in Sec. 30 as the cutoff procedure. We modify it here by thinking of the electrons as being coupled to several kinds of "photons" of different masses. In principle, we should recalculate the functions $\Sigma_2^{(0)}$, $R^{(0)}$, and $S^{(0)}$ for various photon masses in order to form such linear combinations that the integrals for the "barred" functions (31.18), (31.35), (31.36) converge. Such a calculation can probably be done; however, it is a little more efficient to use (31.8) and (31.19) as follows:

$$\left.\begin{aligned}[\overline{\Sigma}_1^{(0)}(q^2) + (i\gamma q + m)\,\overline{\Sigma}_2^{(0)}(q^2)]^{\text{reg}} &= \frac{e^2}{8\pi^3}\int dk\,[i\gamma(q+k)+2m]\times \\ &\times\sum_i C_i\left[\frac{\delta((q+k)^2+m^2)}{k^2+\mu_i^2}+\frac{\delta(k^2+\mu_i^2)}{(q+k)^2+m^2}\right].\end{aligned}\right\}\quad(34.4)$$

In the final result we shall take

$$C_1\to 1,\qquad \sum_i C_i = 0,\qquad \mu_1\to 0,\qquad \mu_i\to\infty\quad(i\neq 1),$$

where μ_i $(i\neq 1)$ are the other "photon masses". Using (30.6) and a few simple transformations, we obtain from (34.4),

$$\begin{aligned}&[\overline{\Sigma}_2^{(0)}(-m^2) + 2m\,\overline{\Sigma}_1^{(0)\prime}(-m^2)]^{\text{reg}} = \\ &= -\frac{e^2}{8\pi^3}\int_0^1 d\alpha\left\{(1-\alpha)\frac{i}{2\pi}\int_{-\infty}^{+\infty} w\,dw\int dk\sum C_i\,e^{iw[k^2+m^2\alpha^2+\mu_i^2(1-\alpha)]} + \right. \\ &\left.\quad + 2\pi\sum_i C_i\frac{\alpha(1-\alpha^2)}{\alpha^2+\frac{\mu_i^2}{m^2}(1-\alpha)}\right\} = \qquad\qquad (34.5) \\ &= -\frac{e^2}{8\pi^2}\sum_i C_i\int_0^1 d\alpha\left\{(1-\alpha)\log\left(\alpha^2+\frac{\mu_i^2}{m^2}(1-\alpha)\right)+\frac{2\alpha(1-\alpha^2)}{\alpha^2+\frac{\mu_i^2}{m^2}(1-\alpha)}\right\}.\end{aligned}$$

In the first term of (34.5) we integrate by parts with respect to α and make use of

$$\sum_i C_i\int_0^1\frac{d\alpha\,\frac{\mu_i^2}{m^2}(1-\alpha)f(\alpha)}{\alpha^2+\frac{\mu_i^2}{m^2}(1-\alpha)} = -\sum_i C_i\int_0^1\frac{d\alpha\cdot\alpha^2 f(\alpha)}{\alpha^2+\frac{\mu_i^2}{m^2}(1-\alpha)}\qquad(34.6)$$

to obtain

$$\left.\begin{aligned}&[\overline{\Sigma}_2^{(0)}(-m^2) + 2m\,\overline{\Sigma}_1^{(0)\prime}(-m^2)]^{\text{reg}} = \\ &= -\frac{e^2}{4\pi^2}\sum_i C_i\int_0^1\frac{\alpha\,d\alpha}{1-\alpha}\left[\frac{1-2\alpha+\frac{3}{4}\alpha^3}{\alpha^2+\frac{\mu_i^2}{m^2}(1-\alpha)}+\frac{1}{4}\right].\end{aligned}\right\}\quad(34.7)$$

The integral (34.7) can be done by elementary methods; however, we shall leave it in its present form until later. In a similar way, we find for the constants $\bar{R}^{(0)}(0)$ and $\bar{S}^{(0)}(0)$:

$$\left.\begin{aligned}\gamma_\mu[\bar{R}^{(0)}(0)-\bar{S}^{(0)}(0)]^{\text{reg}} &= -\frac{e^2}{16\pi^3}\int dk\,\gamma_\lambda(i\gamma(q-k)-m)\gamma_\mu(i\gamma(q-k)-m)\gamma_\lambda\times\\ &\times\sum_i C_i\left\{\frac{\delta(k^2+\mu_i^2)}{[(q-k)^2+m^2][(q'-k)^2+m^2]}+\right.\\ &\left.+\frac{\delta((q-k)^2+m^2)}{(k^2+\mu_i^2)[(q'-k)^2+m^2]}+\frac{\delta((q'-k)^2+m^2)}{(k^2+\mu_i^2)[(q-k)^2+m^2]}\right\}_{q=q'}.\end{aligned}\right\}\quad(34.8)$$

The expression (34.8) is to be evaluated under the restrictions made after (31.20). By means of[1]

$$\frac{\delta(a)}{b\cdot c}+\frac{\delta(b)}{a\cdot c}+\frac{\delta(c)}{a\cdot b}=\int_0^1 d\alpha\int_0^\alpha d\beta\,\delta''(a\beta+b(\alpha-\beta)+c(1-\alpha))\quad(34.9)$$

we can combine the terms inside the curly brackets of (34.8):

$$\left.\begin{aligned}\{\ldots\}_{q=q'} &= \int_0^1 d\alpha\int_0^\alpha d\beta\,\delta''(\alpha((k-q)^2+m^2)+(k^2+\mu_i^2)(1-\alpha)) =\\ &= \int_0^1 \alpha\,d\alpha\,\delta''((k-q\alpha)^2+m^2\alpha^2+\mu_i^2(1-\alpha)).\end{aligned}\right\}\quad(34.10)$$

Substitution of (34.10) into Eq. (34.8) and use of the symmetry in $k'=k-q\alpha$ gives

$$\left.\begin{aligned}\gamma_\mu[\bar{R}^{(0)}(0)-\bar{S}^{(0)}(0)]^{\text{reg}} &= -\frac{e^2}{16\pi^3}\int_0^1\alpha\,d\alpha\int dk'\times\\ &\times\sum_i C_i\,\delta''(k'^2+m^2\alpha^2+\mu_i^2(1-\alpha))\left[-k'^2\gamma_\mu+4m^2\gamma_\mu\left(1-\alpha-\frac{\alpha^2}{2}\right)\right].\end{aligned}\right\}\quad(34.11)$$

The k'-integration is now convergent, can be done, and gives the result

$$\left.\begin{aligned}[\bar{R}^{(0)}(0)-\bar{S}^{(0)}(0)]^{\text{reg}} &= -\frac{e^2}{8\pi^2}\sum_i C_i\int_0^1\alpha\,d\alpha\times\\ &\times\left\{\log\left(\alpha^2+\frac{\mu_i^2}{m^2}(1-\alpha)\right)+\frac{2\left(1-\alpha-\frac{\alpha^2}{2}\right)}{\alpha^2+\frac{\mu_i^2}{m^2}(1-\alpha)}\right\}.\end{aligned}\right\}\quad(34.12)$$

Here we can transform (34.12) in a manner similar to (34.5) [by applying (34.6)] and we obtain

$$\left.\begin{aligned}[\bar{R}^{(0)}(0)-\bar{S}^{(0)}(0)]^{\text{reg}} &= \frac{-e^2}{4\pi^2}\sum_i C_i\int_0^1\frac{\alpha\,d\alpha}{1-\alpha}\left[\frac{1-2\alpha+\frac{3}{4}\alpha^3}{\alpha^2+\frac{\mu_i^2}{m^2}(1-\alpha)}+\frac{1}{4}\right]=\\ &= [\bar{\Sigma}_2^{(0)}(-m^2)+2m\,\bar{\Sigma}_1^{(0)\prime}(-m^2)]^{\text{reg}}.\end{aligned}\right\}\quad(34.13)$$

Equation (34.3) now vanishes by the use of the identity (34.13) and, in this approximation, we have the same charge renormalization for an electron as for an external field. In a later section,

1. J. Schwinger, Phys. Rev. 76, 790 (1949).

we shall see that this result is true not only in this order but can be proved quite generally. The relation between infinite constants expressed in (34.13) was first used by Schwinger[1] and later shown by Ward[2] to be true in general. The equation to which (34.13) is the first approximation is often referred to in the literature as the "Ward identity".

The second approximation to the current operator now takes the form

$$\left.\begin{aligned}\langle q|\, j_\mu^{(2)}(x)\,|q'\rangle = \langle q|\, j_\mu^{(0)}(x)\,|q'\rangle [-\overline{\Pi}^{(0)}(Q^2) + \overline{\Pi}^{(0)}(0) + \overline{R}^{(0)}(Q^2) -\\ - \overline{R}^{(0)}(0) - \overline{S}^{(0)}(Q^2) + \overline{S}^{(0)}(0) - i\pi\,\varepsilon(Q)\,(\Pi^{(0)}(Q^2) -\\ - R^{(0)}(Q^2) + S^{(0)}(Q^2))] + i\,Q_\nu \langle q|\, m_{\mu\nu}^{(0)}(x)|\, q'\rangle \times\\ \times [\overline{S}^{(0)}(Q^2) + i\pi\,\varepsilon(Q)\, S^{(0)}(Q^2)].\end{aligned}\right\} \quad (34.14)$$

All expressions occurring in (34.14) are convergent. From Eqs. (24.35), (31.32) and (31.33) we have

$$\Pi^{(0)}(Q^2) = \frac{e^2}{12\pi^2}\left(1 - \frac{2m^2}{Q^2}\right)\sqrt{1 + \frac{4m^2}{Q^2}}\;\Theta(-Q^2 - 4m^2), \quad (34.15)$$

$$\left.\begin{aligned}&-\overline{\Pi}^{(0)}(Q^2) + \overline{\Pi}^{(0)}(0) =\\ &= \frac{e^2}{12\pi^2}\left[\frac{4m^2}{Q^2} - \frac{5}{3} + \left(1 - \frac{2m^2}{Q^2}\right)\sqrt{1 + \frac{4m^2}{Q^2}}\,\log\frac{1 + \sqrt{1 + \frac{4m^2}{Q^2}}}{\left|1 - \sqrt{1 + \frac{4m^2}{Q^2}}\right|}\right],\end{aligned}\right\} \quad (34.16)$$

$$R^{(0)}(Q^2) = \frac{e^2}{8\pi^2}\left[\frac{1 + \frac{2m^2}{Q^2}}{1 + \frac{4m^2}{Q^2}}\log\left[1 - \frac{Q^2 + 4m^2}{\mu^2}\right] - \frac{3}{2}\right]\sqrt{1 + \frac{4m^2}{Q^2}}\;\Theta(-Q^2 - 4m^2), \quad (34.17)$$

$$\left.\begin{aligned}&\overline{R}^{(0)}(Q^2) - \overline{R}^{(0)}(0) =\\ &= \frac{e^2}{4\pi^2}\left\{\frac{1 + \frac{2m^2}{Q^2}}{\sqrt{1 + \frac{4m^2}{Q^2}}}\left[\Phi\left(\frac{-2\sqrt{1 + \frac{4m^2}{Q^2}}}{1 + \sqrt{1 + \frac{4m^2}{Q^2}}}\right) + \frac{\pi^2}{3} + \frac{\pi^2}{4}\Theta\left(1 - \sqrt{1 + \frac{4m^2}{Q^2}}\right) -\right.\right.\\ &\left.- \frac{1}{4}\log^2\frac{1 + \sqrt{1 + \frac{4m^2}{Q^2}}}{\left|1 - \sqrt{1 + \frac{4m^2}{Q^2}}\right|}\right] + \frac{3}{4}\left[\sqrt{1 + \frac{4m^2}{Q^2}}\log\frac{1 + \sqrt{1 + \frac{4m^2}{Q^2}}}{\left|1 - \sqrt{1 + \frac{4m^2}{Q^2}}\right|} - 2\right] +\\ &\left.+ \log\frac{m}{\mu}\left[1 - \frac{1 + \frac{2m^2}{Q^2}}{\sqrt{1 + \frac{4m^2}{Q^2}}}\log\frac{1 + \sqrt{1 + \frac{4m^2}{Q^2}}}{\left|1 - \sqrt{1 + \frac{4m^2}{Q^2}}\right|}\right]\right\},\end{aligned}\right\} \quad (34.18)$$

$$S^{(0)}(Q^2) = -\frac{e^2}{4\pi^2}\frac{m^2}{Q^2}\frac{\Theta(-Q^2 - 4m^2)}{\sqrt{1 + \frac{4m^2}{Q^2}}}, \quad (34.19)$$

---

1. J. Schwinger, Phys. Rev. 76, 790 (1949); see eqs. (1.95), (1.97) and (1.100).
2. J. C. Ward, Phys. Rev. 78, 182 (1950).

$$\overline{S}^{(0)}(Q^2) = \frac{e^2}{8\pi^2} \frac{2\frac{m^2}{Q^2}}{\sqrt{1+\frac{4m^2}{Q^2}}} \log \frac{1+\sqrt{1+\frac{4m^2}{Q^2}}}{\left|1-\sqrt{1+\frac{4m^2}{Q^2}}\right|}. \qquad (34.20)$$

In (34.18), $\Phi(x)$ stands for the function

$$\Phi(x) = \int_1^x \frac{dt}{t} \log|1+t|. \qquad (34.21)$$

It satisfies many functional equations of which we give the following examples here[1]

$$\Phi(x) + \Phi\left(\frac{1}{x}\right) = \frac{1}{2}\log^2|x| - \frac{\pi^2}{2}\Theta(-x), \qquad (34.22)$$

$$\Phi(x) + \Phi(-1-x) = -\frac{\pi^2}{3} + \log|x|\log|1+x|, \qquad (34.23)$$

$$\Phi(x) - \Phi(-x) = \Phi\left(-\frac{2}{1+x}\right) - \Phi\left(-\frac{2x}{1+x}\right) + \frac{\pi^2}{4} - \pi^2\Theta(-1-x). \qquad (34.24)$$

In (34.16), (34.18) and (34.20), it is assumed that $1+\frac{4m^2}{Q^2}$ is positive. If this is not the case, the logarithms in these equations and in (34.21) must be replaced by arctan functions. For small values of Q^2/m^2 we find

$$\overline{R}^{(0)}(Q^2) - \overline{R}^{(0)}(0) = -\frac{e^2}{12\pi^2}\frac{Q^2}{m^2}\left[\log\frac{m}{\mu} - \frac{1}{8}\right] + \cdots, \qquad (34.25)$$

$$\overline{S}^{(0)}(Q^2) = \frac{e^2}{8\pi^2} - \frac{e^2}{48\pi^2}\frac{Q^2}{m^2} + \cdots. \qquad (34.26)$$

---

1. See, for example, B. K. Mitchell, Phil. Mag. 40, 351 (1949) and W. Gröbner and N. Hofreiter, Integraltafeln, Vienna and Innsbruck, 1950. Equation (34.24), which is not given by these authors, can be shown in the following way:

$$\Phi(x) - \Phi(-x) = \int_{-x}^{x} \frac{dt}{t}\log|1+t| = \int_1^x \frac{dz}{z}\log\left|\frac{1+z}{1-z}\right| - \Phi(-1).$$

The substitution $1+y = \frac{1-z}{1+z}$ in the integral gives

$\Phi(x) - \Phi(-x) =$

$$= -\int_{-1}^{-\frac{2x}{1+x}} dy\left[\frac{1}{y} - \frac{1}{2+y}\right]\log|1+y| - \Phi(-1) - \Theta(-1-x)\int_{-\infty}^{+\infty} dy\left[\frac{1}{y} - \frac{1}{2+y}\right]\log|1+y| =$$

$$= -\Phi\left(\frac{-2x}{1+x}\right) + \int_1^{\frac{2}{1+x}} \frac{dt}{t}\log|t-1| - \pi^2\Theta(-1-x) =$$

$$= \Phi\left(-\frac{2}{1+x}\right) - \Phi\left(\frac{-2x}{1+x}\right) + \frac{\pi^2}{4} - \pi^2\Theta(-1-x),$$

since $\Phi(-1)$ is equal to $-\frac{\pi^2}{4}$ according to (34.22).

From this and (29.36) follows

$$\left.\begin{aligned}\langle q|j_\mu^{(2)}(x)|q'\rangle = \frac{e^2}{12\pi^2}\left[\log\frac{m}{\mu} - \frac{23}{40}\right]\frac{\square}{m^2}\langle q|j_\mu^{(0)}(x)|q'\rangle + \\ + \frac{e^2}{8\pi^2}\frac{\partial}{\partial x_\nu}\langle q|m_{\mu\nu}^{(0)}(x)|q'\rangle + \cdots .\end{aligned}\right\} \quad (34.27)$$

The expressions obtained so far contain only integrals which are convergent for large values of the integration variable a and which can be calculated. Equation (34.18) and hence (34.27) also depend upon the quantity μ. The limit $\mu \to 0$ cannot be taken here without encountering new divergences. Therefore our results cannot be compared directly with experiment. What we must do is to discuss the experimental conditions more carefully in order to decide how μ is to be interpreted in each special case.

35. Radiative Corrections to the Scattering in an External Field. The Infrared Catastrophe

As an application of the results of Sec. 34, we now examine the radiative corrections to the scattering cross section for an electron in an external field. The lowest order approximation to this cross section has been worked out in Sec. 24. With the assumption that the external field can be treated[1] in Born approximation, we can write the S-matrix as follows:

$$S = i\int dx\, j_\mu(x)\, A_\mu^{\text{äuss}}(x). \quad (35.1)$$

The quantity $j_\mu(x)$ in (35.1) is to be understood as the complete current operator. We obtain the result of Sec. 24 if we consider only the lowest order approximation to the current. The radiative corrections can be obtained by using higher order approximations for the current. Using the quantities defined in Sec. 34, the next higher approximation to the S-matrix becomes

$$\langle q|S|q'\rangle = i\langle q|j_\mu|q'\rangle\, A_\mu^{\text{äuss}}(Q)\, 2\pi\delta(Q_0)\,, \quad (35.2)$$

where

$$\left.\begin{aligned}\langle q|j_\mu|q'\rangle = \langle q|j_\mu^{(0)}|q'\rangle\,[1 - \overline{\Pi}^{(0)}(Q^2) + \overline{\Pi}^{(0)}(0) + \overline{R}^{(0)}(Q^2) - \overline{R}^{(0)}(0) + \\ + \overline{S}^{(0)}(0)] - \overline{S}^{(0)}(Q^2)\frac{e}{2m}(q_\mu + q'_\mu)\frac{\bar{u}(q)\,u(q')}{V}\,,\end{aligned}\right\} \quad (35.3)$$

$$Q = q' - q. \quad (35.4)$$

In Eq. (35.3) we have used the fact that the vector Q is space-like, so that the "unbarred" functions R, S, and Π vanish. Using the methods explained in previous sections, we find the cross section for unpolarized electron scattering in an external field:

---

1. The second Born approximation in the external field has been calculated by R. G. Newton, Phys. Rev. 97, 1162 (1955). See also R. G. Newton, Phys. Rev. 98, 1514 (1955); M. Chrétien, Phys. Rev. 98, 1515 (1955); H. Suura, Phys. Rev. 99, 1020 (1955).

$$\left.\begin{aligned}\frac{d\sigma}{d\Omega}=\frac{e^2}{4\pi^2}\Big\{&q_\mu q_\nu\left[1-2\left(\overline{\Pi}^{(0)}(Q^2)-\overline{\Pi}^{(0)}(0)-\overline{R}^{(0)}(Q^2)+\overline{R}^{(0)}(0)+\overline{S}^{(0)}(Q^2)-\overline{S}^{(0)}(0)\right)\right]+\\ &+\delta_{\mu\nu}\frac{Q^2}{4}\left[1-2\left(\overline{\Pi}^{(0)}(Q^2)-\overline{\Pi}^{(0)}(0)-\overline{R}^{(0)}(Q^2)+\overline{R}^{(0)}(0)-\overline{S}^{(0)}(0)\right)\right]\Big\}\times\\ &\times A_\mu^{\text{äuss}}(\boldsymbol{Q})\,A_\nu^{\text{äuss}}(-\boldsymbol{Q}).\end{aligned}\right\}\quad(35.5)$$

In (35.5), terms of order e^6 have been dropped, since (35.3) should be improved if this order is to be retained. Also, it has been assumed in (35.5) that the gauge of the external field satisfies the Lorentz condition $Q_\mu A_\mu^{\text{äuss}}(\boldsymbol{Q})=0$. In particular, if the external field is a Coulomb field, then

$$A_\mu(\boldsymbol{Q})=-i\,\delta_{\mu 4}\frac{Ze}{4p^2\sin^2\frac{\Theta}{2}}. \quad (35.6)$$

Using the lowest order cross section of Sec. 24,

$$\frac{d\sigma^{(0)}}{d\Omega}=\left(\frac{Ze^2}{8\pi p^2\sin^2\frac{\Theta}{2}}\right)^2\left(E^2-p^2\sin^2\frac{\Theta}{2}\right) \quad (35.7)$$

we can rewrite (35.5) as follows:

$$\left.\begin{aligned}\frac{d\sigma}{d\Omega}=\frac{d\sigma^{(0)}}{d\Omega}\Big[&1-2\left(\overline{\Pi}^{(0)}(Q^2)-\overline{\Pi}^{(0)}(0)-\overline{R}^{(0)}(Q^2)+\overline{R}^{(0)}(0)+\overline{S}^{(0)}(Q^2)-\right.\\ &\left.-\overline{S}^{(0)}(0)\right)-\frac{\frac{p^2}{m^2}\sin^2\frac{\Theta}{2}}{1+\frac{p^2}{m^2}\cos^2\frac{\Theta}{2}}\,2\overline{S}^{(0)}(Q^2)\Big],\end{aligned}\right\}\quad(35.8)$$

$$Q^2=4p^2\sin^2\frac{\Theta}{2}. \quad (35.9)$$

In (35.8) the function $\overline{R}^{(0)}(Q^2)-\overline{R}^{(0)}(0)$ is still present; by (34.18) this expression contains the small photon mass μ. In order to interpret this quantity, we note that in any actual experiment it is never possible to measure only the elastic scattering (35.8). One can never be sure that the electron has not radiated a photon of very low energy; the detector always has a finite resolving power and electrons of energy only slightly less than the incident energy will be counted as having been elastically scattered. This effect is one form of bremsstrahlung, and so we can take the cross section of the process from Sec. 26. If the energy of the light quantum in Eqs. (26.3), (26.10), and (26.13) is taken as very low with respect to the energy of the electrons, we obtain

$$\frac{d\sigma^{\text{Strahl}}}{d\Omega}=\frac{d\sigma^{(0)}}{d\Omega}\frac{e^2}{(2\pi)^3}\int\limits_{|\boldsymbol{k}|<\Delta E}\frac{d^3k}{2\omega}\left[\frac{2m^2+4p^2\sin^2\frac{\Theta}{2}}{kq\cdot kq'}-\frac{m^2}{(kq)^2}-\frac{m^2}{(kq')^2}\right], \quad (35.10)$$

Here we have integrated over all emitted photons of energy smaller than the resolution ΔE of the measuring apparatus. The integral (35.10) over the emitted photons diverges for small energies and again we shall cut it off by giving the photon a small mass. Thus, instead of (35.10), we have to examine

$$\frac{d\sigma_{\mathrm{Strahl}}}{d\Omega}=\frac{d\sigma^{(0)}}{d\Omega}\frac{e^2}{(2\pi)^3}\int\limits_{|\boldsymbol{k}|<\Delta E} dk\,\delta(k^2+\mu^2)\,\Theta(k)\left[\frac{2m^2+4p^2\sin^2\frac{\Theta}{2}}{kq\cdot kq'}-\frac{m^2}{(kq)^2}-\frac{m^2}{(kq')^2}\right]. \quad (35.11)$$

Since we now have cut off (35.11) and (35.8) in the same way, we can add both cross sections and see if the limit $\mu\to 0$ can be taken in the total cross section.

In order to compute (35.11), we have to work out the integral

$$\left.\begin{aligned} J &= \int\limits_{|\boldsymbol{k}|<\Delta E} dk\,\delta(k^2+p^2)\,\Theta(k)\,\frac{1}{kq\cdot kq'}=\\ &=\frac{1}{2}\int\limits_0^{\Delta E}\frac{\boldsymbol{k}^2\,d|\boldsymbol{k}|}{\sqrt{\boldsymbol{k}^2+\mu^2}}\int\frac{d\Omega_k}{(\boldsymbol{p}\boldsymbol{k}-E\sqrt{\boldsymbol{k}^2+\mu^2})(\boldsymbol{p}'\boldsymbol{k}-E\sqrt{\boldsymbol{k}^2+\mu^2})}=\\ &=\frac{1}{2}\int\limits_0^{\Delta E}\frac{\boldsymbol{k}^2\,d|\boldsymbol{k}|}{\sqrt{\boldsymbol{k}^2+\mu^2}}\int\limits_0^1 d\alpha\int\frac{d\Omega_k}{[(\boldsymbol{p}+\boldsymbol{Q}\alpha)\,\boldsymbol{k}-E\sqrt{\boldsymbol{k}^2+\mu^2}]^2}. \end{aligned}\right\} \quad (35.12)$$

For the last transformation in (35.12) we have used the identity[1]

$$\frac{1}{a\cdot b}=\int\limits_0^1\frac{d\alpha}{[a\alpha+b(1-\alpha)]^2}. \quad (35.13)$$

The k-integration in (35.12) can be done by elementary methods and gives

$$J=2\pi\int\limits_0^1\frac{d\alpha}{A}\left\{\log\frac{2\Delta E}{\mu}-\frac{E}{2\sqrt{E^2-A}}\log\frac{E+\sqrt{E^2-A}}{E-\sqrt{E^2-A}}\right\}, \quad (35.14)$$

$$A=m^2+Q^2\alpha(1-\alpha). \quad (35.15)$$

By the use of (35.14) and (35.15), we can write (35.11) as follows:

$$\frac{d\sigma^{\mathrm{Strahl}}}{d\Omega}=\frac{d\sigma^{(0)}}{d\Omega}\frac{e^2}{2\pi^2}\left[\log\frac{2\Delta E}{\mu}F_1+F_2\right], \quad (35.16)$$

$$F_1=\int\limits_0^1 d\alpha\left[\frac{m^2+\frac{Q^2}{2}}{m^2+Q^2\alpha(1-\alpha)}-1\right], \quad (35.17)$$

$$\left.\begin{aligned} F_2=\int\limits_0^1 d\alpha\Bigg[&\frac{m^2+\frac{Q^2}{2}}{m^2+Q^2\alpha(1-\alpha)}\cdot\frac{E}{2\sqrt{p^2-Q^2\alpha(1-\alpha)}}\log\frac{E-\sqrt{p^2-Q^2\alpha(1-\alpha)}}{E+\sqrt{p^2-Q^2\alpha(1-\alpha)}}-\\ &-\frac{E}{2p}\log\frac{E-p}{E+p}\Bigg]. \end{aligned}\right\} \quad (35.18)$$

The α-integration in (35.17) can be done easily, although the integration in (35.18) is quite tedious. We give only the results:

1. R. P. Feynman, Phys. Rev. 76, 785 (1949).

$$F_1 = \frac{1 + \frac{2m^2}{Q^2}}{\sqrt{1 + \frac{4m^2}{Q^2}}} \log \frac{\sqrt{1 + \frac{4m^2}{Q^2}} + 1}{\sqrt{1 + \frac{4m^2}{Q^2}} - 1} - 1 , \tag{35.19}$$

$$\left.\begin{aligned} F_2 = & -\frac{1 + \frac{2m^2}{Q^2}}{\sqrt{1 + \frac{4m^2}{Q^2}}} \left[\Phi\left(\frac{-2\sqrt{1 + \frac{4m^2}{Q^2}}}{1 + \sqrt{1 + \frac{4m^2}{Q^2}}} \right) + \frac{\pi^2}{3} - \frac{1}{4}\left(\log \frac{\sqrt{1 + \frac{4m^2}{Q^2}} + 1}{\sqrt{1 + \frac{4m^2}{Q^2}} - 1} \right)^2 \right] + \\ & + \frac{E}{2p} \log \frac{E+p}{E-p} - \frac{1 + \frac{2m^2}{Q^2}}{\sqrt{1 + \frac{4m^2}{Q^2}}} \log \frac{2E}{m} \log \frac{\sqrt{1 + \frac{4m^2}{Q^2}} + 1}{\sqrt{1 + \frac{4m^2}{Q^2}} - 1} + \frac{Q^2}{4E^2}\left(1 + \frac{2m^2}{Q^2}\right) I , \end{aligned}\right\} \tag{35.20}$$

$$\left.\begin{aligned} I = & \frac{1}{2\beta\delta\sin\frac{\Theta}{2}} \left[\Phi\left(-\left(\frac{x-1}{x+1}\right)^2 \right) - \Phi\left(-\left(\frac{1-y}{1+y}\right)^2 \right) + \right. \\ & \left. + \log\frac{x-1}{x+1} \log \frac{(x+y)^2 (x+1)^2}{4x^2(1-y^2)} - \log\frac{1-y}{1+y} \log\frac{(x+y)^2(1+y)^2}{4y^2(x^2-1)} \right] , \end{aligned}\right\} \tag{35.21}$$

$$x = \frac{\beta\left(1 + \sin\frac{\Theta}{2}\right)}{1 - \delta} \geq 1 , \tag{35.22a}$$

$$y = \frac{\beta\left(1 + \sin\frac{\Theta}{2}\right)}{1 + \delta} \leq 1 , \tag{35.22b}$$

$$\delta = \sqrt{1 - \beta^2 \cos^2\frac{\Theta}{2}} \leq 1 , \tag{35.22c}$$

$$\beta = \frac{p}{E} < 1 . \tag{35.22d}$$

The function $\Phi(x)$ used in (35.20) and (35.21) has been defined in (34.21).

Adding (35.8) and (35.16) in order to get the total cross section, the mass μ drops out of the result by the use of (34.18) and (35.19). The observable quantity $\frac{d\sigma^{tot}}{d\Omega}$ is therefore independent of the photon mass and the role of the cutoff has now been assumed by the resolution of the measuring apparatus.[1] For very small values of ΔE, the probability of emission of many photons has to be considered also, as Bloch and Nordsieck and Jauch and Rohrlich[1] have

1. The first discussion of the infrared problem was given by F. Bloch and A. Nordsieck, Phys. Rev. 52, 54 (1937). See also W. Pauli and M. Fierz, Nuovo Cim. 15, 167 (1938). The treatment used here is due to J. Schwinger, Phys. Rev. 76, 790 (1949). A similar discussion of the higher orders of the problem has been given by J. M. Jauch and F. Rohrlich, Helv. Phys. Acta 27, 613 (1954). See also p. 258 of the book by these authors.

shown. By these considerations, it can be shown that $\frac{d\sigma^{\text{tot}}}{d\Omega}$ remains finite even in the limit $\Delta E \to 0$. In this problem we can regard the motion of the electron as given, so that this is really a problem of the interaction of photons with a classical current. Just this problem was already discussed in Sec. 11 and we shall not repeat the details here. At least for the accuracy of measurement presently attainable, such a refinement of the theory is not necessary.[1]

The complete expression for the total cross section can certainly be obtained from the above equations, but it is rather complicated. We therefore restrict ourselves to the extreme relativistic limit. Under these conditions, the cross section simplifies to

$$\frac{d\sigma^{\text{tot}}}{d\Omega} = \frac{d\sigma^{(0)}}{d\Omega}(1-\delta), \tag{35.23}$$

with

$$\delta = \frac{e^2}{\pi^2}\left\{\left[\log\left(\frac{2E}{m}\left|\sin\frac{\Theta}{2}\right|\right) - \frac{1}{2}\right]\left[\log\frac{E}{\Delta E} - \frac{13}{12}\right] + \frac{17}{72} - \frac{1}{2}\sin^2\frac{\Theta}{2}\,I\right\} \tag{35.24}$$

and

$$I = \frac{1}{2\sin^2\frac{\Theta}{2}}\left[\Phi\left(-\sin^2\frac{\Theta}{2}\right) + \frac{\pi^2}{12} - \log\left(\sin^2\frac{\Theta}{2}\right)\log\left(\cos^2\frac{\Theta}{2}\right)\right]. \tag{35.25}$$

For special angles, for example, $\Theta = \pi$ or $\pi/2$, the function I can be evaluated exactly by means of the formulas given in (34.22) through (34.24). For arbitrary angles there are detailed tables[2] of the function Φ. That a finite cross section results after the renormalization of charge and mass was noted by several authors[3] in 1948. The first careful calculation of the cross section was carried out by J. Schwinger[4] and we have used what is essentially his method.

A more exact experimental verification of Eqs. (35.23) through (35.25) is still lacking. Certainly many measurements have been made of the scattering of high-energy electrons by nuclei.[5] First, the accuracy of these measurements is not high enough so that the results can be said to confirm the theory. Furthermore, composite nuclei are usually involved, and for these the spatial extension of the nuclear charge distribution has sizable effects. Indeed, the deviation from Eq. (35.7) caused by the nuclear size is much

1. See, however, E. Lomon, Nuclear Phys. 1, 101 (1955); D. R. Yennie and H. Suura, Phys. Rev. 105, 1378 (1957).

2. See, for example, B. K. Mitchell, Phil. Mag. 40, 351 (1949).

3. Z. Koba and S. Tomonaga, Progr. Theor. Phys. 3, 290 (1948); H. W. Lewis, Phys. Rev. 73, 173 (1948); J. Schwinger, Phys. Rev. 73, 416 (1948).

4. J. Schwinger, Phys. Rev. 76, 790 (1949). See also L. R. B. Elton and H. H. Robertson, Proc. Phys. Soc. Lond. , A65, 145 (1952).

5. See, for example, the comprehensive review by R. Hofstadter, Rev. Mod. Phys. 28, 214 (1956).

larger than the radiative corrections of (35.23). These measurements are therefore more a determination of the nuclear charge distribution than a verification of the theory. More recently, there have been experiments[1,2] measuring the scattering of electrons by protons and here the charge distribution is not such a big effect. The experiments have been done so as to determine first the structure of the proton from the angular distribution of the scattered electrons. Then the results can be used for a comparison[2] of the theoretical and experimental values for the total cross section. With an electron energy of about 140 MeV, Tautfest and Panofsky obtained

$$\sigma_{\text{exp}}/\sigma_{\text{theor}} = 0.988 \pm 0.021 \,. \tag{35.26}$$

Since the theoretically important part of the radiative corrections is of order of magnitude $\frac{e^2}{\pi^2} \log \frac{E}{m}$, i.e., about 4% at 140 MeV, the experimental result (35.26) cannot be viewed as an exact verification of the theory. At least these results are not in contradiction with Eqs. (35.23) through (35.25).

36. The Hyperfine Structure of the Hydrogen Atom

As the next application of the results derived in Sec. 35, and as preparation for the discussion of the Lamb shift of Sec. 37, we shall now study the hyperfine structure in the $1s$ state of the hydrogen atom. We are therefore considering a system in which the electromagnetic field is the sum of the external Coulomb field of the proton, the field of the magnetic dipole of the proton, and the radiation field. We take the electron in the electrostatic field as the unperturbed system with energy levels given by the Dirac theory of the hydrogen atom. We are concerned with the splitting of the $1s$ level under the influence of the magnetic dipole of the proton, and corrections arising from the coupling of the electron to the electromagnetic radiation field will have to be considered. First we shall consider only the magnetic dipole of the proton as the perturbation and, from the elementary principles of quantum mechanics, we obtain the following expression for the splitting of the energy level:

$$\delta E = - \int d^3x \, \langle n | \, j_\mu(x) \, | n \rangle \, A_\mu^{\text{mag}}(\boldsymbol{x}) \,, \tag{36.1}$$

$$A_\mu^{\text{mag}}(\boldsymbol{x}) = \frac{\boldsymbol{\mu} \times \boldsymbol{x}}{4\pi r^3} \,. \tag{36.2}$$

Here μ is the magnetic moment of the proton and $\langle n | \, j_\mu(x) \, | n \rangle$ is

1. R. W. McAllister and R. Hofstadter, Phys. Rev. 102, 851 (1956); E. E. Chambers and R. Hofstadter, Phys. Rev. 103, 1454 (1956).

2. G. W. Tautfest and W. K. H. Panofsky, Phys. Rev. 105, 1356 (1957).

the expectation value of the current operator in the unperturbed state under consideration. For the "unperturbed state" we must consider the electron and its radiation field. Since an exact expression for this expectation value is clearly not possible, as an additional approximation, we shall develop it in powers of the radiation field. We cannot use the results of Sec. 34 directly because the electrostatic field is strong enough so that it must be treated exactly in every step of the calculation. In principle, we can do this by transforming the differential equations for the field operators,

$$\left(\gamma \frac{\partial}{\partial x} + m\right) \psi(x) = i e \gamma \left(A^{\text{Strahl}}(x) + A^{\text{Coulomb}}(x)\right) \psi(x), \quad (36.3)$$

$$\Box A_\mu^{\text{Strahl}}(x) = -\frac{ie}{2} [\overline{\psi}(x), \gamma_\mu \psi(x)], \quad (36.4)$$

into integral equations:

$$\psi(x) = \psi^C(x) - ie \int S_R(x, x') \gamma A^{\text{Strahl}}(x') \psi(x') \, dx', \quad (36.5)$$

$$A_\mu^{\text{Strahl}}(x) = A_\mu^{(0)}(x) + \frac{ie}{2} \int D_R(x - x') [\overline{\psi}(x'), \gamma_\mu \psi(x')] \, dx'. \quad (36.6)$$

In (36.5), $\psi^C(x)$ is the solution of the equation

$$\left(\gamma \frac{\partial}{\partial x} + m\right) \psi^C(x) = i e \gamma A^{\text{Coulomb}}(x) \psi^C(x), \quad (36.7)$$

and $S_R(x, x')$ is given by

$$\left(\gamma \frac{\partial}{\partial x} + m\right) S_R(x, x') - i e \gamma A^{\text{Coulomb}}(x) S_R(x, x') = -\delta(x - x'), \quad (36.8)$$

$$S_R(x, x') = 0, \qquad \text{for} \quad x_0 < x_0'. \quad (36.9)$$

This function is a singular function of the same kind as was studied in Sec. 16. In principle, an explicit calculation of this function is possible, since the eigenvalues and eigenfunctions of (36.7) are known. Up to now, this has not been done. Next, if we consider the radiation field in (36.5) as a small quantity, we can iterate these equations in the usual way and thus obtain the first non-zero correction to the current operator. This correction is exactly what results from (31.2) through (31.4) if the singular functions of the free electron, $S_R(x-x')$, $S_A(x-x')$, and $S^{(1)}(x-x')$ are replaced everywhere by the corresponding functions for the bound electron, $S_R(x, x')$, $S_A(x, x')$, and $S^{(1)}(x, x')$. In principle, the expression obtained in this way takes account of the Coulomb field exactly. This procedure leads to

$$\delta E = \delta E^{(0)} + \delta E^{(1)}, \quad (36.10)$$

$$\delta E^{(0)} = -\frac{ie}{2} \int d^3x \langle n | [\overline{\psi}^C(x), \gamma_\mu \psi^C(x)] | n \rangle A_\mu^{\text{mag}}(\mathbf{x}), \quad (36.11)$$

$$\delta E^{(1)} = -\int d^3x \langle n | \delta j_\mu(x) | n \rangle A_\mu^{\text{mag}}(\mathbf{x}), \quad (36.12)$$

$$\left.\begin{aligned}\delta j_\mu(x) &= \frac{ie}{2}\int dx'\,[\overline{\psi}^C(x),\gamma_\mu S_R(x,x')\,\Phi^C(x')] + \frac{ie}{2}\int dx'\,[\overline{\Phi}^C(x')S_A(x',x),\gamma_\mu\psi^C(x)]\\ &-\frac{ie^3}{4}\iint dx'\,dx''\,[\overline{\psi}^C(x'),\gamma_\lambda(S^{(1)}(x',x)\gamma_\mu S_R(x,x'')\,D_R(x''-x')+\\ &+S_A(x',x)\gamma_\mu S^{(1)}(x,x'')\,D_R(x'-x'')+S_A(x',x)\gamma_\mu S_R(x,x'')\,D^{(1)}(x'-x''))\,\gamma_\lambda\psi^C(x'')]\\ &+\frac{ie^3}{4}\iint dx'\,dx''\,(\mathrm{Sp}\,[\gamma_\mu S_R(x,x')\gamma_\nu S^{(1)}(x',x)]+\\ &+\mathrm{Sp}\,[\gamma_\mu S^{(1)}(x,x')\gamma_\nu S_A(x',x)])\,D_R(x'-x'')\,[\overline{\psi}^C(x''),\gamma_\nu\psi^C(x'')]\,,\end{aligned}\right\}\quad(36.13)$$

$$\Phi^C(x) = -\frac{e^2}{2}\int \gamma_\lambda\,[S^{(1)}(x,x')\,D_A(x'-x)+S_R(x,x')\,D^{(1)}(x'-x)]\,\gamma_\lambda\psi^C(x')\,dx'. \quad (36.14)$$

We begin by evaluating (36.11). From (36.2) we find

$$\delta E^{(0)} = -\frac{ie}{4\pi}\int d^3x\,\bar{u}_n(\boldsymbol{x})\,\gamma_k\,u_n(\boldsymbol{x})\,\frac{[\boldsymbol{\mu}\times\boldsymbol{x}]_k}{r^3}. \quad (36.15)$$

Here $u_n(\boldsymbol{x})$ is the eigenfunction of the time-independent Dirac equation corresponding to the state $|n\rangle$:

$$\langle 0|\,\psi^C(x)\,|n\rangle = u_n(\boldsymbol{x})\,e^{-iE_n x_0}. \quad (36.16)$$

For the $1s$ state of the hydrogen atom with nuclear charge Ze we therefore have

$$u_n(\boldsymbol{x}) = \frac{f(r)}{\sqrt{4\pi}}\left[1, 0, i\frac{z}{r}\frac{Z\alpha}{1+\varrho}, i\frac{x+iy}{r}\frac{Z\alpha}{1+\varrho}\right] \quad \text{for } s_z = +\frac{1}{2}, \quad (36.17a)$$

$$u_n(\boldsymbol{x}) = \frac{f(r)}{\sqrt{4\pi}}\left[0, 1, i\frac{x-iy}{r}\frac{Z\alpha}{1+\varrho}, -i\frac{z}{r}\frac{Z\alpha}{1+\varrho}\right] \text{ for } s_z = -\frac{1}{2}, \quad (36.17b)$$

$$f(r) = \frac{(2mZ\alpha r)^\varrho}{r}\sqrt{\frac{(1+\varrho)\,mZ\alpha}{\Gamma(2\varrho+1)}}\,e^{-Zm\alpha r}, \quad (36.18)$$

$$\varrho = \sqrt{1-Z^2\alpha^2}. \quad (36.19)$$

From these, we find after a simple calculation

$$i\bar{u}_n(\boldsymbol{x})\,\gamma_k\,u_n(\boldsymbol{x}) = -\frac{f^2(r)}{\pi}\frac{\alpha Z}{1+\varrho}\frac{[\boldsymbol{x}\times\boldsymbol{s}]_k}{r}, \quad (36.20)$$

where $\mathbf{s}$ is the vector

$$\mathbf{s} = (0, 0, s_z). \quad (36.20a)$$

Substitution into (36.15) gives

$$\left.\begin{aligned}\delta E^{(0)} &= -\frac{e}{4\pi^2}\frac{\alpha Z}{1+\varrho}\int d^3x\,f^2(r)\,\frac{1}{r^4}\left((\boldsymbol{\mu}\,\mathbf{s})\,r^2-(\boldsymbol{\mu}\,\boldsymbol{x})\cdot(\mathbf{s}\,\boldsymbol{x})\right) =\\ &= -\frac{e}{\pi}\frac{\alpha Z}{1+\varrho}(\boldsymbol{\mu}\,\mathbf{s})\frac{2}{3}\int_0^\infty dr\,f^2(r) = -\frac{4}{3}\frac{e}{2m}(\boldsymbol{\mu}\,\mathbf{s})\frac{m^3Z^3\alpha^3}{\pi}\frac{1}{\varrho(2\varrho-1)}.\end{aligned}\right\}\quad(36.21)$$

We can expand (36.19) in powers of $Z\alpha$ for small values of this quantity. Accordingly, (36.21) simplifies to

$$\delta E^{(0)} = -\frac{4}{3}\frac{e}{2m}(\boldsymbol{\mu}\,\mathbf{s})\frac{m^3Z^3\alpha^3}{\pi}\left[1+\frac{3}{2}Z^2\alpha^2+\cdots\right]. \quad (36.22)$$

Moreover, since the magnetic moment of the nucleus is proportional to its spin I, we get

$$(\mu\, \mathbf{s}) = \frac{\mu_k}{2I}\left(J(J+1) - I(I+1) - \frac{3}{4}\right) \tag{36.23}$$

from the well known rules for the composition of angular momentum. In (36.23), J stands for the total angular momentum of the whole atom. For $J = I + \frac{1}{2}$, we have

$$(\mu\, \mathbf{s}) = \frac{\mu_k}{2}, \tag{36.24}$$

and for $J = I - \frac{1}{2}$,

$$(\mu\, \mathbf{s}) = -\frac{\mu_k}{2}\left(1 + \frac{1}{I}\right). \tag{36.25}$$

Hence the 1s state is split into two states with an energy difference[1]

$$\Delta E = \frac{2}{3}\mu_e \mu_k \frac{2I+1}{I} \frac{m^3 Z^3 \alpha^3}{\pi}\left[1 + \frac{3}{2} Z^2 \alpha^2 + \cdots\right], \tag{36.26}$$

$$\mu_e = \frac{e}{2m}. \tag{36.26a}$$

The energy difference (36.22) is of the order $m^3 \mu_e \mu_k Z^3 \alpha^3 \equiv E_m$, with a correction term of the order $Z^2 \alpha^2 \cdot E_m$. If we were not interested in this last term, we could have derived the first approximation to (36.22) in a simpler way. By the use of the Gordon decomposition of the current operator [c.f. Eq. (33.2)], we have

$$i\,\bar{u}(x)\,\gamma_k\,u(x) = \frac{i}{2m}\left(\frac{\partial \bar{u}}{\partial x_k} u - \bar{u}\frac{\partial u}{\partial x_k}\right) + \frac{1}{2m}\frac{\partial}{\partial x_l}(\bar{u}\,\sigma_{kl}\,u). \tag{36.27}$$

Now by the use of (36.17) it follows that

$$\frac{i}{2m}\left(\frac{\partial \bar{u}}{\partial x_k} u - \bar{u}\frac{\partial u}{\partial x_k}\right) = \frac{f^2(r)}{2m\pi}\frac{\alpha^2 Z^2}{(1+\varrho)^2}\frac{[\mathbf{x}\times \mathbf{s}]_k}{r^2}, \tag{36.28}$$

$$\frac{1}{2m}\bar{u}\,\sigma_{kl}\,u = \frac{f^2(r)}{2m\pi(1+\varrho)}\left[\mathbf{s} - \frac{\alpha^2 Z^2}{1+\varrho}\frac{(\mathbf{s}\,\mathbf{x})\cdot \mathbf{x}}{r^2}\right]_m, \tag{36.29}$$

and cyclic permutations of k, l, m. If all terms with a factor $Z^2\alpha^2$ can be dropped, then we can write approximately

$$\frac{i}{2m}\left(\frac{\partial \bar{u}}{\partial x_k} u - \bar{u}\frac{\partial u}{\partial x_k}\right) = 0, \tag{36.30}$$

$$\frac{1}{2m}\bar{u}\,\sigma_{kl}\,u = \frac{\varphi^2(r)}{4\pi m}(\mathbf{s})_m, \tag{36.31}$$

$$\varphi(r) = \frac{1}{\sqrt{2}}(2m Z\alpha)^{\frac{3}{2}} e^{-Zm\alpha r}. \tag{36.32}$$

The function $\varphi(r)$ in (36.32) is the radial wave function of the nonrelativistic Schroedinger equation for the hydrogen atom. In this approximation the current of the electron is just the current of the

1. E. Fermi, Z. Physik 60, 320 (1930); G. Breit, Phys. Rev. 35, 1447 (1930).

magnetic dipole of the electron spin.[1] Integration by parts in (36.15) then gives

$$\left.\begin{aligned}\delta E^{(0)}&=-\frac{e}{16\pi^2 m}\int d^3x\,\varphi^2(r)\left(\mathbf{s}\,\mathrm{rot}\,\frac{[\boldsymbol{\mu}\times\boldsymbol{x}]}{r^3}\right)=-\frac{e}{16\pi^2 m}\int d^3x\,\varphi^2(r)\left(\mathbf{s}\,\mathrm{rot}\left(\mathrm{rot}\frac{\boldsymbol{\mu}}{r}\right)\right)=\\&=-\frac{e}{16\pi^2 m}\int d^3x\,\varphi^2(r)\,\mathbf{s}\left(\mathrm{grad}\left(\mathrm{grad}\,\frac{\boldsymbol{\mu}}{r}\right)-\boldsymbol{\mu}\,\Delta\left(\frac{1}{r}\right)\right)=\\&=\frac{e(\mathbf{s}\,\boldsymbol{\mu})}{24\pi^2 m}\int d^3x\,\varphi^2(r)\,\Delta\left(\frac{1}{r}\right)=-\frac{e}{6\pi m}(\mathbf{s}\,\boldsymbol{\mu})\,|\varphi(0)|^2=-\frac{4}{3}\mu_e(\mathbf{s}\,\boldsymbol{\mu})\,\frac{|\varphi(0)|^2}{4\pi}.\end{aligned}\right\}\quad(36.33)$$

Because of (36.32), there is complete agreement between (36.33) and the first approximation to (36.22).

From (36.12) we obtain the corrections to this result which are of the order of $E_m\alpha$. That is, the corrections to (36.22) evidently are not to be dropped even if Z is of the order of 1 (for example, in hydrogen), since further terms of the same order or even larger order result from (36.12). We shall now evaluate the term of order $E_m\alpha$. Here it is clearly sufficient to use the simplified theory of the spin given above. Moreover, we can replace the singular functions for the Coulomb field by those for a free particle throughout (36.13) and (36.14) because the terms neglected by this all contain a further factor of $Z\alpha$. The quantity δj_u in (36.13) then becomes the same as those expressions which were treated in Sec. 31 to 34, except that we have to use solutions of the non-relativistic Schroedinger equation for initial and final states, rather than plane waves. Certainly the Dirac equation for free particles was used for initial and final states at several places in Sec. 31. This is not strictly correct for the present calculation; however, the error involved is again of order $E_m Z\alpha^2$, and thus not disturbing. In this approximation we obtain

$$\left.\begin{aligned}\delta E^{(1)}&=-\frac{1}{4\pi}\int d^3x\,\frac{[\boldsymbol{\mu}\times\boldsymbol{x}]_k}{r^3}\iint\frac{d^3q\,d^3x'}{(2\pi)^3}\,e^{i\boldsymbol{q}(\boldsymbol{x}-\boldsymbol{x}')}\Big[i\,e\,\bar{u}_n(\boldsymbol{x}')\,\gamma_k\,u_n(\boldsymbol{x}')\times\\&\quad\times\left(-\bar{\Pi}^{(0)}(\boldsymbol{q}^2)+\bar{\Pi}^{(0)}(0)+\bar{R}^{(0)}(\boldsymbol{q}^2)-\bar{R}^{(0)}(0)-\bar{S}^{(0)}(\boldsymbol{q}^2)+\bar{S}^{(0)}(0)\right)+\\&\quad+\mu_e\,\bar{S}^{(0)}(\boldsymbol{q}^2)\,\frac{\partial}{\partial x_l'}\left(\bar{u}_n(\boldsymbol{x}')\,\sigma_{kl}\,u_n(\boldsymbol{x}')\right)\Big]\approx\\&\approx\frac{\mu_e}{4\pi}\int d^3x\,\frac{\partial}{\partial x_l}\left(\frac{[\boldsymbol{\mu}\times\boldsymbol{x}]_k}{r^3}\right)\iint\frac{d^3q\,d^3x'}{(2\pi)^3}\,e^{i\boldsymbol{q}(\boldsymbol{x}-\boldsymbol{x}')}\,\frac{\varphi^2(r')}{2\pi}\,(\mathbf{s})_m\times\\&\quad\times\left[\bar{S}^{(0)}(0)+\bar{R}^{(0)}(\boldsymbol{q}^2)-\bar{R}^{(0)}(0)-\bar{\Pi}^{(0)}(\boldsymbol{q}^2)+\bar{\Pi}^{(0)}(0)\right]=\\&=-\frac{\mu_e}{8\pi^2}\int d^3x\left(\mathbf{s}\,\mathrm{rot}\,\frac{[\boldsymbol{\mu}\times\boldsymbol{x}]}{r^3}\right)\iint\frac{d^3q\,d^3x'}{(2\pi)^3}\,e^{i\boldsymbol{q}(\boldsymbol{x}-\boldsymbol{x}')}\,\varphi^2(r')\times\\&\quad\times\left[\frac{\alpha}{2\pi}+\bar{R}^{(0)}(\boldsymbol{q}^2)-\bar{R}^{(0)}(0)-\bar{\Pi}^{(0)}(\boldsymbol{q}^2)+\bar{\Pi}^{(0)}(0)\right].\end{aligned}\right\}\quad(36.34)$$

1. This is not true in the exact theory, since the spin and orbital angular momentum are not (separately) good quantum numbers. In the 1s state with total angular momentum 1/2, the electron is principally in the state with $l=0$ and $s_z=1/2$; however, there is a small probability of order $Z^2\alpha^2$ that the electron can be found in the state with $l=1$, $m=1$, $s_z=-1/2$. This is the origin of the expression (36.28) and the last term in (36.29). There is an accidental cancellation among the contributions to (36.27), so that the actual difference between (36.20) and the result of the approximations made here is given by replacing $\frac{f^2}{1+\varrho}$ by $\frac{\varphi^2}{2}$.

Now we have

$$\int \frac{d^3x'}{(2\pi)^3} e^{-iqx'} \varphi^2(r') = \frac{|\varphi(0)|^2}{\pi^2} \frac{2mZ\alpha}{[4m^2Z^2\alpha^2+q^2]^2}, \tag{36.35}$$

and

$$\int d^3x\, e^{iqx}\left(s\cdot \mathrm{rot}\,\frac{[\mu\times x]}{r^3}\right) = 4\pi\left[(\mu\, s) - \frac{(\mu\, q)\,(s\, q)}{q^2}\right]. \tag{36.36}$$

Therefore, from (36.34) we find

$$\left.\begin{aligned} \delta E^{(1)} &= -\frac{4}{3}\mu_e(\mu\, s)\frac{|\varphi(0)|^2}{\pi^2}\int_0^\infty \frac{2mZ\alpha\, q^2 d|q|}{[4m^2Z^2\alpha^2+q^2]^2}\left[\frac{\alpha}{2\pi}+\cdots\right] = \\ &= \delta E^{(0)}\cdot\frac{4}{\pi}\int_0^\infty \frac{2mZ\alpha\, q^2 d|q|}{[4m^2Z^2\alpha^2+q^2]^2}\left[\frac{\alpha}{2\pi}+\bar{R}^{(0)}(q^2)-\bar{R}^{(0)}(0)-\bar{\Pi}^{(0)}(q^2)+\bar{\Pi}^{(0)}(0)\right]. \end{aligned}\right\} \tag{36.37}$$

The square brackets in (36.37) vary slowly with q^2 because the dimensionless variable in this expression is q^2/m^2. The first factor is almost zero for $q^2/m^2 \gg 4Z^2\alpha^2$. Its effect is therefore almost the same as a delta function and, up to terms of order $E_m Z^2\alpha^3$, we obtain

$$\delta E^{(1)} = \delta E^{(0)} \frac{\alpha}{2\pi}, \tag{36.38}$$

$$\delta E^{(0)} + \delta E^{(1)} = -\frac{4}{3}\mu_e(\mu\, s)\frac{m^3Z^3\alpha^3}{\pi}\left(1+\frac{\alpha}{2\pi}\right). \tag{36.39}$$

In this fashion, the anomalous magnetic moment of the electron enters directly into the expression for the hyperfine structure splitting, as would be expected naively. From the above discussion, it follows that (36.39) actually should contain a further correction of the order $E_m Z\alpha^2$. This term can be calculated from the preceding expressions if the Coulomb field of the nucleus is included in the next order. We shall not go further into these rather complicated calculations. The result is[1]

$$\delta E = -\frac{4}{3}\mu_e(\mu\, s)\frac{m^3Z^3\alpha^3}{\pi}\left[1+\frac{\alpha}{2\pi}-\frac{0.328\,\alpha^2}{\pi^2}-Z\alpha^2\left(\frac{5}{2}-\log 2\right)\right]\left[1+\frac{3}{2}Z^2\alpha^2\right]. \tag{36.40}$$

In (36.40) the fourth order magnetic moment has been included since, for small Z, this contribution is of the same order as the other new terms in (36.40).

Exact measurements of the hyperfine structure splitting in hydrogen have been carried out by Prodell and Kusch[2] and by Wittke and Dicke.[2] Because the ratio of the magnetic moments of the

1. R. Karplus and A. Klein, Phys. Rev. 85, 972 (1952); N. M. Kroll and F. Pollock, Phys. Rev. 86, 876 (1952).

2. A. G. Prodell and P. Kusch, Phys. Rev. 88, 184 (1952); P. Kusch, Phys. Rev. 100, 1188 (1955); J. P. Wittke and R. H. Dicke, Phys. Rev. 103, 620 (1956).

proton and electron and the quantity $\frac{1}{2} m\alpha^2$ (the Rydberg constant) are known to greater accuracy,[1] Eq. (36.40) is suitable for an exact determination of the fine structure constant[2] α. It must be remembered that the mass m in (36.40) comes from the non-relativistic wave function of the hydrogen atom, and therefore should be taken as the reduced mass in hydrogen. In this way the result

$$\frac{1}{\alpha} = 137.0384 \pm 0.0003. \tag{36.41}$$

is obtained. The limits of error given here come from the paper of Kroll and Pollock, which was mentioned above, and contain only the uncertainty in the measured values. In addition, there are the terms neglected in (36.40). These are, first, the contributions of order α^3, which are probably very small. Second, the motion of the nucleus must be considered. In (36.40) this motion has been described only by means of the reduced mass, but it can give rise to terms of the order $\alpha \frac{m_e}{m_P}$. Third, it is possible that the assumption made here to regard the proton as a point particle is not sufficient.[3] The corrections arising from this cannot be evaluated without special assumptions; however, it is quite probable that they affect the last two figures in the above result. Finally, it should be mentioned that the hyperfine structure splitting of the $2s$ state in hydrogen has also been measured.[4] Theoretically, it would be expected that this energy difference would also be given by (36.40), if only the last factor were modified[5] to $(1 + \frac{17}{8} Z^2\alpha^2)$ and a factor 8 introduced into the denominator, so that

$$R = \frac{\delta E(2s)}{\delta E(1s)} = \frac{1}{8}\left[1 + \frac{5}{8} Z^2\alpha^2\right] = \frac{1}{8} \cdot 1.0000333. \tag{36.42}$$

The experimental result of Reich, Heberle, and Kusch is

$$R = \tfrac{1}{8}[1.0000346 \pm 0.0000003] \tag{36.43}$$

for hydrogen[4] and

$$R = \tfrac{1}{8}[1.0000342 \pm 0.0000006] \tag{36.44}$$

for deuterium.[6] The equality of the two results (36.43) and (36.44) indicates that the deviation between them and (36.42) cannot be

---

1. S. Koenig, A. G. Prodell and P. Kusch, Phys. Rev. 88, 191 (1952); J. W. M. Du Mond and E. R. Cohen, Phys. Rev. 82, 555 (1951).
2. H. A. Bethe and C. Longmire, Phys. Rev. 75, 306 (1949), and the references cited in footnote 1, p. 172.
3. E. E. Salpeter and W. A. Newcomb, Phys. Rev. 87, 150 (1952); R. Arnowitt, Phys. Rev. 92, 1002 (1953); A. C. Zemach, Phys. Rev. 104, 1771 (1956).
4. H. Reich, J. Heberle and P. Kusch, Phys. Rev. 98, 1194 (1955), 101, 612 (1956).
5. See G. Breit, Phys. Rev. 35, 1447 (1930).
6. H. Reich, J. Heberle and P. Kusch, Phys. Rev. 104, 1585 (1956).

explained by nuclear effects such as a finite extension of the nucleus or its recoil. A theoretical evaluation of the radiative corrections of order α^3 by Mittleman[1] gives the result

$$R = \tfrac{1}{8}\left[1 + \tfrac{5}{8} Z^2 \alpha^2 + 5.28\,\alpha^3\right] = \tfrac{1}{8} \cdot 1.0000354. \tag{36.45}$$

in place of the relation (36.42). The difference between theory and experiment has not yet been explained.

37. Level Shifts in the Hydrogen Atom: The Lamb Shift

After these preliminaries, we are ready to discuss the level shift in the hydrogen atom which was discovered in 1947 by Lamb and Retherford[2] and which was a major stimulus in the development of modern quantum electrodynamics. We are going to use the methods of Sec. 36; i.e., we attempt to treat the Coulomb field of the nucleus as a small perturbation except in the wave functions of the initial and final states. In order to be consistent, we must use the approximate theory of the electron spin given in Sec. 36. From this we obtain immediately the first order shift in the state $|n\rangle$:

$$\delta E = -\int \langle n|\,\delta j_\mu(x)\,|n\rangle\, A_\mu^{\text{Coulomb}}(\boldsymbol{x})\, d^3x, \tag{37.1}$$

$$\left.\begin{aligned}\langle n|\,\delta j_\mu(x)\,|n\rangle &= \frac{i e}{(2\pi)^3}\int d^3 q\, e^{i\boldsymbol{q}(\boldsymbol{x}-\boldsymbol{x}')}\,\bar{u}_n(\boldsymbol{x}')\,\gamma_\mu\, u_n(\boldsymbol{x}') \times \\ &\times\left[-\overline{\Pi}^{(0)}(\boldsymbol{q}^2) + \overline{\Pi}^{(0)}(0) + \overline{R}^{(0)}(\boldsymbol{q}^2) - \overline{R}^{(0)}(0) - \overline{S}^{(0)}(\boldsymbol{q}^2) + \overline{S}^{(0)}(0)\right] + \\ &+ \frac{\mu_e}{(2\pi)^3}\int d^3 q\, e^{i\boldsymbol{q}(\boldsymbol{x}-\boldsymbol{x}')}\,\overline{S}^{(0)}(\boldsymbol{q}^2)\,\frac{\partial}{\partial x'_\nu}\left(\bar{u}_n(\boldsymbol{x}')\,\sigma_{\mu\nu}\, u_n(\boldsymbol{x}')\right).\end{aligned}\right\} \tag{37.2}$$

For simplicity, we have completely neglected[3] the effect of the magnetic dipole of the nucleus in (37.1) and (37.2). We again make the approximation used in (36.37) and expand the functions $\overline{\Pi}^{(0)}(\boldsymbol{q}^2)$, $\overline{R}^{(0)}(\boldsymbol{q}^2)$, and $\overline{S}^{(0)}(\boldsymbol{q}^2)$ in powers of $\boldsymbol{q}^2$. From (34.27) we find

$$\left.\begin{aligned}\langle n|\,\delta j_\mu(x)\,|n\rangle \approx \frac{i\alpha e}{3\pi}\left[\log\frac{m}{\mu} - \frac{23}{40}\right]\frac{\Delta}{m^2}\left(\bar{u}_n(\boldsymbol{x})\,\gamma_\mu\, u_n(\boldsymbol{x})\right) + \\ + \frac{\alpha}{2\pi}\,\mu_e\,\frac{\partial}{\partial x_\nu}\left(\bar{u}_n(\boldsymbol{x})\,\sigma_{\mu\nu}\, u_n(\boldsymbol{x})\right).\end{aligned}\right\} \tag{37.3}$$

Now, using

$$A_\mu^{\text{Coulomb}}(\boldsymbol{x}) = -i\,\delta_{\mu 4}\,\frac{Z e}{4\pi r}, \tag{37.4}$$

we find

$$\delta E = \delta E^{(1)} + \delta E^{(2)}, \tag{37.5}$$

$$\left.\begin{aligned}\delta E^{(1)} &= -\frac{Z\alpha^2}{3\pi}\left[\log\frac{m}{\mu} - \frac{23}{40}\right]\frac{1}{m^2}\int d^3x\,\bar{u}_n(\boldsymbol{x})\,\gamma_4\, u_n(\boldsymbol{x})\,\Delta\left(\frac{1}{r}\right) = \\ &= \frac{4}{3}\,\frac{Z\alpha^2}{m^2}\left[\log\frac{m}{\mu} - \frac{23}{40}\right]|\varphi_n(0)|^2,\end{aligned}\right\} \tag{37.6}$$

1. M. H. Mittleman, Phys. Rev. 107, 1170 (1957).
2. W. E. Lamb and R. C. Retherford, Phys. Rev. 72, 241 (1947).
3. In the final comparison with experiment, clearly both effects have to be included together.

$$\delta E^{(2)} = \frac{\alpha}{2\pi}\mu_e \frac{Ze}{4\pi}\int d^3x \frac{1}{r}\frac{\partial}{\partial x_k}(\bar{u}_n(x)\gamma_4\gamma_k u_n(x)). \tag{37.7}$$

By the use of the Dirac equation, we have

$$\left.\begin{aligned} &\frac{\partial}{\partial x_k}(\bar{u}_n(x)\gamma_4\gamma_k u_n(x)) = \frac{\partial}{\partial x_k}(u_n^*(x)\gamma_k u_n(x)) = \\ &\qquad = 2\left[E_n + \frac{Z\alpha}{r}\right]\bar{u}_n(x)\,u_n(x) - 2m\,u_n^*(x)\,u_n(x). \end{aligned}\right\} \tag{37.8}$$

For our purposes it is sufficiently accurate to express the small components of the function $u_n(x)$ in terms of the large ones by

$$u_{\text{small}}(x) = -\frac{i}{2m}\sigma\cdot\operatorname{grad} u_{\text{large}}(x). \tag{37.9}$$

From (37.8) we then find

$$\left.\begin{aligned} &\frac{\partial}{\partial x_k}(u_n^*(x)\gamma_k u_n(x)) \approx 2\left[E_n - m + \frac{Z\alpha}{r}\right]|\varphi_n(x)|^2 - \\ &- \frac{1}{m}\{\operatorname{grad}\varphi_n^*(x)\cdot\operatorname{grad}\varphi_n(x) + 2i\,\mathbf{s}\,[\operatorname{grad}\varphi_n^*(x)\times\operatorname{grad}\varphi_n(x)]\} = \\ &= \frac{-1}{2m}\Delta|\varphi_n(x)|^2 - \frac{2i\mathbf{s}}{m}[\operatorname{grad}\varphi_n^*(x)\times\operatorname{grad}\varphi_n(x)], \end{aligned}\right\} \tag{37.10}$$

where $\varphi_n(x)$ is again the non-relativistic Schroedinger wave function. In (37.10) we have consistently dropped all terms with a factor $Z^2\alpha^2$ and have used the wave equation for the function $\varphi_n(x)$ in obtaining the last result. Integrating (37.10) by parts, it follows that

$$\left.\begin{aligned} &\int\frac{d^3x}{r}\frac{\partial}{\partial x_k}(u_n^*(x)\gamma_k u_n(x)) = \frac{2\pi}{m}|\varphi_n(0)|^2 + \\ &\qquad + \frac{j(j+1) - l(l+1) - \frac{3}{4}}{m}\int\frac{d^3x\,|\varphi_n(x)|^2}{r^3}, \end{aligned}\right\} \tag{37.11}$$

where the last term is defined to be zero for $l=0$. Using the familiar expressions for the non-relativistic wave functions, we obtain from (37.6), (37.7), and (37.11),

$$\delta E^{(1)} = \frac{4}{3\pi}\frac{mZ^4\alpha^5}{n^3}\left[\log\frac{m}{\mu} - \frac{23}{40}\right]\delta_{l,0}, \tag{37.12}$$

$$\delta E^{(2)} = \frac{1}{2\pi}\frac{mZ^4\alpha^5}{n^3}\frac{C_{l,j}}{2l+1}, \tag{37.13}$$

$$C_{l,j} = \begin{cases} \dfrac{1}{l+1} & \text{for} \quad j = l + \dfrac{1}{2}, \\ -\dfrac{1}{l} & \text{for} \quad j = l - \dfrac{1}{2}. \end{cases} \tag{37.14}$$

For states of non-zero orbital angular momentum, therefore, (37.12) vanishes, and only the term (37.13) coming from the anomalous magnetic moment of the electron gives a shift to the state. For

s-states (37.12) does not vanish; however, it contains the small photon mass μ. Just as in Sec. 35, the occurrence of this quantity in our result means that we have neglected something essential in our calculation. In this case, the mistake is that the formal development in powers of $Z\alpha$ is actually not possible for the bound electron. In a virtual state, if a photon enters with an energy which is small compared to the binding energy of the electron, then the electron can no longer be regarded as almost free and the singular S-functions in Eqs. (36.13) and (36.14) cannot be replaced by those of the free electron. In order to take this into account, we first note that the binding energy of the electron is of the order of magnitude $mZ^2\alpha^2$, i.e., much smaller than the rest mass of the electron. It is therefore possible to break our calculation into two parts. First we consider only virtual states in which the photon has an energy which is much smaller than some limiting energy K. For K we choose an energy which is much larger than the binding energy but much smaller than the rest mass of the electron. By this device, it is possible to treat the electrons in the virtual states by means of the non-relativistic theory. This means a considerable simplification in the formal calculations. Finally, for the other states, for which the energy of the photon is larger than K, we can neglect the effect of the binding and use the preceding calculation. In this part we can set μ equal to zero; however, we are not to integrate over the whole of k-space. Rather than the small mass μ, the energy K enters as the cutoff quantity in (37.12). The desired connection between μ and K is obtained most simply from the calculation of Sec. 35. If we introduce the quantity K rather than μ into Eq. (35.12), we get

$$\left.\begin{aligned} \int\limits_{K<|\boldsymbol{k}|<\Delta E} \frac{dk\,\delta(k^2)}{kq\cdot kq'}\,\Theta(k) &= \frac{1}{2}\int\limits_{K}^{\Delta E} k\,dk \int\limits_0^1 d\alpha \int \frac{d\Omega_k}{[(\boldsymbol{p}+\boldsymbol{Q}\alpha)\,\boldsymbol{k} - E\,|\boldsymbol{k}|]^2} = \\ &= 2\pi \log\frac{\Delta E}{K}\int\limits_0^1 \frac{d\alpha}{A}. \end{aligned}\right\} \quad (37.15)$$

We are interested only in the case $\left|\frac{Q^2}{m^2}\right| \ll 1$, and so we have

$$\frac{d\sigma^{\text{Strahl}}}{d\Omega} = \frac{d\sigma^{(0)}}{d\Omega}\,\frac{e^2}{2\pi^2}\log\frac{\Delta E}{K}\,\frac{1}{3}\,\frac{Q^2}{m^2}. \quad (37.16)$$

The two functions F_1 and F_2 in Eq. (35.16) have the power series

$$F_1 = \frac{1}{3}\,\frac{Q^2}{m^2} + \cdots\,, \quad (37.17)$$

$$F_2 = -\frac{5}{18}\,\frac{Q^2}{m^2} + \cdots\,. \quad (37.18)$$

as is readily verified. By comparison of (37.16) and (35.16), it follows by the use of (37.17) and (37.18) that

$$\log \frac{2K}{\mu} - \frac{5}{6} = 0. \tag{37.19}$$

From this we have[1]

$$\delta E^{(1)} = \frac{4}{3\pi} \frac{m Z^4 \alpha^5}{n^3} \left[\log \frac{m}{2K} + \frac{31}{120} \right] \delta_{l,0}, \tag{37.20}$$

rather than (37.12).

For discussion of the non-relativistic intermediate states, we begin with the Schroedinger equation

$$-\frac{1}{2m} \Delta \varphi + \frac{ie}{m} A_l^{\text{Strahl}}(x) \frac{\partial \varphi}{\partial x_l} - \frac{Z\alpha}{r} \varphi = E \varphi . \tag{37.21}$$

The term with the radiation field is taken as a small perturbation and in the lowest non-vanishing order it gives[2]

$$(\Delta E)_n = \sum_{m,k,\lambda} \frac{|H^{(1)}_{n,m+k}|^2}{E_n - E_m - \omega}, \tag{37.22}$$

$$H^{(1)}_{n,m+k} = -e_l^{(\lambda)} \langle m | v_l | n \rangle \frac{e}{\sqrt{2V\omega}}, \tag{37.23}$$

$$\langle m | v_l | n \rangle = -\frac{i}{m} \int d^3x \, \varphi_m^*(x) \frac{\partial \varphi_n(x)}{\partial x_l} e^{ikx}. \tag{37.24}$$

As before, in (37.22) through (37.24) $\varphi_n(x)$ stands for the eigenfunction of the state $|n\rangle$ of the unperturbed problem. The corresponding energy is E_n . The limit $V \to \infty$ and the summation over the directions of polarization λ of the photon give

$$(\Delta E)_n = \frac{2}{3} \frac{\alpha}{\pi} \int_0^K \omega \, d\omega \sum_m \frac{\langle n | v_k | m \rangle \langle m | v_k | n \rangle}{E_n - E_m - \omega}. \tag{37.25}$$

We have set the exponential factor equal to 1 in (37.24), as is allowed by the restrictions made on K.

The expression (37.25) contains the total energy change of the state $|n\rangle$,i.e., the change in the non-relativistic kinetic energy which is due to the self-mass. If the external field is zero, then (37.25) contains only this last contribution. In this case the inte-

---

1. Justification for the use of the result (37.19) in (37.12) comes about as follows: As can be seen from (31.23), the infrared term in (37.12) arises from an integral of the type (35.12). Thus in both cases it is a question of one cutoff in the same k-space.

2. In this treatment of the non-relativistic problem, we regard the longitudinal and scalar degrees of freedom of the radiation field as having been replaced by the electrostatic interaction between the particles (c.f. Sec. 38a). Since there is only one electron present here, and since the electrostatic self-energy is the same for a free particle or a bound state and can therefore be neglected, only the transverse electromagnetic field enters (37.21). The summation over λ in (37.22) is hence only over 1 and 2.

gral (37.24) vanishes for $n \neq m$, and we have

$$(\Delta E)_n^{\text{free}} = -\frac{2}{3}\frac{\alpha}{\pi} K \langle n | \bar{v}^2 | n \rangle = \frac{1}{2} \delta m \langle n | \bar{v}^2 | n \rangle, \quad (37.26)$$

with

$$\delta m = -\frac{4}{3}\frac{\alpha}{\pi} K. \quad (37.27)$$

The actual level shift is therefore

$$\left.\begin{aligned} \delta E^{(3)} &= (\Delta E)_n - (\Delta E)_n^{\text{free}} = \\ &= \frac{2}{3}\frac{\alpha}{\pi} \sum_m \langle n | v_k | m \rangle \langle m | v_k | n \rangle \int_0^K \left[\frac{\omega}{E_n - E_m - \omega} + 1 \right] d\omega = \\ &= \frac{2}{3}\frac{\alpha}{\pi} \sum_m |\langle n | \bar{v} | m \rangle|^2 (E_m - E_n) \log \frac{K}{|E_m - E_n|}. \end{aligned}\right\} \quad (37.28)$$

The sum in (37.28) has to be worked out numerically for each state $|n\rangle$. In order to get a feeling for the order of magnitude, we replace the energy in the logarithm by an average value $\langle E \rangle$, and find

$$\delta E^{(3)} = \frac{2}{3}\frac{\alpha}{\pi} \log \frac{K}{\langle E \rangle} \sum_m |\langle n | \bar{v} | m \rangle|^2 (E_m - E_n). \quad (37.29)$$

Equation (37.29) can be regarded as a definition of the mean value $\langle E \rangle$. This quantity will probably be of the order of the binding energy. The sum in (37.29) can be worked out exactly:

$$\left.\begin{aligned} &\sum |\langle n | \bar{v} | m \rangle|^2 (E_m - E_n) = \langle n | [\bar{v}, H^{(0)}] \bar{v} | n \rangle = \langle n | \bar{v} [H^{(0)}, \bar{v}] | n \rangle = \\ &= -\frac{1}{2m^2} \left(\langle n | \left[\mathbf{p}, \frac{Z\alpha}{r} \right] \mathbf{p} | n \rangle + \langle n | \mathbf{p} \left[\frac{Z\alpha}{r}, \mathbf{p} \right] | n \rangle \right) = \\ &= \frac{Z\alpha}{2m^2} \int d^3x \operatorname{grad}\left(\frac{1}{r}\right) \operatorname{grad} |\varphi_n(\mathbf{x})|^2 = \frac{2\pi Z\alpha}{m^2} |\varphi_n(0)|^2 = 2 \frac{Z^4 \alpha^4}{n^3} m \, \delta_{l,0}, \end{aligned}\right\} \quad (37.30)$$

$$\delta E^{(3)} = \frac{4}{3\pi} \frac{m Z^4 \alpha^5}{n^3} \log \frac{K}{\langle E \rangle} \delta_{l,0}. \quad (37.31)$$

From (37.31), (37.20), and (37.13), the total level shift becomes

$$\delta E = \frac{4}{3\pi} \frac{m Z^4 \alpha^5}{n^3} \left\{ \log \frac{m}{2 \langle E_s \rangle} + \frac{19}{30} \right\} \quad \text{for } l = 0, \quad (37.32a)$$

$$\delta E = \frac{4}{3\pi} \frac{m Z^4 \alpha^5}{n^3} \left\{ \log \frac{\text{Ry}}{\langle E_l \rangle} + \frac{3}{8} \frac{C_{l,j}}{2l+1} \right\} \quad \text{for } l \neq 0. \quad (37.32b)$$

The first term in (37.32b) has been added in order to take account of the fact that the sum (37.28) does not exactly vanish for $l \neq 0$, as it should do from (37.30). In this case, it is independent of K.

The non-relativistic part of the above calculation was first performed by Bethe.[1] The last terms in (37.32) were given[2] independently by Kroll and Lamb, and by French and Weisskopf. The

1. H. A. Bethe, Phys. Rev. 72, 339 (1947).

2. N. M. Kroll and W. E. Lamb, Phys. Rev. 75, 388 (1949); J. B. French and V. F. Weisskopf, Phys. Rev. 75, 1240 (1949).

method used here follows a calculation of Feynman.[1] Independently, Tomonaga and coworkers[2] have derived the result (37.32) by similar methods. Finally, it must be mentioned that Kramers gave[3] a semi-classical discussion of the level shift from the electromagnetic interaction prior to the authors mentioned above.

For comparison with experiment, the mean value $\langle E \rangle$ must first be calculated. This has been done by Bethe, Brown, and Stehn,[4] as well as by Harriman.[5] We write their result as

$$\langle E_s \rangle = 16.640 \text{ Ry} \quad \text{for the } 2s \text{ state,} \tag{37.33a}$$

$$\langle E_p \rangle = 0.9704 \text{ Ry} \quad \text{for the } 2p \text{ state.} \tag{37.33b}$$

For the two states $2p_{\frac{1}{2}}$ and $2p_{\frac{3}{2}}$ in hydrogen, (37.32b) gives an energy difference of

$$\delta E = \frac{\alpha^5}{32\pi} m. \tag{37.34}$$

From the Dirac theory, there is also an energy difference

$$\delta' E = m \frac{\alpha^4}{32} \left[1 + \frac{5}{8} \alpha^2 + \cdots \right], \tag{37.35}$$

so that the total energy difference is given by

$$\delta E + \delta' E \equiv \Delta E = m \left[\frac{\alpha^4}{32} + \frac{5}{256} \alpha^6 + \frac{\alpha^5}{32\pi} \left(1 - \frac{0.656}{\pi} \alpha \right) \right]. \tag{37.36}$$

In this equation we have replaced $\frac{\alpha}{2\pi}$ from (37.7) [and hence also from (37.34)] by the fourth order magnetic moment (c.f. Sec. 33), so as to take account of the next term in the series expansion in the radiation field. Using the value (36.41) for the fine structure constant, we obtain for deuterium

$$\Delta E = 10971.6 \text{ Mc/sec.} \tag{37.37}$$

This agrees very well with the measurement of Dayhoff, Triebwasser, and Lamb, who obtained[6] the value

$$\Delta E = (10971.6 \pm 0.2) \text{ Mc/sec.} \tag{37.38}$$

1. R. P. Feynman, Phys. Rev. 74, 1430 (1948); 76, 769 (1949).

2. H. Fukuda, Y. Miyamoto and S. Tomonaga, Progr. Theor. Phys. 4, 47, 121 (1949). See also Y. Nambu, Progr. Theor. Phys. 4, 82 (1949), as well as O. Hara and T. Tokano, Progr. Theor. Phys. 4, 103 (1949).

3. H. A. Kramers, Report Solvay Conference 1948, Brussels, 1950.

4. H. A. Bethe, L. M. Brown and J. R. Stehn, Phys. Rev. 77, 370 (1950).

5. J. M. Harriman, Phys. Rev. 101, 594 (1956).

6. E. S. Dayhoff, S. Triebwasser and W. E. Lamb, Phys. Rev. 89, 106 (1953).

Alternatively, one can take these measurements as an independent determination of the fine structure constant, and this is justified in view of the uncertainty in the interpretation of the hyperfine structure. In this way one obtains[1]

$$\frac{1}{\alpha} = 137.0383 \pm 0.0012. \tag{37.39}$$

The words "Lamb shift" refer not to the energy difference (37.36), but to the splitting of the $2s_{\frac{1}{2}}$ and $2p_{\frac{1}{2}}$ levels of the hydrogen atom. According to the Dirac theory these two levels should have exactly the same energy, but according to (37.32) the $2s_{\frac{1}{2}}$ level is a trifle higher in energy than the other state. We find

$$L = \delta E\,(2s_{\frac{1}{2}}) - \delta E\,(2p_{\frac{1}{2}}) = \frac{m\,\alpha^5}{6\pi}\left[\log\frac{m\langle E_p\rangle}{2\,\mathrm{Ry}\,\langle E_s\rangle} + \frac{91}{120}\right]. \tag{37.40}$$

With the numerical results (37.33) and the value (37.39) [or (36.41)] for α one finds

$$L = 1052.1\ \mathrm{Mc/sec}. \tag{37.41}$$

The experimental result is[2]

$$L = (1057.8 \pm 0.1)\ \mathrm{Mc/sec}. \tag{37.42}$$

The difference of (37.42) and (37.41) is far outside the experimental limits of error. As several authors[3] have shown, this discrepancy can be explained to a large extent by the next term in the series in $Z\alpha$. The corresponding contribution is of the order of 7 Mc/sec. If additional small effects, such as terms of order α^2 from the functions $\overline{\Pi}(p^2)$, $\overline{R}(p^2)$, and $\overline{S}(p^2)$ and the finite extension and recoil of the nucleus are considered,[4] the theoretical value is changed to

$$L = (1057.9 \pm 0.2)\ \mathrm{Mc/sec}. \tag{37.43}$$

To within experimental error, (37.43) and (37.42) agree completely. Similar measurements of the level shifts have been done in deuterium[2] (for $n=2$), in ordinary hydrogen[5] (for $n=3$), in ionized helium[6] (He^+ for $n=2$), and in neutral helium[7] (He). The results

1. See footnote 6, p. 179

2. E. S. Dayhoff, S. Triebwasser and W. E. Lamb, Phys. Rev. 89, 98 (1953).

3. M. Baranger, H. A. Bethe and R. P. Feynman, Phys. Rev. 92, 482 (1953); R. Karplus, A. Klein and J. Schwinger, Phys. Rev. 86, 288 (1952).

4. E. E. Salpeter, Phys. Rev. 89, 92 (1953); C. M. Sommerfield, Phys. Rev. 107, 328 (1957); A. Petermann, Helv. Phys. Acta 30, 407 (1957).

5. W. E. Lamb and T. M. Saunders, Phys. Rev. 103, 313 (1956).

6. R. Novick, E. Lipworth and P. F. Yergin, Phys. Rev. 100, 1153 (1955).

7. I. Wieder and W. E. Lamb, Phys. Rev. 107, 125 (1957).

are

$$\delta E(2s_{\frac{1}{2}}) - \delta E(2p_{\frac{1}{2}}) = (1059.0 \pm 0.1) \text{ Mc/sec in Deuterium,} \quad (37.44)$$

$$\delta E(3s_{\frac{1}{2}}) - \delta E(3p_{\frac{1}{2}}) = (315 \pm 10) \text{ Mc/sec in Hydrogen,} \quad (37.45)$$

$$\delta E(2s_{\frac{1}{2}}) - \delta E(2p_{\frac{1}{2}}) = (14043 \pm 13) \text{ Mc/sec in He}^+, \quad (37.46)$$

$$\delta E(2p_2) - \delta E(2p_1) = (2291.7 \pm 0.4) \text{ Mc/sec in Helium,} \quad (37.47)$$

$$\delta E(3p_1) - \delta E(3p_0) = (8113.8 \pm 0.2) \text{ Mc/sec in Helium,} \quad (37.48)$$

$$\delta E(3p_2) - \delta E(3p_1) = (658.6 \pm 0.2) \text{ Mc/sec in Helium,} \quad (37.49)$$

The corresponding theoretical values are

$$\delta E(2s_{\frac{1}{2}}) - \delta E(2p_{\frac{1}{2}}) = (1059.3 \pm 0.2) \text{ Mc/sec in Deuterium,}^{1} \quad (37.50)$$

$$\delta E(3s_{\frac{1}{2}}) - \delta E(3p_{\frac{1}{2}}) = (314.95 \pm 0.05) \text{ Mc/sec in Hydrogen,}^{2} \quad (37.51)$$

$$\delta E(2s_{\frac{1}{2}}) - \delta E(2p_{\frac{1}{2}}) = (14043 \pm 3) \text{ Mc/sec in He}^+,^{3} \quad (37.52)$$

Within the limits of error, all these values agree very well with the measurements. For the two-electron problem in helium, the mathematical difficulties are so great that up to now there has been no sufficiently accurate theory with which to compare (37.47) through (37.49). By optical methods, G. Herzberg has measured[4] the energy difference between the $2p_{\frac{1}{2}}$ and $1s_{\frac{1}{2}}$ states of deuterium with sufficient accuracy so that a value for the shift of the 1 s level has been obtained. The accuracy of the measured shift obtained is much lower than for the previously cited measurements. The result is (0.26 ± 0.04) cm^{-1} and agrees with the theoretical value of 0.2726 cm^{-1}. Finally, it should also be noted that optical measurements of the x-ray spectra of a few heavy elements have shown the influence of the radiative shifts[5] of the energy levels. In conclusion, it must be said that the good agreement between theory and experiment indicated here can be regarded as a confirmation of the validity of the general ideas of quantum electrodynamics and the renormalization program. As was said in Sec. 29, the function $\bar{\Pi}^{(0)}(p^2) - \bar{\Pi}^{(0)}(0)$ in (37.2) gives a contribution (of about 27 Mc/sec) which is much larger than the experimental uncertainty, so that the good agreement of calculated and measured values for the Lamb shift may also be considered as proof of the existence of the vacuum polarization.

---

1. See footnote 4, p. 180.

2. J. M. Harriman, Phys. Rev. 101, 594 (1956). In going to (37.51) we have changed the theoretical value of the level shift given in this paper so as to include the corrected (theoretical) value for the fourth order anomalous moment of the electron.

3. See footnote 6, p. 180.

4. G. Herzberg, Proc. Roy. Soc. Lond., A234, 526 (1956).

5. R. L. Shacklett and J. W. M. Du Mond, Phys. Rev. 106, 501 (1957).

38. Positronium[1]

For completeness we shall now discuss the bound states of an electron and a positron, called "positronium". In the lowest non-relativistic approximation the binding energy is determined by the instantaneous electrostatic interaction of the two particles, just as in the hydrogen atom. The sole difference is that here the reduced mass is half the electron mass, so that the energy eigenvalues are given by

$$E_n^{(0)} = -\frac{m}{4}\frac{\alpha^2}{n^2}. \tag{38.1}$$

The spectrum of positronium is therefore strongly degenerate in this approximation, as is the corresponding spectrum for the non-relativistic hydrogen atom. A more exact treatment gives a fine structure to this spectrum in which the level splitting is of the order of $m\alpha^4$. Just as in hydrogen, this splitting comes partly from relativistic terms in the kinetic energy, partly from spin effects, and partly from corrections to the electrostatic interaction due to the finite speed of propagation of electromagnetic effects. Our discussion of positronium will be done so that we first isolate the electrostatic interaction of the particles and treat this exactly, and afterwards consider the other terms by means of a perturbation calculation.

a. The Electrostatic Interaction

The complete Hamiltonian for a system of electrons and the electromagnetic field can be written in the following form:

$$H = H_\psi^{(0)} + H_{Tr}^{(0)} + H_l^{(0)} + H_{sk}^{(0)} + H_{Tr}^W + H_l^W + H_{sk}^W. \tag{38.2}$$

Here the first term is the Hamiltonian of the free electrons, and the second, third, and fourth terms are the Hamiltonian of free photons with transverse, longitudinal, and scalar polarizations, respectively (c.f. Sec. 6). Finally, the last three terms give the interaction of the electrons with the three kinds of photons. Accordingly, in x-space we shall split up the electromagnetic field into three parts as follows:

$$A_k(x) = \mathscr{A}_k(x) + \frac{\partial \Lambda(x)}{\partial x_k}, \tag{38.3}$$

$$\frac{\partial \mathscr{A}_k(x)}{\partial x_k} = 0, \tag{38.4}$$

$$\Lambda(x) = -\frac{1}{4\pi}\int\limits_{x_0'=x_0} \frac{d^3x'}{r_{xx'}}\frac{\partial A_k(x')}{\partial x_k'} \equiv \Delta^{-1}\frac{\partial A_k(x)}{\partial x_k}, \tag{38.5}$$

$$A_4(x) = i\,V(x). \tag{38.6}$$

1. A more detailed discussion of positronium will be found in the article by L. Simons in Vol. XXXIV of this handbook (Handbuch der Physik, edited by S. Flügge, Springer-Verlag, Heidelberg).

We define

$$\chi(x) = \frac{\partial A_\mu(x)}{\partial x_\mu} = \Delta \Lambda(x) + \frac{\partial V(x)}{\partial x_0}\,, \tag{38.7}$$

$$\dot{\chi}(x) = \frac{\partial^2 A_\mu(x)}{\partial x_0 \partial x_\mu} = \frac{\partial^2 V(x)}{\partial x_0^2} + \Delta \frac{\partial \Lambda(x)}{\partial x_0} = \Delta\left(\frac{\partial \Lambda(x)}{\partial x_0} + V(x)\right) + \varrho(x). \tag{38.8}$$

The last form in (38.8) follows from the equations of motion for $V(x)$; $\varrho(x)$ is the charge density. Using these equations and integrating by parts a few times, we can write the sum of the terms $H_l^{(0)}$, $H_{sk}^{(0)}$, and H_{sk}^W as

$$\left.\begin{aligned} H_l^{(0)} + H_{sk}^{(0)} + H_{sk}^W &= \frac{1}{2}\int d^3x \left[\frac{\partial^2\Lambda(x)}{\partial x_k \partial x_l}\frac{\partial^2\Lambda(x)}{\partial x_k \partial x_l} + \frac{\partial^2\Lambda(x)}{\partial x_k \partial x_0}\frac{\partial^2\Lambda(x)}{\partial x_k \partial x_0} - \right.\\ &\left. - \frac{\partial V(x)}{\partial x_k}\frac{\partial V(x)}{\partial x_k} - \frac{\partial V(x)}{\partial x_0}\frac{\partial V(x)}{\partial x_0}\right] + \int d^3x\, \varrho(x)\, V(x) = \\ &= H_C + \frac{1}{2}\int d^3x \left[\left(\Delta\Lambda(x) - \frac{\partial V(x)}{\partial x_0}\right)\chi(x) + \left(V(x) + \Delta^{-1}\varrho(x) - \frac{\partial \Lambda(x)}{\partial x_0}\right)\dot{\chi}(x)\right] \end{aligned}\right\} \tag{38.9}$$

with

$$H_C = \frac{1}{8\pi}\iint\limits_{x_0 = x_0'} \frac{d^3x\, d^3x'}{r_{xx'}}\,\varrho(x)\,\varrho(x'). \tag{38.9a}$$

Ignoring the terms which contain the subsidiary condition (38.7) and its time derivative, the sum of the three terms in (38.9) is just the instantaneous electrostatic interaction of the electrons. We now make the gauge transformation

$$\psi(x) \to \psi(x)\, e^{ie\Lambda(x)}\,, \qquad H_\psi^{(0)} + H_l^W \to H_\psi^{(0)}\,, \tag{38.10}$$

and obtain

$$H = H_\psi^{(0)} + H_{Tr}^{(0)} + H_C + H_{Tr}^W + \frac{1}{2}\int d^3x\,[(\ldots)\,\chi(x) + (\ldots)\,\dot{\chi}(x)], \tag{38.11}$$

$$H_{Tr}^{(0)} = \frac{1}{2}\int d^3x \left[\frac{\partial \mathscr{A}_k(x)}{\partial x_l}\frac{\partial \mathscr{A}_k(x)}{\partial x_l} + \frac{\partial \mathscr{A}_k(x)}{\partial x_0}\frac{\partial \mathscr{A}_k(x)}{\partial x_0}\right], \tag{38.11a}$$

$$H_\psi^{(0)} = \frac{1}{2}\int d^3x \left[\overline{\psi}(x), \left(\gamma_k \frac{\partial}{\partial x_k} + m\right)\psi(x)\right], \tag{38.11b}$$

$$H_{Tr}^W = -\int d^3x\, \mathscr{A}_k(x)\, j_k(x) = -\frac{ie}{2}\int d^3x\, \mathscr{A}_k(x)\,[\overline{\psi}(x), \gamma_k\, \psi(x)]. \tag{38.11c}$$

instead of (38.2). For physically interesting states, for which the subsidiary condition must be satisfied, the last two terms in (38.11) have no significance and can be dropped. In this way the effect of the longitudinal and scalar photons has been replaced[1] by the Coulomb energy (38.9a). By this procedure, the explicit relativistic

1. See Sec. 8 where it was shown that these degrees of freedom of the field can have no effect for free photons. The method given there for treating the state vector by means of a limit must actually be used here also in order to eliminate the last two terms in a consistent manner.

covariance has been lost, but for our applications this is not troublesome.

b. The Unperturbed Problem

For the unperturbed state of positronium we write the state vector in the form

$$|z\rangle = \sum_{q,q'} \Phi(q, q') \, |q, q'\rangle \ , \tag{38.12}$$

and drop the perturbation H_{Tr}^{W} and the last integral in (38.11). The symbol $|q, q'\rangle$ stands for an eigenstate of $H_{\psi}^{(0)}$ containing one electron-positron pair.[1] The Schroedinger equation

$$\left.\begin{aligned} \sum_{q,q'} (E_q + E_{q'}) \, \Phi(q, q') \, |q, q'\rangle + \sum |q, q'\rangle \langle q, q'| \, H_C \, |q_1, q_1'\rangle \, \Phi(q_1, q_1') = \\ = E \sum \Phi(q, q') \, |q, q'\rangle \end{aligned}\right\} \tag{38.13}$$

follows from this. In x-space the non-relativistic approximation is

$$\left[-\frac{1}{2m} \frac{\partial^2}{\partial \boldsymbol{x}_1^2} - \frac{1}{2m} \frac{\partial^2}{\partial \boldsymbol{x}_2^2} - \frac{e^2}{4\pi r} \right] \psi(\boldsymbol{x}_1, \boldsymbol{x}_2) = (E - 2m) \, \psi(\boldsymbol{x}_1, \boldsymbol{x}_2), \tag{38.14}$$

with

$$\psi(\boldsymbol{x}_1, \boldsymbol{x}_2) = \sum_{q,q'} e^{i\boldsymbol{q}\boldsymbol{x}_1 + i\boldsymbol{q}'\boldsymbol{x}_2} \, \Phi(\boldsymbol{q}, \boldsymbol{q}') . \tag{38.15}$$

The binding energy $E - 2m$ takes the values given in (38.1).

c. Treatment of the Transverse Photons by Perturbation Theory

For a more complete treatment of the problem, we must also consider states with transverse photons in (38.12). Thus, in the next approximation we write

$$|z\rangle = \sum_{q,q'} \Phi(q, q') \, |q, q'\rangle + \sum_n a_n \, |n, \boldsymbol{k}\rangle . \tag{38.16}$$

The second term in (38.16) contains only states which can be produced by the operator H_{Tr}^{W} from a state with one pair. Thus, except for the photon, the states contain zero, one, or two pairs. Formally we write the vector $|z\rangle$ in (38.16) as

$$|z\rangle = e^{iS} \, |\varphi\rangle . \tag{38.16a}$$

Here the vector $|\varphi\rangle$ is again of the form (38.12) and the transformation matrix S is determined from H_{Tr}^{W} in the following way:

$$H' \, |\varphi\rangle = E \, |\varphi\rangle \ , \tag{38.17}$$

$$\left.\begin{aligned} H' = e^{-iS} H e^{iS} \approx H + i[H, S] - \tfrac{1}{2}[[H, S], S] \approx \\ \approx H_0 + H_{Tr}^{W} + i[H_0, S] + i[H_{Tr}^{W}, S] - \tfrac{1}{2}[[H_0, S], S] \ , \end{aligned}\right\} \tag{38.17a}$$

$$H_0 = H_{\psi}^{(0)} + H_A^{(0)} + H_C . \tag{38.17b}$$

---

1. States of this kind must be clearly differentiated from states with a given number of incoming particles. C.f. Sec. 11, especially Eqs. (11.47) through (11.53).

By means of the choice

$$i\,[H_0, S] = -H_{Tr}^W, \tag{38.18}$$

all terms which are linear in the creation or annihilation operators of the photons are eliminated in this order. Since all diagonal elements of the operator H_{Tr}^W vanish, Eq. (38.18) can be satisfied by a non-singular matrix S. The commutator on the left side of (38.18) is formally the time derivative of the operator S in an interaction picture which is associated with the operator H_0 [c.f. Eq. (2.9)]. Equation (38.18) is therefore satisfied by

$$S = -\int_{-\infty}^{x_0} H_{Tr}^W(x_0')\,dx_0\,. \tag{38.18a}$$

Here $H_{Tr}^W(x_0)$ is the time-dependent operator represented in this picture. We substitute (38.18a) into (38.17a) and the new Hamiltonian becomes

$$H' = H_\varphi^{(0)} + H_{Tr}^{(0)} + H_C - \frac{i}{2}\int_{-\infty}^{x_0} [H_{Tr}^W(x_0), H_{Tr}^W(x_0')]\,dx_0'. \tag{38.19}$$

Since the last term in (38.19) contains the commutator of the photon operators (i.e., is a c-number) the photon variables of the problem have been formally eliminated. Using the equation

$$[\mathscr{A}_k(x), \mathscr{A}_l(x')] = -\frac{1}{(2\pi)^3}\int dk\, e^{ik(x'-x)}\,\delta(k^2)\,\varepsilon(k)\left[\delta_{kl} - \frac{k_k k_l}{\boldsymbol{k}^2}\right], \tag{38.20}$$

we then obtain

$$\left.\begin{aligned}&[H_{Tr}^W(x_0), H_{Tr}^W(x_0')] = \\ &= \iint d^3x\, d^3x'\, j_k(x)\, j_l(x')\,\frac{-1}{(2\pi)^3}\int dk\, e^{ik(x'-x)}\,\delta(k^2)\,\varepsilon(k)\left[\delta_{kl} - \frac{k_k k_l}{\boldsymbol{k}^2}\right].\end{aligned}\right\} \tag{38.21}$$

For the transition to a Schroedinger picture, all operators present in H' have to be taken at the same time. We therefore write the current operator $j_l(x')$ in the form

$$j_l(x') = -\frac{ie}{2}\iint_{x_0''=x_0'''=x_0} d^3x''\, d^3x'''\,[\overline{\psi}(x''')\,\gamma_4 S(x'''-x'), \gamma_l\, S(x'-x'')\,\gamma_4\psi(x'')]. \tag{38.22}$$

In (38.22) the singular functions of the complete, unperturbed problem ought to be used. That is, they should include the influence of the term H_C in the unperturbed Hamiltonian. As in the previous sections, we shall approximate these functions by the singular functions for free particles. Then the time integral (38.19) can be done and, upon going over to the Schroedinger picture, we have

$$\left.\begin{aligned}&\delta H_{Tr} = \frac{i}{2}\int_{-\infty}^{0} dx_0'\,[H_{Tr}^W(0), H_{Tr}^W(x_0')] = \frac{e^2}{8}\frac{1}{(2\pi)^{13}}\int\cdots\int\frac{d^3q\, d^3q_1\, d^3q'\, d^3q_1'}{(q-q_1)^2}\times\\ &\times\delta(\boldsymbol{q}+\boldsymbol{q}_1'-\boldsymbol{q}_1-\boldsymbol{q}')\left[\delta_{kl} - \frac{(\boldsymbol{q}-\boldsymbol{q}_1)_k(\boldsymbol{q}-\boldsymbol{q}_1)_l}{(\boldsymbol{q}-\boldsymbol{q}_1)^2}\right][\overline{\varphi}(\boldsymbol{q}_1), \gamma_k\varphi(\boldsymbol{q})]\times\\ &\times[\overline{\varphi}(\boldsymbol{q}'), \gamma_l\varphi(\boldsymbol{q}_1')]\end{aligned}\right\} \tag{38.23}$$

with[1]

$$\varphi(\boldsymbol{q}) = \int d^3x\, e^{-i\boldsymbol{q}\boldsymbol{x}} \psi(x)|_{x_0=0}. \tag{38.23a}$$

d. The Fine Structure of Positronium

The remainder of the calculation is simple in principle, but somewhat tedious. In the barycentric system the total perturbation energy is

$$\delta H = \delta H_r + \delta H_C + \delta H_{Tr} + \delta H^{\text{exchange}}, \tag{38.24}$$

$$\langle q, -q|\delta H_r|q_1, -q_1\rangle = -\frac{(\boldsymbol{q}^2)^2}{4m^3}\,\delta(\boldsymbol{q}-\boldsymbol{q}_1), \tag{38.24a}$$

$$\left.\begin{aligned}\langle q, -q|\delta H_C|q_1, -q_1\rangle = -\frac{e^2}{(2\pi)^3}\frac{1}{(\boldsymbol{q}-\boldsymbol{q}_1)^2}\times \\ \times\left[u^{*(+)}(q)\,u^{(+)}(q_1)\,u^{*(-)}(q)\,u^{(-)}(q_1) - 1\right],\end{aligned}\right\} \tag{38.24b}$$

$$\left.\begin{aligned}\langle q, -q|\delta H_{Tr}|q_1, -q_1\rangle = -\frac{e^2}{(2\pi)^3}\frac{1}{(q-q_1)^2}\bar{u}^{(+)}(q)\,\gamma_k\,u^{(+)}(q_1)\times \\ \times\,\bar{u}^{(-)}(q)\,\gamma_l\,u^{(-)}(q_1)\left[\delta_{kl} - \frac{(\boldsymbol{q}-\boldsymbol{q}_1)_k(\boldsymbol{q}-\boldsymbol{q}_1)_l}{(\boldsymbol{q}-\boldsymbol{q}_1)^2}\right],\end{aligned}\right\} \tag{38.24c}$$

$$\left.\begin{aligned}\langle q, -q|\delta H^{\text{exchange}}|q_1, -q_1\rangle = \frac{e^2}{(2\pi)^3}\frac{-1}{4(m^2+\boldsymbol{q}^2)}\times \\ \times\,\bar{u}^{(+)}(q)\,\gamma_\mu\,u^{(-)}(q)\,\bar{u}^{(-)}(q_1)\,\gamma_\mu\,u^{(+)}(q_1).\end{aligned}\right\} \tag{38.24d}$$

In (38.24), δH_r contains the relativistic corrections to the kinetic energy; δH_C, the corresponding terms for the electrostatic energy; and δH_{Tr}, the terms coming from (38.23). To this order we must also include exchange effects for the electrons, and this is the origin of the term (38.24d). In order to obtain this expression, it is simplest to start[1] with the sum $H_C + \delta H_{Tr}$. Using the approximations

$$u^{(+)}_{\text{large}}(q) = \left[1 - \frac{\boldsymbol{q}^2}{8m^2}\right]|s\rangle, \tag{38.25a}$$

$$u^{(+)}_{\text{small}}(q) = \frac{\boldsymbol{q}\boldsymbol{\sigma}}{2m}|s\rangle, \tag{38.25b}$$

and the exact relation (c.f. Sec. 14)

$$u^{(-)}(-q) = -C\,\bar{u}^{(+)}(q), \qquad C = \gamma_2\gamma_4, \tag{38.26}$$

---

1. If (38.23) and (38.9a) are added, we obtain the result

$$H_C + \delta H_{Tr} = \frac{e^2}{8}\frac{1}{(2\pi)^{13}}\times$$
$$\times\int\cdots\int\frac{d^3q\ldots d^3q_1'}{(q-q_1)^2}\,\delta(\boldsymbol{q}+\boldsymbol{q}_1-\boldsymbol{q}_1-\boldsymbol{q}')\,[\bar{\varphi}(\boldsymbol{q}_1),\gamma_\mu\,\varphi(\boldsymbol{q})]\,[\bar{\varphi}(\boldsymbol{q}'),\gamma_\mu\,\varphi(\boldsymbol{q}_1')].$$

This is just the interaction operator of Sec. 27 for the scattering of two electrons from each other. There, however, the whole expression was treated as a small perturbation; this is not allowed here.

where $|s\rangle$ is the non-relativistic spin function for the unperturbed problem, we can develop the expression (38.24) in powers of q^2/m^2 and $(q-q')^2/m^2$. After some rather lengthy computations, we get

$$\delta H = \delta H_r + \delta H_l + \delta H_{ls} + \delta H_s + \delta H^{\text{exchange}}, \qquad (38.27)$$

$$\langle \delta H_l \rangle = -\frac{e^2}{(2\pi)^3}\left[\frac{1}{m^2 Q^2}\left(q^2 - \frac{(q\,Q)^2}{Q^2}\right) - \frac{1}{4m^2}\right], \quad Q = q - q_1, \qquad (38.27a)$$

$$\langle \delta H_{ls} \rangle = -\frac{e^2}{(2\pi)^3}\frac{3i}{2m^2}\frac{([q\times Q]\cdot S)}{Q^2}, \qquad (38.27b)$$

$$\langle \delta H_S \rangle = \frac{e^2}{(2\pi)^3}\frac{1}{2m^2}\left[1 - S^2 - \frac{(Q\,S)^2}{Q^2}\right], \qquad (38.27c)$$

$$\langle \delta H^{\text{exchange}} \rangle = \frac{e^2}{(2\pi)^3}\frac{S^2}{4m^2}. \qquad (38.27d)$$

Here S is the total spin of the positronium and the terms (38.27) have been grouped according to their dependence on the spin. The expectation values of the expression (38.27) can then be calculated for the solutions of Eq. (38.14). In this way we obtain the perturbed energy levels of positronium:

$s = 0$:

$$\delta E = \frac{m\alpha^4}{16}\left[\frac{11}{4n^4} - \frac{4}{n^3(l+\frac{1}{2})}\right], \qquad (38.28a)$$

$s = 1$:

$$\delta E = \frac{m\alpha^4}{16}\left[\frac{11}{4n^4} - \frac{4}{n^3(l+\frac{1}{2})}\right] + \frac{7}{12}\frac{m\alpha^4}{n^3}\delta_{l,0} + \frac{m\alpha^4(1-\delta_{l,0})}{8n^3(l+\frac{1}{2})}A_{l,j} \qquad (38.28b)$$

with

$$A_{l,j} = \begin{cases} \dfrac{3l+4}{(l+1)(2l+3)} & \text{for} \quad j = l+1 \\ -\dfrac{1}{l(l+1)} & \text{for} \quad j = l \\ -\dfrac{3l-1}{l(2l-1)} & \text{for} \quad j = l-1. \end{cases} \qquad (38.28c)$$

From this, we find the splitting of the states 1^3S_1 and 1^1S_0,

$$\Delta E = \tfrac{7}{12}\, m\alpha^4 = 2.044 \cdot 10^5 \text{ Mc/sec}. \qquad (38.29)$$

Formulas for the fine structure of positronium have been given by Pirenne,[1] Landau and Berestetski,[2] and by Ferrell.[3] Experimental investigations of the energy difference (38.29) have been carried

---

1. J. Pirenne, Arch. Sci. Phys. Nat. 28, 233 (1946); 29, 121, 207, 265 (1947).

2. L. D. Landau and V. B. Berestetski, J. Exp. Theor. Phys. USSR 19, 673 (1949); V. B. Berestetski, J. Exp. Theor. Phys. USSR 19, 1130 (1949).

3. R. Ferrell, Phys. Rev. 84, 858 (1951); thesis, Princeton, 1951.

out by Deutsch and coworkers.[1] They have obtained the value

$$\Delta E = (2.0338 \pm 0.0004)\, 10^5 \text{ Mc/sec}. \quad (38.30)$$

The difference between (38.29) and (38.30) is fully explained by the next higher term in α. Karplus and Klein[2] have calculated the energy difference between $1\,^3S_1$ and $1\,^1S_0$ states, taking into account terms of order $m\alpha^5$. They find the result

$$\Delta E = \frac{m\alpha^4}{4}\left[\frac{7}{3} - \frac{\alpha}{\pi}\left(\frac{32}{9} + 2\log 2\right)\right] = 2.0337 \cdot 10^5 \text{ Mc/sec}. \quad (38.31)$$

Similar results for the 2S and 2P states have been given by Fulton and Martin.[3]

e. The Lifetime of Positronium

Whether they are bound in positronium or are free particles, an electron and a positron can annihilate each other and radiate two or more photons. This means that positronium is not a stable form, but that it has a finite lifetime. We shall now compute this lifetime.

The element of the S-matrix for a transition from a state $|q, q'\rangle$ with one pair to a state with two photons is given[4] by the rules of Sec. 21 and 22 as

$$\left.\begin{aligned} \langle q, q' | S | k_1, k_2\rangle &= \delta(q + q' - k_1 - k_2)\, \frac{(2\pi)^4}{V^2}\, \frac{e^2}{2\sqrt{\omega_1\omega_2}}\, \bar{u}^{(+)}(q) \times \\ &\times \left[\frac{i\gamma e^{(1)}(i\gamma(q-k_1) - m)\, i\gamma e^{(2)}}{-2qk_1} + \frac{i\gamma e^{(2)}(i\gamma(q-k_2) - m)\, i\gamma e^{(1)}}{-2qk_2}\right] \bar{u}^{(-)}(-q'). \end{aligned}\right\} \quad (38.32)$$

The vectors $e^{(1)}$ and $e^{(2)}$ are the polarization vectors of the two photons k_1 and k_2. By means of the methods used repeatedly in Chap. V, we obtain the transition probability per unit time w for the annihilation of a state $|z\rangle$:

$$|z\rangle = \sum_q \Phi(q)\, |q, q'\rangle|_{q+q'=0}\,, \quad (38.33)$$

1. M. Deutsch and S. Brown, Phys. Rev. 85, 1047 (1952); R. Weinstein, M. Deutsch and S. Brown, Phys. Rev. 98, 223 (1955). See also V. W. Hughes, S. Marder and C. S. Wu, Phys. Rev. 106, 934 (1957).

2. R. Karplus and A. Klein, Phys. Rev. 87, 848 (1952).

3. T. Fulton and P. C. Martin, Phys. Rev. 95, 811 (1954).

4. As emphasized before, the states with incoming particles in (38.32) are quite different from the states with "particles at time zero" in (38.12) and (38.16). Only if there is no interaction are these two kinds of states the same. Despite this, we can identify them with each other here because we are concerned with the first non-vanishing approximation in a perturbation calculation. In higher orders this identification is not allowed.

$$w = \frac{\omega^2}{2(2\pi)^2} \iint d^3x\, d^3x'\, \varphi^*(\boldsymbol{x})\, \varphi(\boldsymbol{x}') \frac{1}{V^2} \sum_{\boldsymbol{q},\boldsymbol{q}'} e^{i\boldsymbol{q}\boldsymbol{x} - i\boldsymbol{q}'\boldsymbol{x}'} \times \\ \times \int d\Omega_k\, U^*(\boldsymbol{q}', \boldsymbol{k})\, U(\boldsymbol{q}, \boldsymbol{k}), \tag{38.34}$$

$$U(\boldsymbol{q}, \boldsymbol{k}) = \frac{e^2}{2\omega} \bar{u}^{(+)}(q) \times \\ \times \left[\frac{i\gamma e^{(1)} [i\gamma(q - k_1) - m]\, i\gamma e^{(2)}}{-2q k_1} + \frac{i\gamma e^{(2)} [i\gamma(q - k_2) - m]\, i\gamma e^{(1)}}{-2q k_2} \right] u^{(-)}(q) \Big|_{-\boldsymbol{k}_2 = \boldsymbol{k}_1 = \boldsymbol{k}}, \tag{38.35}$$

$$\varphi(\boldsymbol{x}) = \frac{1}{\sqrt{V}} \sum_{\boldsymbol{q}} \Phi(\boldsymbol{q})\, e^{i\boldsymbol{q}\boldsymbol{x}}. \tag{38.36}$$

In (38.34) we can go to the non-relativistic limit by the use of (38.25) and (38.26) (i.e., where $|\boldsymbol{q}| \ll m$ and $\omega \approx m$). If the two photons are polarized parallel to each other, we obtain

$$U(\boldsymbol{q}, \boldsymbol{k}) = 0 \qquad \text{for} \quad \boldsymbol{e}^{(1)} \parallel \boldsymbol{e}^{(2)}. \tag{38.37a}$$

If the two photons are polarized perpendicular to each other, we have instead

$$U(\boldsymbol{q}, \boldsymbol{k}) = \frac{-i e^2}{\sqrt{2} m^2} \left(1 - \frac{\boldsymbol{S}^2}{2}\right) \quad \text{for} \quad \boldsymbol{e}^{(1)} \perp \boldsymbol{e}^{(2)}. \tag{38.37b}$$

In (38.37b), $\boldsymbol{S}$ again stands for the total spin of the positronium. Consequently, for triplet states (38.37b) vanishes. Hence these states cannot decay by two photons.[1] For singlet states,[2] however, we can use (38.34) and (38.37b) to obtain

$$w^{\text{singlet}} = \frac{4\pi\alpha^2}{m^2} |\varphi(0)|^2 = \frac{m\alpha^5}{2n^3} \delta_{l,0} = \frac{0.80 \cdot 10^{10}}{n^3} \delta_{l,0} \sec^{-1}, \tag{38.38}$$

$$w^{\text{triplet}} = 0. \tag{38.39}$$

The result (38.39) does not mean that the triplet state is stable, but only that it must decay by at least three photons. The transition probability of this decay must be of the order of $m\alpha^6$ and a detailed calculation using the previous methods gives[3] the result

---

1. This has been shown only for the order being considered. By the use of conservation of angular momentum, the statement can be shown to be an exact selection rule for a ^3S-state: Because of momentum conservation, the two photons must be emitted in opposite directions. Either they have total angular momentum zero (mutually perpendicular polarization directions), or they have total angular momentum 2 (mutually parallel polarization directions). Neither case can originate from a ^3S-state. See also L. Michel, Nuovo Cim. 10, 319 (1953).

2. J. A. Wheeler, Ann. N. Y. Acad. Sci. 46, 221 (1946).

3. A. Ore and J. L. Powell, Phys. Rev. 75, 1696 (1949); R. Ferrell, thesis, Princeton, 1951.

$$w^{\text{triplet}} = \frac{2}{9\pi}(\pi^2 - 9)\frac{m\alpha^6}{n^3}\delta_{l,0} = \frac{0.72 \cdot 10^7}{n^3}\delta_{l,0}\,\text{sec}^{-1}. \tag{38.40}$$

The difference between (38.38) and (38.40) has been used experimentally in verifying the existence of positronium.[1] A direct measurement of w^{triplet} for $n=1$, $l=0$ has also been made by Deutsch.[1] The result,

$$w^{\text{triplet}}_{\text{exp}} = (0.68 \pm 0.07) \cdot 10^7\,\text{sec}^{-1}, \tag{38.41}$$

is in good agreement with (38.40).

39. A Survey of Radiative Corrections in Other Processes

a. Compton Scattering

Radiative corrections for the Klein-Nishina formula (25.23) have been worked out by several authors. Jost and Corinaldesi[2] have treated the corresponding problem for particles of spin 0. In the non-relativistic limit, their result is[3]

$$\left.\begin{aligned} d\sigma = \frac{1}{2} r_0^2 (1+\cos^2\Theta)\, d\Omega \Big[1 - \frac{4\alpha}{3\pi}\frac{\omega^2}{m^2} \times \\ \times \left(\frac{-3(1+\cos\Theta+\cos^2\Theta)+\cos^3\Theta}{1+\cos^2\Theta}\log\frac{m}{\omega} + (1-\cos\Theta)\log\frac{m}{2\Delta E}\right) + \cdots\Big],\end{aligned}\right\} \tag{39.1}$$

where

$$r_0 = \frac{e^2}{4\pi m} = \frac{\alpha}{m}, \tag{39.2}$$

and

$$\omega \ll m. \tag{39.3}$$

Here ω is the frequency of the incident photon and ΔE is the resolution of the measuring apparatus (as in Sec. 35). As before, the term with ΔE is present because of the infrared divergence which appears in an integration over virtual photons. Just as in the scattering of electrons in an external field, this divergence is to be understood in terms of properties of the measuring apparatus.

For particles with spin 1/2 one also obtains[4] (39.1) in the non-relativistic limit. For $\omega \gg m$, the cross section has a rather complicated form; we shall forego reproducing it here.[4] In this case also there is an infrared divergence which is treated in a similar way. The correction factor in (39.1) has the form $1 + O\left(\alpha\frac{\omega^2}{m^2}\right)$, i.e., the radiative corrections vanish for $\omega \ll m$. This is not only

1. M. Deutsch, Phys. Rev. 82, 455 (1951); 83, 866 (1951).
2. E. Corinaldesi and R. Jost, Helv. Phys. Acta 21, 183 (1948).
3. In this limit the cross section for Compton scattering goes over into the classical Thompson formula both for particles of spin zero and for the particles of spin 1/2 considered previously. At higher energies the formula for spin zero differs from the Klein-Nishina formula.
4. L. M. Brown and R. P. Feynman, Phys. Rev. 85, 231 (1952). An earlier discussion of the same problem was given by M. R. Schafroth, Helv. Phys. Acta. 22, 501 (1949); 23, 542 (1950).

true in lowest order, but can be proved in general.[1]

b. Electron-Electron Scattering

The corrections to Eq. (27.20) have been evaluated by Redhead.[2] Here also the complete result is very complicated and we shall restrict ourselves to the non-relativistic limit. If one electron is at rest while the other is incident with velocity v and is scattered through the angle Θ, then we can write the formula with corrections as follows:

$$\left.\begin{aligned} d\sigma = 4\pi r_0^2 \frac{\sin 2\Theta\, d\Theta}{v^4}\Big[&\frac{1}{\sin^4\Theta}\left(1 - \pi v \alpha \sin\Theta\,(1-\sin\Theta)\right) + \\ &+ \frac{1}{\cos^4\Theta}\left(1 - \pi v \alpha \cos\Theta\,(1-\cos\Theta)\right) - \\ &- \frac{1}{\sin^2\Theta\cos^2\Theta}\left(1 + \frac{\pi v \alpha}{2}(1 - \sin\Theta - \cos\Theta)\right)\Big], \end{aligned}\right\} \quad (39.4)$$

for

$$v \ll 1. \quad (39.4a)$$

In the general case when (39.4a) is not satisfied, the radiative correction terms to this cross section also contain infrared terms which must be treated as in Sec. 35. These terms have been dropped from (39.4), since they are multiplied by v^2 and this is neglected here.

c. The Self-Stress of the Electron

Another problem which can be treated simply by the methods developed here is the so-called "self-stress" of the electron. To do this, we examine the energy-momentum tensor of our system:

$$T_{\mu\nu}(x) = T^{(1)}_{\mu\nu}(x) + T^{(2)}_{\mu\nu}(x) \ , \quad (39.5)$$

$$\left.\begin{aligned} T^{(1)}_{\mu\nu}(x) = \tfrac{1}{8}\big(&[\overline{\psi}(x), \gamma_\mu \partial_\nu \psi(x)] + [\overline{\psi}(x), \gamma_\nu \partial_\mu \psi(x)] - \\ &- [\partial^*_\mu \overline{\psi}(x), \gamma_\nu \psi(x)] - [\partial^*_\nu \overline{\psi}(x), \gamma_\mu \psi(x)]\big) \ , \end{aligned}\right\} \quad (39.6)$$

$$T^{(2)}_{\mu\nu}(x) = -\tfrac{1}{2}\{F_{\mu\lambda}, F_{\nu\lambda}\} + \tfrac{1}{4}\delta_{\mu\nu} F_{\lambda\varrho} F_{\lambda\varrho} \ , \quad (38.7)$$

$$\partial_\mu = \frac{\partial}{\partial x_\mu} - i e A_\mu(x) \ , \quad (39.8)$$

$$F_{\mu\nu} = \frac{\partial A_\nu(x)}{\partial x_\mu} - \frac{\partial A_\mu(x)}{\partial x_\nu}. \quad (39.9)$$

Consider the expectation value of the tensor $T_{\mu\nu}$ for a state with one electron. In the rest frame of this electron, the expectation value of all components of $T_{\mu\nu}$ vanishes, except for the diagonal terms. This follows directly from symmetry considerations. We now define

1. W. Thirring, Phil. Mag. 41, 1193 (1950); F. E. Low, Phys. Rev. 96, 1428 (1954); M. Gell-Mann and M. L. Goldberger, Phys. Rev. 96, 1433 (1954).

2. M. L. G. Redhead, Proc. Roy. Soc. Lond., A220, 219 (1953). Another treatment of the extreme relativistic limit is given by Akhiezer and Polovin, Akad. Nauk USSR 90, 55 (1953).

$$E(0) = -\int \langle q | T_{44} | q \rangle |_{q=0} \, d^3x \,, \qquad (39.10)$$

$$S(0) = \int \langle q | T_{11} | q \rangle |_{q=0} \, d^3x \,. \qquad (39.11)$$

Clearly $E(0)$ is the total energy of an electron at rest and we shall call $S(0)$ the "self-stress". Since Eqs. (39.5) through (39.7) have been written for the unrenormalized theory, we have

$$E(0) = m_{\text{exp}} = m_0 + \delta m \,, \qquad (39.12)$$

where m_0 is the mass of the bare electron and δm is the electromagnetic self-energy. From

$$T^{(1)}_{\mu\mu} = -\frac{m_0}{2} [\overline{\psi}(x), \psi(x)] \,, \qquad (39.13\text{a})$$

$$T^{(2)}_{\mu\mu} = 0 \,, \qquad (39.13\text{b})$$

it follows that

$$\left. \begin{aligned} 3\,S(0) - E(0) &= \int d^3x \langle q | T_{\mu\mu}(x) | q \rangle |_{q=0} = \\ &= -\frac{m_0}{2} \int d^3x \langle q | [\overline{\psi}(x), \psi(x)] | q \rangle |_{q=0} \,, \end{aligned} \right\} \qquad (39.14)$$

or

$$m_0 + \delta m - \frac{m_0}{2} \int d^3x \langle q | [\overline{\psi}(x), \psi(x)] | q \rangle |_{q=0} = 3\,S(0) \,. \qquad (39.15)$$

If H is the Hamiltonian of the unrenormalized theory [Eq. (17.18) through (17.22)], then

$$\frac{1}{2} \int d^3x \langle q | [\overline{\psi}(x), \psi(x)] | q \rangle |_{q=0} = \langle q | \frac{\partial H}{\partial m_0} | q \rangle \Big|_{q=0} = \frac{\partial}{\partial m_0} (m_0 + \delta m) \,. \qquad (39.16)$$

From (39.15) and (39.16) we conclude[1]

$$S(0) = -\frac{1}{3} \left[m_0 \frac{\partial \, \delta m}{\partial m_0} - \delta m \right] = -\frac{m_0^2}{3} \frac{\partial}{\partial m_0} \left(\frac{\delta m}{m_0} \right) . \qquad (39.17)$$

Thus it is possible in principle to obtain the self-stress from the self-mass by differentiation. In fact the self-mass is infinite [at least in lowest order perturbation theory; c.f. (32.6)], so that Eq. (39.17) actually has no meaning. We will now show from considerations of invariance that the self-stress (39.17) must vanish identically. In particular, (39.17) can then be used <u>to define</u> a cutoff Λ in (31.18) as a function of the mass m_0.

In order to show this, we consider the electron in a system where it is moving with velocity v in the x-direction. We denote all quantities in the new coordinate system by a prime, so that the following transformations hold:

$$\langle T'_{44} \rangle = \frac{1}{1 - v^2} \left(\langle T_{44} \rangle - v^2 \langle T_{11} \rangle \right) \,, \qquad (39.18)$$

1. A. Pais and S. T. Epstein, Rev. Mod. Phys. 21, 445 (1949).

$$i\langle T'_{41}\rangle = \frac{v}{1-v^2}\left(\langle T_{44}\rangle - \langle T_{11}\rangle\right), \tag{39.19}$$

$$d^3x' = \sqrt{1-v^2}\, d^3x. \tag{39.20}$$

For the energy $E(v)$ and the momentum $P_x(v)$ in the new coordinate frame, we therefore obtain

$$E(v) = \frac{1}{\sqrt{1-v^2}}\left(E(0) + v^2 S(0)\right), \tag{39.21}$$

$$P_x(v) = \frac{v}{\sqrt{1-v^2}}\left(E(0) + S(0)\right). \tag{39.22}$$

Hence, in order that $E(v)$ and $P_x(v)$ have the transformation properties of a four-vector, it is necessary that $S(0)$ vanish identically. Therefore we must have

$$\frac{\partial}{\partial m_0}\left(\frac{\delta m}{m_0}\right) = 0. \tag{39.23}$$

Consequently, the dimensionless quantity $\delta m/m_0$ cannot depend on m_0. In a convergent theory this would also be impossible for reasons of dimensionality, since the only dimensionless variable which could occur here would be e^2. However, in a theory with a cutoff, the mass m_0 can occur in the combination A/m_0^2. (Here A is the cutoff.) Our result shows that A must be chosen[1] proportional to m_0^2. Finally, we remark that a cutoff using the regularization procedure of Sec. 30 also gives[2] the result zero for the self-stress.

d. Scattering of Light Quanta by an External Field (Delbrück Scattering) and Photon-Photon Scattering

In Sec. 29 we have seen that an external field can produce virtual pairs and that the vacuum behaves like a polarizable medium. An indirect confirmation of this effect is obtained by the exact measurement of the Lamb shift. For a free photon with energy-momentum vector satisfying $k^2 = 0$, the effect vanishes, as demonstrated by the charge renormalization of Eq. (29.23). However, if the photon passes through an external field, then it can produce virtual pairs which can be scattered[3] by the external field, and then be annihilated with the emission of a different photon. The result is a scattering of the photon by the external field. This effect was first predicted by Delbrück.[4] A complete evaluation of the cross section for this process has not yet been made. For

1. S. Borowitz and W. Kohn, Phys. Rev. 86, 985 (1952); Y. Takahashi and H. Umezawa, Progr. Theor. Phys. 8, 193 (1952).

2. F. Rohrlich, Phys. Rev. 77, 357 (1950); F. Villars, Phys. Rev. 79, 122 (1950).

3. From the charge symmetry of the theory it can be shown that the virtual particles must be scattered at least twice by the external field if a non-zero result is to be obtained.

4. M. Delbrück, Z. Physik 84, 144 (1933).

a scattering angle of zero, Rohrlich and Gluckstern[1] have calculated the scattering amplitude and Bethe and Rohrlich[2] have evaluated the cross section for small angles and large energies. For a Coulomb field with charge Ze the cross section is of the order of magnitude

$$\sigma \sim (Z\alpha)^4 r_0^2. \tag{39.24}$$

The total cross section is much smaller than the cross section for Compton scattering ($\sigma_{\text{Compton}} \sim r_0^2$); however, the angular distribution is quite different from that of Compton scattering. The Delbrück scattering has a very sharp maximum for small angles and this may make possible its experimental detection.[3]

If the external field of Delbrück scattering is produced by a second photon, then one has a process in which two photons scatter from each other by means of virtual pairs. A careful evaluation of the cross section for this process has been made by Karplus and Neumann.[4] Here the cross section is of the order of magnitude

$$\sigma \sim \alpha^2 r_0^2, \tag{39.25}$$

and the whole effect is scarcely observable with the present experimental techniques. In principle this effect represents a most interesting departure from the superposition principle of classical electrodynamics.

---

1. F. Rohrlich and R. L. Gluckstern, Phys. Rev. 86, 1 (1952).
2. H. A. Bethe and F. Rohrlich, Phys. Rev. 86, 10 (1952).
3. See R. R. Wilson, Phys. Rev. 90, 720 (1953).
4. R. Karplus and M. Neuman, Phys. Rev. 83, 776 (1950). Certain special cases were investigated previously by A. I. Akhiezer, Phys. Z. Sowjt. 11, 263 (1937) and H. Euler, Ann. Phys. 26, 398 (1936).

CHAPTER VII

GENERAL THEORY OF RENORMALIZATION*

40. General Definition of Particle Numbers

In the last chapter we worked out various examples of radiative corrections. We proved that the infinite quantities, appearing in the lowest approximation, may be interpreted physically by means of the renormalization of the charge and mass of the electron and so may be eliminated. The remaining finite terms can be compared with experimental measurements. The calculated and measured values show remarkably good agreement. In a few cases the results of calculations of terms of higher order were given also, but this was done without further discussion of the treatment of the infinite quantities which would occur. It should be noted, though, that the idea of renormalization is not necessarily tied to the occurrence of infinite quantities. Even if the self-mass and self-charge were finite, it would be necessary first to isolate them, before a comparison with experiment could be made. In this chapter we shall attempt to give as general a discussion of the renormalization procedure as possible, and so we shall not rely on perturbation theory. Also, we will not be preoccupied with whether or not the intermediate results are finite. In this discussion we shall often speak of states of a given number of particles. In a theory with interactions, the definition of the particle number is not trivial. We therefore begin with a discussion of this concept.

In a special example in Sec. 11, we saw that a theory with interactions allows various possibilities for the definition of states with a given number of particles. We recall the most important point: particle numbers were defined by the use of the Fourier de-

---

* (Translator's note) A survey of recent developments in the formal theory of quantum fields has been given by G. Källén in Fundamental Problems in Elementary Particle Physics (Proceedings of the 14th Physics Conference of the Solvay Institute), Wiley, New York, 1968. Other articles by Källén will be found in The Quantum Theory of Fields (Proceedings of the 12th Physics Conference of the Solvay Institute), edited by R. Stoops, Interscience-Wiley, New York, 1961; and Proceedings of the 1967 International Conference on Particles and Fields, edited by C.R. Hagen, G. Guralnik and V. S. Mathur, Interscience-Wiley, New York, 1967. For an introduction to some other viewpoints, see P.C.T., Spin, Statistics and All That, R.F. Streater and A. S. Wightman, Benjamin, New York, 1964; and Stephen Gasiorowicz, Elementary Particle Physics, Wiley, New York, 1966; and references cited there.

composition of a free field in which the coefficients were taken as the creation or annihilation operators. Here the canonical commutation relations played a decisive role and determined the eigenvalues of the particle numbers. We have seen this repeatedly in Sec. 6, 12, and 13. From the fact that the free and coupled fields have the same canonical commutation relations for equal times, it follows that for every time T, free fields $A_\mu^{(0)}(x, T)$ and $\psi^{(0)}(x, T)$ [see (11.47)] can be introduced which are defined by the equations

$$\psi^{(0)}(x, T) = -i \int_{x_0'=T} S(x-x')\gamma_4\psi(x')\, d^3x', \tag{40.1}$$

$$A_\mu^{(0)}(x, T) = -\int_{x_0'=T} \left[\frac{\partial A_\mu(x')}{\partial x_0'} D(x-x') + A_\mu(x') \frac{\partial D(x-x')}{\partial x_0}\right] d^3x'. \tag{40.2}$$

These quantities obviously satisfy the equations for the free fields and are identical with the coupled fields for $x_0 = T$. From this we can now construct a system of state vectors in which every vector is an eigenvector of the Hamiltonian

$$H^{(0)}(T) = H^{(0)}(A_\mu^{(0)}(x, T), \psi^{(0)}(x, T)). \tag{40.3}$$

Thus the operator $H^{(0)}$ is a sum of the operators (17.20) and (17.21). By means of the equations of motion (40.1) and (40.2), it can easily be shown that the operator (40.3) is independent of x_0. It depends only on the "initial time" T. From this it does not follow that the operator (40.3) commutes with the complete Hamiltonian operator of the system $H(A_\mu(x), \psi(x))$, because the operators (40.1) and (40.2) and thus (40.3) contain the time coordinate explicitly in the singular functions. In fact, we see that $H^{(0)}(T)$ is identical with $H^{(0)}(A_\mu(x), \psi(x))$ for $x_0 = T$ and, since the latter quantity does not commute with $H^{(1)}(A_\mu(x), \psi(x))$ for $x_0 = T$, neither does (40.3) commute with the complete Hamiltonian of the system. We conclude that the state vectors introduced above for an arbitrary time T are not eigenstates of the (complete) Hamiltonian. The particle numbers introduced in this way therefore cannot describe the actual "physical" particles. Despite this, the corresponding states were often employed, especially in the older literature, and they were called[1] "states of free particles". The physical states have to be formed from linear combinations of states of free particles. From the usual formalism of quantum theory (conservation of probability and completeness relation), it follows that the transformation from the "free" to the "physical" states is described by a unitary matrix. In this formulation the principal goal of the theory is to determine this matrix. By the example of Sec. 11 and

1. Previously we have used states with free particles of this type in Sec. 38, Eq. (38.12), etc., in treating positronium. In particular, we put $T = 0$ and went over from a Heisenberg picture to a Schroedinger picture at this time. There the physical states were made up of linear combinations of states of free particles.

also by the discussion of Sec. 17 and its subsequent applications, we have seen that the theory can be formulated in another way and that this alternative has proven very useful in the application of perturbation theory. With this in mind, we let the time T in (40.1) and (40.2) tend toward $-\infty$. By this, the corresponding free fields become formally the "in-fields" $A_\mu^{(0)}(x)$ and $\psi^{(0)}(x)$ which were used previously. As equations of motion for the operators in the Heisenberg picture, we obtain the Eqs. (17.14) and (17.15), which we write here as

$$\psi(x) = \psi^{(0)}(x) - \int S_R(x-x')\, f(x')\, dx' \,, \tag{40.4}$$

$$f(x) = i\, e\, \gamma\, A(x)\, \psi(x) \,, \tag{40.4a}$$

$$A_\mu(x) = A_\mu^{(0)}(x) + \int D_R(x-x')\, j_\mu(x')\, dx' \,, \tag{40.5}$$

$$j_\mu(x) = \frac{i\,e}{2}\,[\overline{\psi}(x), \gamma_\mu \psi(x)]. \tag{40.5a}$$

By means of these free fields, we can again introduce a system of state vectors with definite numbers of particles. From now on we shall call these particles the "incoming" or "in-particles". The corresponding state vectors $|\mathrm{Ein}\rangle$ ($|\mathrm{in}\rangle$) are therefore eigenvalues of the operator $H^{(0)}(A_\mu^{(0)}(x), \psi^{(0)}(x))$, i.e.,

$$H^{(0)}(A_\mu^{(0)}(x), \psi^{(0)}(x))\, |\mathrm{Ein}\rangle = E_{\mathrm{Ein}}\, |\mathrm{Ein}\rangle \,. \tag{40.6}$$

The essential difference between these states and those which are eigenstates of (40.3) for finite T is the following: the particles defined by means of (40.1) and (40.2) would appear if the interaction of the system <u>suddenly</u> vanished at time T. Because the interaction changes the number of these particles continuously, one cannot ascribe any actual physical significance to them. However, the particles of Eq. (40.6) are physical because they have existed in the system for a very long time. These particles have a physical significance because in a collision problem we are concerned with a certain number of incoming particles which scatter from each other and then move away from each other as different free particles. In the process, the number of final particles can certainly be larger or smaller than the original number. The mathematical advantage of the in-particles is that we have at our disposal an infinite length of time in which to "switch on" the interaction of the particles. In contrast to the description by means of the particles defined by Eq. (40.3), we can now switch on[1] the interaction (charge) <u>adiabatically</u>. By the adiabatic theorem of quantum mechanics, the state $|\mathrm{Ein}\rangle$ ($|\mathrm{in}\rangle$) is also an eigen-

1. As we have seen in perturbation theory, such a switching is necessary so that certain oscillating integrals will assume a well-defined value.

state of the complete Hamiltonian of the system. Up to the present, the most complete proof of the adiabatic theorem has been given[1] by Born and Fock; however, as was noted in Sec. 11, there exists no really satisfactory proof for a system with an infinite number of degrees of freedom. We have indicated in Sec. 18 how a proof can be given[2] for each term in a power series in the charge. Without further discussion, we shall now assume that the adiabatic theorem is valid for quantum electrodynamics, even if the solution is not represented by a perturbation series. Alternatively, we can say that the existence (or any other property) of a solution which satisfies the adiabatic theorem is meant if we speak of the existence (or of any other property) of a solution. We can make no mathematical statements about other, more singular solutions. Furthermore, we shall understand that we always refer to a state with in-particles if we speak of a state with a definite number of particles. Such a state is therefore an eigenstate of the operator (40.6) as well as of the complete Hamiltonian for the interacting fields.[3]

Perhaps it should be explained that we have had to make no assumption about the completeness of our system of states. It is completely acceptable mathematically if there exist other eigenstates of the complete Hamiltonian, and therefore if the in-states form only a subspace in the Hilbert space. The other states would then correspond to bound states of two or more particles. Certainly in quantum electrodynamics there is no reason to suspect that such states exist;[4] however, if the corresponding argument is used in a meson theory, one should be prepared for the emergence of such states (the deuteron, etc.).

41. Mass Renormalization of the Electron

We are now ready to discuss the mass renormalization of the electron (Sec. 32) in greater generality. We require that a state with a physical electron (i.e., an in-state) should have mass equal to m, the mass appearing on the left side of the Dirac equation. In order to do this, we add a term δm to the right side and obtain in place of (40.4) and (40.4a),

$$\left(\gamma \frac{\partial}{\partial x} + m\right) \psi(x) = f(x) \,, \tag{41.1}$$

1. See footnote 2, p. 54.

2. In the renormalized theory, a rather complete discussion for the perturbation-theoretic treatment of the problem has been given by F. J. Dyson, Phys. Rev. 82, 428 (1951).

3. For an alternative formulation of the "adiabatic hypothesis", see H. Lehmann, K. Symanzik and W. Zimmermann, Nuovo Cim. 1, 205 (1955).

4. The positronium discussed in Sec. 38 is not stable and therefore is not an actual eigenstate of the complete Hamiltonian operator. In this connection, see W. Glaser and G. Källén, Nucl. Phys. 2, 706 (1957).

$$f(x) = i e \gamma A(x) \psi(x) + \delta m \psi(x). \qquad (41.2)$$

The object of this section is to give a definition, even if it is only an implicit one, for the term δm. This definition should not be tied to any explicit perturbation calculation. We start by examining the matrix elements of the operator $\psi(x)$ between the vacuum (i.e., the physical vacuum, which is the same as the vacuum for in-states) and a state $|q\rangle$ of an electron with momentum $\boldsymbol{q}$. Thus we seek the quantity $\langle 0|\psi(x)|q\rangle$. From Eq. (3.21) the x-dependence is given by

$$\langle 0|\psi_\alpha(x)|q\rangle = \langle 0|\psi_\alpha|q\rangle e^{iqx}. \qquad (41.3)$$

After carrying out the mass renormalization, the time component q_0 in (41.3) is to be given by $\sqrt{\boldsymbol{q}^2 + m^2}$. This is the definition of mass renormalization. From this we conclude that the x-dependence of the matrix element of the <u>in-field</u> is just the same as the x-dependence of the matrix element (41.3), since in the former quantity the m of (41.1) is the mass by definition. Thus we can write

$$\langle 0|\psi_\alpha^{(0)}(x)|q\rangle = \langle 0|\psi_\alpha^{(0)}|q\rangle e^{iqx}, \qquad (41.4)$$

where [c.f. Eq. (12.22)] the quantities $\langle 0|\psi_\alpha^{(0)}|q\rangle$ are given by the functions $\frac{1}{\sqrt{V}} u_\alpha^{(r)}(\boldsymbol{q})$ of Sec. 12.

In order to make a connection between the x-independent factors in (41.3) and (41.4) and to obtain an equation for the self-mass δm in (41.2), we need to introduce the following device: From the equation of motion for $\overline{\psi}(x)$,

$$\overline{\psi}(x) = \overline{\psi}^{(0)}(x) - \int \overline{f}(x'') S_A(x'' - x)\, dx'', \qquad (41.5)$$

and from the integral equation

$$\overline{\psi}(x) = -\int_{x'}^{x} \overline{f}(x'') S(x'' - x)\, dx'' - i \int_{x_0''=x_0'} \overline{\psi}(x'') \gamma_4 S(x'' - x)\, d^3x'', \qquad (41.6)$$

it follows that

$$\overline{\psi}^{(0)}(x) = \int_{-\infty}^{x'} \overline{f}(x'') S(x'' - x)\, dx'' - i \int_{x_0''=x_0'} \overline{\psi}(x'') \gamma_4 S(x'' - x)\, d^3x''. \qquad (41.7)$$

In (41.6) and hence in (41.7) also, the time x_0' is arbitrary. Equation (41.7) allows us to express the in-field $\overline{\psi}^{(0)}(x)$ as a surface integral over the field $\overline{\psi}(x)$ at an arbitrary time x_0' and a four-dimensional volume integral from $-\infty$ to x_0'. In these equations only the limits for the time integrations have been shown explicitly; the spatial integrals always run over the whole volume V. If we set the arbitrary time in (41.7) equal to the time in some arbitrary operator $F(x)$, then we obtain

$$\left.\begin{aligned} \{F(x), \overline{\psi}^{(0)}(x')\} = \int_{-\infty}^{x} \{F(x), \overline{f}(x'')\} S(x'' - x')\, dx'' - \\ - i \int_{x''=x_0} \{F(x), \overline{\psi}(x'')\} \gamma_4 S(x'' - x')\, d^3x''. \end{aligned}\right\} \qquad (41.8)$$

The last integral in (41.8) contains only operators for which the time coordinates are equal and can therefore be calculated from the canonical commutation relations.

We now examine the vacuum expectation value on the left of (41.8):

$$\left.\begin{aligned}&\langle 0|\{F(x),\bar{\psi}^{(0)}(x')\}|0\rangle =\\ &=\sum_{|z\rangle}\langle 0|F(x)|z\rangle\langle z|\bar{\psi}^{(0)}(x')|0\rangle+\sum_{|z\rangle}\langle 0|\bar{\psi}^{(0)}(x')|z\rangle\langle z|F(x)|0\rangle .\end{aligned}\right\}\quad(41.9)$$

In principle, the sum over states $|z\rangle$ in (41.9) has a contribution from each state in a complete set of state vectors. In particular, though, if we choose this set as the states of the physical particles (the in-states), supplemented, if necessary, by the bound states), then it follows by definition that the states $|z\rangle$ between the operators must contain just one electron for the first term and just one positron for the second term. Therefore if we compute the vacuum expectation value of the quantity (41.8), we obtain the matrix elements of the operator $F(x)$ between the vacuum and the one-electron or one-positron states. Moreover, we recall Eqs. (15.1) through (15.3), according to which the function $S(x''-x')$ on the right side of Eq. (41.8), can be written as a sum over the matrix elements (41.4). Hence, by comparison of the coefficients of $\langle q|\bar{\psi}^{(0)}(x')|0\rangle$, we find

$$\left.\begin{aligned}\langle 0|F(x)|q\rangle = i\int_{-\infty}^{x}\langle 0|\{F(x),\bar{f}(x'')\}|0\rangle\langle 0|\psi^{(0)}(x'')|q\rangle dx'' +\\ +\langle 0|\frac{\partial F(x)}{\partial\psi(x)}|0\rangle\langle 0|\psi^{(0)}(x)|q\rangle .\end{aligned}\right\}\quad(41.10)$$

Equation (41.10) allows us to express the matrix element between the vacuum and a one-electron state for an arbitrary operator $F(x)$ in terms of the vacuum expectation value of the anticommutator of the same operator with the right side of the Dirac equation. The advantage of this procedure is that the vacuum is invariant under a Lorentz transformation (this is not true for the one-particle states) and so we can use simple considerations of invariance. As an example, we take the operator $F(x)$ to be $\psi(x)$. We find

$$\left.\begin{aligned}&\langle 0|\psi_\alpha(x)|q\rangle =\\ &= i\int_{-\infty}^{x}\langle 0|\{\psi_\alpha(x),\bar{f}_\beta(x'')\}|0\rangle\langle 0|\psi_\beta^{(0)}(x'')|q\rangle dx''+\langle 0|\psi_\alpha^{(0)}(x)|q\rangle .\end{aligned}\right\}\quad(41.11)$$

The spinor indices have been shown explicitly, for clarity. By (3.21) the expression $\langle 0|\{\psi(x),\bar{f}(x'')\}|0\rangle$ can depend only on the difference of coordinates $x-x''$, and, by using invariance considerations, it can be written as

$$\langle 0|\{\psi_\alpha(x),\bar{f}_\beta(x'')\}|0\rangle = \int dp\, e^{ip(x-x'')}\{\delta_{\alpha\beta}A(p)+i(\gamma_\mu)_{\alpha\beta}p_\mu B(p)\}. \quad (41.12)$$

We now make an assumption which we shall frequently use later: We assume that the energy — momentum vector of every physical state is time-like and that the vacuum is the state of lowest energy. The energy of the vacuum is taken as zero. Equation (41.12) then follows from the fact that from γ_μ and p_μ only the two combinations $\delta_{\alpha\beta}$ and $(\gamma p)_{\alpha\beta}$ have the correct transformation properties. The two invariant functions A and B depend only on the vector p, and, of course, in an invariant way.[1]

Since the quantity (41.12) vanishes for space-like $x-x''$,

$$\Theta(x-x'')\langle 0|\{\psi(x),\bar{f}(x'')\}|0\rangle$$

is also an invariant. We can therefore write

$$\left.\begin{aligned} &i\Theta(x-x'')\langle 0|\{\psi(x),\bar{f}(x'')\}|0\rangle = \\ &\quad = \frac{1}{(2\pi)^4}\int dp\, e^{ip(x-x'')}[A'(p)+i\gamma p\, B'(p)], \end{aligned}\right\} \quad (41.13)$$

where $A'(p)$ and $B'(p)$ are different functions. Then from (41.11) it follows that

$$\left.\begin{aligned} \langle 0|\psi(x)|q\rangle &= [1+A'(q)+i\gamma q\, B'(q)]\langle 0|\psi^{(0)}(x)|q\rangle = \\ &= [1+A'(q)-m\,B'(q)]\langle 0|\psi^{(0)}(x)|q\rangle. \end{aligned}\right\} \quad (41.14)$$

The equation of motion for $\psi^{(0)}(x)$ has been used in the last step in (41.14). For each one-particle state $|q\rangle$ we have $q^2=-m^2$ and $q_0>0$. From the general form of the functions $A'(q)$ and $B'(q)$ we then conclude that the factor $1-A'(q)-mB'(q)$ is independent of the spatial part $\mathbf{q}$ of q. Accordingly, we can write (41.14) as follows:

$$\langle 0|\psi(x)|q\rangle = N\langle 0|\psi^{(0)}(x)|q\rangle, \quad (41.15)$$

where N is a "universal" constant (i.e., independent of x and q). From (41.15) it now follows that

$$\langle 0|f(x)|q\rangle = \left(\gamma\frac{\partial}{\partial x}+m\right)\langle 0|\psi(x)|q\rangle = 0. \quad (41.16)$$

The two equations (41.10) and (41.16) can now be used to obtain an implicit equation for the self-mass δm in (41.2). To do this we take $F(x)$ in (41.10) equal to $f(x)$. This gives

$$\left.\begin{aligned} 0 = i\int\Theta(x-x')\langle 0|\{f(x),\bar{f}(x')\}|0\rangle\langle 0|\psi^{(0)}(x')|q\rangle dx' + \\ +[ie\gamma_\nu\langle 0|A_\nu(x)|0\rangle+\delta m]\langle 0|\psi^{(0)}(x)|q\rangle \end{aligned}\right\} \quad (41.17)$$

or

$$\left.\begin{aligned} &\delta m\langle 0|\psi^{(0)}(x)|q\rangle = \\ &\quad = -i\int\Theta(x-x')\langle 0|\{f(x),\bar{f}(x')\}|0\rangle\langle 0|\psi^{(0)}(x')|q\rangle dx' \end{aligned}\right\} \quad (41.18)$$

since the vacuum expectation value of the potential vanishes in our gauge. To reduce the right side of (41.18) we return to Eq. (3.21).

---

1. The most general possible form for A is therefore $A_1(p^2)+A_2(p^2)\,\varepsilon(p)$, where $A_i(p^2)$ is non-zero only for time-like p. It can be shown (see below) that in this case $A_1(p^2)=0$. The corresponding statements also hold for the function $B(p)$.

Using it, we can write

$$\left.\begin{aligned}\langle 0|\{f_\alpha(x), \bar{f}_\beta(x')\}|0\rangle &= \frac{-1}{(2\pi)^3}\int_{p_0>0} dp\, e^{ip(x-x')} \times \\ &\times \left[\Sigma_1^{(+)}(p^2) + (i\gamma p + m)\,\Sigma_2^{(+)}(p^2)\right]_{\alpha\beta} + \frac{1}{(2\pi)^3}\int_{p_0>0} dp\, e^{ip(x'-x)} \times \\ &\times \left[\Sigma_1^{(-)}(p^2) + (-i\gamma p + m)\,\Sigma_2^{(-)}(p^2)\right]_{\alpha\beta}\end{aligned}\right\} \quad (41.19)$$

with

$$\left[\Sigma_1^{(+)}(p^2) + (i\gamma p + m)\,\Sigma_2^{(+)}(p^2)\right]_{\alpha\beta} = -V\sum_{p^{(z)}=p}\langle 0|\bar{f}_\beta|z\rangle\langle z|f_\alpha|0\rangle, \quad (41.19\text{a})$$

$$\left[\Sigma_1^{(-)}(p^2) + (-i\gamma p + m)\,\Sigma_2^{(-)}(p^2)\right]_{\alpha\beta} = V\sum_{p^{(z)}=p}\langle 0|f_\alpha|z\rangle\langle z|\bar{f}_\beta|0\rangle. \quad (41.19\text{b})$$

The four functions $\Sigma_i^{(\pm)}$ are defined by Eqs. (41.19a) and (41.19b). The summations in these equations run over all physical states for which the total energy-momentum vector is p (or $-p$). Again, V is the volume and we have taken the limit $V\to\infty$ in (41.19). From the charge symmetry of the theory, we can now show that only two of the functions $\Sigma_i^{(\pm)}$ are independent. We start with

$$\left[\Sigma_1^{(+)}(p^2) + (i\gamma p + m)\,\Sigma_2^{(+)}(p^2)\right]_{\alpha\beta} = -V\sum_{p^{(z)}=p}\langle 0|\bar{f}'_\beta|z\rangle\langle z|f'_\alpha|0\rangle. \quad (41.20)$$

Now we apply the matrix C of Eq. (14.3) to obtain

$$\left.\begin{aligned}\left[\Sigma_1^{(+)}(p^2) + (i\gamma p+m)\Sigma_2^{(+)}(p^2)\right]_{\alpha\beta} &= -V\sum_{p^{(z)}=p} C_{\alpha\delta}\langle 0|f_\varepsilon|z\rangle\langle z|\bar{f}_\delta|0\rangle (C^{-1})_{\beta\varepsilon} \\ &= C_{\alpha\delta}\left[\Sigma_1^{(-)}(p^2) + (-i\gamma p + m)\,\Sigma_2^{(-)}(p^2)\right]_{\varepsilon\delta}(C^{-1})_{\varepsilon\beta} = \\ &= \left[\Sigma_1^{(-)}(p^2) + (i\gamma p + m)\,\Sigma_2^{(-)}(p^2)\right]_{\alpha\beta}.\end{aligned}\right\} \quad (41.21)$$

In (41.21) we have used (14.4) through (14.6). From this it follows that

$$\Sigma_i^{(+)}(p^2) = \Sigma_i^{(-)}(p^2) \equiv \Sigma_i(p^2)\,. \quad (41.22)$$

Consequently we can write (41.19) as an integral over the entire p-space:

$$\left.\begin{aligned}&\langle 0|\{f_\alpha(x), \bar{f}_\beta(x')\}|0\rangle = \\ &= \frac{-1}{(2\pi)^3}\int dp\, e^{ip(x-x')}\left[\Sigma_1(p^2) + (i\gamma p + m)\,\Sigma_2(p^2)\right]_{\alpha\beta}\,\varepsilon(p)\,.\end{aligned}\right\} \quad (41.23)$$

Now we shall consider the expression $\Theta(x-x')\langle 0|\{f(x), \bar{f}(x')\}|0\rangle$. We write $\Theta(x-x') = \frac{1}{2}(1+\varepsilon(x-x'))$ and use the Fourier representation for $\varepsilon(x-x')$,

$$\varepsilon(x-x') = \frac{1}{i\pi}P\int_{-\infty}^{+\infty}\frac{d\tau}{\tau}\,e^{i\tau(x_0-x_0')}. \quad (41.24)$$

This gives

$$\left.\begin{aligned}&\frac{1}{(2\pi)^3}\,\varepsilon(x-x')\int dp\, e^{ip(x-x')}\varepsilon(p)\,\Sigma_1(p^2)=\\&=\frac{-2i}{(2\pi)^4}\int dp\, e^{ip(x-x')}P\int\limits_{-\infty}^{+\infty}\frac{d\tau}{\tau}\,\frac{p_0+\tau}{|p_0+\tau|}\,\Sigma_1\left(\mathbf{p}^2-(p_0+\tau)^2\right)=\\&=\frac{-2i}{(2\pi)^4}\int dp\, e^{ip(x-x')}P\int\limits_0^{\infty}\frac{da\,\Sigma_1(-a)}{a+p^2}\end{aligned}\right\}\tag{41.25}$$

and

$$\left.\begin{aligned}&\frac{1}{(2\pi)^3}\,\varepsilon(x-x')\int dp\, e^{ip(x-x')}(i\gamma p+m)\,\varepsilon(p)\,\Sigma_2(p^2)=\\&=\frac{-2i}{(2\pi)^4}\int dp\, e^{ip(x-x')}(i\gamma p+m)\,P\int\limits_0^{\infty}\frac{da\,\Sigma_2(-a)}{a+p^2}-\\&-\frac{2i}{(2\pi)^4}\int dp\, e^{ip(x-x')}P\int\limits_{-\infty}^{+\infty}d\tau\,\frac{i\gamma_4(p_0+\tau)}{|p_0+\tau|}\,\Sigma_2\left(\mathbf{p}^2-(p_0+\tau)^2\right).\end{aligned}\right\}\tag{41.26}$$

The last integral in (41.26) vanishes by reason of symmetry and we have

$$\left.\begin{aligned}&-i\,\Theta(x-x')\,\langle 0|\{f(x),\bar f(x')\}|0\rangle=\frac{1}{(2\pi)^4}\int dp\, e^{ip(x-x')}\times\\&\times\left[\bar\Sigma_1(p^2)+i\pi\,\varepsilon(p)\,\Sigma_1(p^2)+(i\gamma p+m)\left(\bar\Sigma_2(p^2)+i\pi\,\varepsilon(p)\,\Sigma_2(p^2)\right)\right],\end{aligned}\right\}\tag{41.27}$$

$$\bar\Sigma_i(p^2)=P\int\limits_0^{\infty}\frac{da}{a+p^2}\,\Sigma_i(-a).\tag{41.27a}$$

Substitution of (41.27) into (41.18) gives

$$\delta m\,\langle 0|\,\psi^{(0)}(x)\,|q\rangle=\left[\bar\Sigma_1(-m^2)+i\pi\,\Sigma_1(-m^2)\right]\langle 0|\,\psi^{(0)}(x)\,|q\rangle\,,\tag{41.28}$$

where the equation of motion for $\psi^{(0)}(x)$ has been used. From (41.16) and from the definition (41.19), it follows that $\Sigma_1(-m^2)$ vanishes. Thus from (41.28),

$$\delta m=\bar\Sigma_1(-m^2)=P\int\limits_0^{\infty}\frac{\Sigma_1(-a)}{a-m^2}\,da.\tag{41.29}$$

This result is an implicit equation determining the self-mass. It is implicit because the function $\Sigma_1(p^2)$ has been defined by the use of the right side of the Dirac equation, and from (41.2) this already contains the self-mass.

42. Renormalization of the Dirac Field. Equation for the Constant N

We saw in Eq. (41.15) that the matrix element for the complete operator $\psi(x)$ between the vacuum and the one-electron state is proportional to the corresponding matrix element for the in-field.

We will now rescale the Dirac field so that the constant of proportionality in (41.15) is equal to 1. This is done most simply by dividing each matrix element $\langle a|\psi(x)|b\rangle$ by this constant N. In other words, we introduce a new operator

$$\psi'(x) = \frac{1}{N}\psi(x). \tag{42.1}$$

This new "renormalized" field satisfies the following commutation relation

$$\{\overline{\psi}'(x), \psi'(x')\}_{x_0=x_0'} = \frac{1}{N^2}\gamma_4\,\delta(\boldsymbol{x}-\boldsymbol{x}'). \tag{42.2}$$

Since each term in the Dirac equation contains just one operator $\psi(x)$, the factor N cancels out and the equation for $\psi'(x)$ appears formally the same as that for $\psi(x)$,

$$\left(\gamma\frac{\partial}{\partial x}+m\right)\psi'(x) = f'(x), \tag{42.3}$$

$$f'(x) = ie\gamma A(x)\,\psi'(x) + \delta m\,\psi'(x). \tag{42.4}$$

With this new $f'(x)$ we again define the functions $\Sigma_i'(p^2)$ by (41.23) and obviously we have

$$\Sigma_i'(p^2) = \frac{1}{N^2}\Sigma_i(p^2), \tag{42.5}$$

$$\frac{\delta m}{N^2} = \overline{\Sigma}_1'(-m^2). \tag{42.6}$$

Henceforth we shall be concerned only with the renormalized operator and the old operator $\psi(x)$ will no longer be used. Accordingly, it is not necessary to denote the renormalized operator by a prime and we shall drop it, denoting the renormalized operator simply by $\psi(x)$ from now on. In a similar fashion, $f(x)$ and $\Sigma_i(p^2)$ now denote the renormalized functions. If all the quantities present were finite, this renormalization of the field operators would be without any great significance. As we shall soon see, the constant N^{-1} is not a finite quantity. [See also Eq. (32.20).] This renormalization is therefore quite useful because it allows us to isolate an infinite quantity in the theory and to study this quantity separately.

An equation for the constant N similar to (42.6) can be obtained in the following way: We write the integral equation for the renormalized field as[1]

1. The constant N is dependent upon the charge e and therefore becomes a time-dependent function $N(x_0) = N(e(x_0))$ during the adiabatic switching. Actually the integral in (42.7) should appear as follows:

$$\int S_R(x-x')\frac{N(x_0')}{N(x_0)}f(x')\,dx'.$$

For simplicity, we have not written the details out explicitly.

$$\psi(x) = \frac{1}{N}\psi^{(0)}(x) - \int S_R(x-x')\, f(x')\, dx'. \qquad (42.7)$$

The anticommutator (42.2) can then be written

$$\left.\begin{aligned}\{\overline{\psi}(x), \psi(x')\} &= \frac{1}{N^2}\{\overline{\psi}^{(0)}(x), \psi^{(0)}(x')\} + \frac{1}{N}\left\{\overline{\psi}^{(0)}(x), \psi(x') - \frac{1}{N}\psi^{(0)}(x')\right\} + \\ &+ \frac{1}{N}\left\{\overline{\psi}(x) - \frac{1}{N}\overline{\psi}^{(0)}(x), \psi^{(0)}(x')\right\} + \\ &+ \iint S_R(x'-x'')\{\overline{f}(x'''), f(x'')\}\, S_A(x'''-x)\, dx''\, dx'''.\end{aligned}\right\} (42.8)$$

If we take the vacuum expectation value, the second and third terms in (42.8) contain only the matrix elements $\langle 0|\psi(x')|q\rangle$ and $\langle 0|\overline{\psi}(x)|q\rangle$ [c.f. the discussion following Eq. (41.9)]. By means of the definition

$$\langle 0|\psi(x)|q\rangle = \langle 0|\psi^{(0)}(x)|q\rangle\,, \qquad (42.9)$$

and using the functions $\Sigma_i(p^2)$ we get

$$\left.\begin{aligned}\langle 0|\{\overline{\psi}(x), \psi(x')\}|0\rangle &= -i\, S(x'-x)\left[\frac{1}{N^2} + \frac{2(N-1)}{N^2}\right] - \\ -\frac{1}{(2\pi)^3}\int dp\, e^{ip(x'-x)}\, \varepsilon(p)\, &\frac{i\gamma p - m}{p^2+m^2}\left[\Sigma_1(p^2) + (i\gamma p + m)\Sigma_2(p^2)\right]\frac{i\gamma p - m}{p^2+m^2}.\end{aligned}\right\} (42.10)$$

With the identity

$$\left.\begin{aligned}&\frac{i\gamma p - m}{p^2+m^2}\left[\Sigma_1 + (i\gamma p + m)\,\Sigma_2\right]\frac{i\gamma p - m}{p^2+m^2} = \\ &= (i\gamma p - m)\left[\frac{-\Sigma_2}{p^2+m^2} - \frac{2m\,\Sigma_1}{(p^2+m^2)^2}\right] - \frac{\Sigma_1}{p^2+m^2}\,,\end{aligned}\right\} (42.11)$$

it follows from (42.10) for $x_0 = x_0'$ that

$$\left.\begin{aligned}&\frac{1}{N^2}\gamma_4\,\delta(\boldsymbol{x}-\boldsymbol{x}') = \\ &= \gamma_4\,\delta(\boldsymbol{x}-\boldsymbol{x}')\left[\frac{1+2(N-1)}{N^2} + \int_0^\infty da\left\{\frac{\Sigma_2(-a)}{a-m^2} - \frac{2m\,\Sigma_1(-a)}{(a-m^2)^2}\right\}\right],\end{aligned}\right\} (42.12)$$

or

$$\frac{N-1}{N^2} = -\frac{1}{2}\left(\overline{\Sigma}_2(-m^2) + 2m\,\overline{\Sigma}_1'(-m^2)\right), \qquad (42.13)$$

$$\overline{\Sigma}_1'(-m^2) = -\int_0^\infty \frac{da\,\Sigma_1(-a)}{(a-m^2)^2}\,. \qquad (42.13a)$$

The integral equation (42.7) and the equation following from it, (42.13), require a more detailed discussion, since an uncritical use of these equations can very easily lead to contradictions. Thus, for example, one might readily conclude from (42.7) and (41.16) that the last term of (42.7) vanishes for transitions between the vacuum and a one-particle state, and therefore that

$$\langle 0|\psi(x)|q\rangle = \frac{1}{N}\langle 0|\psi^{(0)}(x)|q\rangle \qquad (42.14)$$

ought to hold. Comparison of (42.14) and (42.9) then shows $N=1$. In a similar way one could conclude from (42.9) and the fact that the functions $\Sigma_i(p^2)$ vanish for $p^2=-m^2$ [see Eq. (41.28) and the remark following it] that the first term on the right side of (42.10) ought to appear multiplied by a factor 1, rather than by

$$\frac{1}{N^2}\left(1+2(N-1)\right).$$

The answer to both these puzzles is that although $\langle 0|f(x)|q\rangle$ does vanish, the convolution integral

$$\frac{N-1}{N}\langle 0|\psi^{(0)}(x)|q\rangle=\int S_R(x-x')\langle 0|f(x')|q\rangle dx' \tag{42.15}$$

is different from zero. In the proof that $\langle 0|f(x)|q\rangle$ vanishes, we neglected the time derivative[1] of the constant N. This is allowed for all finite times because of the adiabatic character of the switching. Neglect of the time derivative is not allowed if integrals from $-\infty$ are involved, as in (42.15). Neglecting the time derivative in such cases can lead to apparent contradictions. In Sec. 32, Eq. (32.10) and following, we worked out in detail the first perturbation-theoretic approximation to this phenomenon. There we saw that after the subtraction of the term with $\Sigma_1^{(0)}(-m^2)$, the matrix element of $f(x)$ [Eq. (32.14)] vanished because of an expression

$$(q^2+m^2)\,\delta(q^2+m^2). \tag{42.16}$$

In the convolution (42.15) there is a factor q^2+m^2 in the denominator, so that the right side actually has the indeterminate form

$$\frac{q^2+m^2}{q^2+m^2}\,\delta(q^2+m^2). \tag{42.17}$$

This holds if the adiabatic switching is not taken into account. However, if the switching is carefully carried through, rather than (42.10), we obtain a term of the order of magnitude α [Eq. (32.16)]. A similar term appears [Eq. (32.13)] in place of the denominator q^2+m^2. The behavior of numerator and denominator is well defined in the limit $\alpha\to 0$ and gives the unambiguous, non-zero result $1-\frac{1}{N}$ in (42.15). For understanding the equations above, it is important to remember that the operator $f(x)$ contains these singular terms. Since the functions $\Sigma_i(p^2)$ have been defined by means of $f(x)$, they contain singular pieces also. This explains the formal contradiction in (42.10). If the adiabatic switching is done carefully, it is evident from the foregoing discussion that the two functions $\Sigma_i(p^2)$ must contain terms of the form

$$\alpha^2\,\delta(p^2+m^2). \tag{42.18}$$

These terms give a non-zero contribution in the integral (42.13a)

1. See footnote 1, p. 204

which just enables us to write (42.10) as

$$\left.\begin{aligned}\langle 0|\{\bar{\psi}(x),\,\psi(x')\}|0\rangle = \frac{-1}{(2\pi)^3}\int dp\, e^{ip(x'-x)}\,\varepsilon(p)\Big[\delta(p^2+m^2)\,(i\gamma p-m)+\\ +\frac{i\gamma p-m}{p^2+m^2}\,[\Sigma_1^{\text{reg}}(p^2)+(i\gamma p+m)\,\Sigma_2^{\text{reg}}(p^2)]\,\frac{i\gamma p-m}{p^2+m^2}\Big].\end{aligned}\right\}\quad(42.19)$$

The two new functions $\Sigma_i^{\text{reg}}(p^2)$ are defined so that they are equal to the original functions $\Sigma_i(p^2)$ everywhere except at the point $p^2=-m^2$. The singular expressions (42.18) at the point $p^2=-m^2$ have been explicitly subtracted from $\Sigma_i^{\text{reg}}(p^2)$, so that not only does $\Sigma_i^{\text{reg}}(-m^2)$ vanish, but also $(p^2+m^2)^{-2}\,\Sigma_i^{\text{reg}}(p^2)$ can be set equal to zero for $p^2=-m^2$. With these new functions, we obtain from (42.19) not (42.13), but rather

$$\frac{1}{N^2}=1+\Sigma_2^{\text{reg}}(-m^2)+2m\,\Sigma_1^{\text{reg}\prime}(-m^2). \qquad (42.20)$$

For the following it will usually be simplest to calculate formally with the original functions $\Sigma_i(p^2)$ and to make the explicit subtraction of the singular parts (42.18) only in the results.

43. The Renormalization of the Charge

We have formally renormalized the mass and the field operator of the electron in Sec. 41 and 42 without the use of perturbation theory. In this section we are going to give a similar treatment for the charge renormalization. Because of the gauge invariance of the theory, the photon has no self-mass and therefore it is not necessary to carry out a renormalization of the energy of the photon. Accordingly, for the matrix elements between the vacuum and a one-photon state $|k\rangle$, there is the same x-dependence for the complete potential $A_\mu(x)$ as for the in-field $A_\mu^{(0)}(x)$. In this case the most general relativistically covariant relation between the two matrix elements is

$$\langle 0|A_\mu(x)|k\rangle = C\left[\delta_{\mu\nu}+M\frac{\partial^2}{\partial x_\mu\,\partial x_\nu}\right]\langle 0|A_\nu^{(0)}(x)|k\rangle. \qquad (43.1)$$

The two constants C and M in (43.1) are independent of k and x. A formal proof for (43.1) can be given following the method of proof of Eq. (41.15) in Sec. 41. Using the equation

$$\left.\begin{aligned}A_\mu^{(0)}(x) &= \int_{-\infty}^{x'} D(x-x'')\,j_\mu(x'')\,dx'' -\\ &-\int_{x_0''=x'}\left[\frac{\partial A_\mu(x'')}{\partial x_0''}\,D(x-x'')+A_\mu(x'')\frac{\partial D(x-x'')}{\partial x_0}\right]d^3x''\end{aligned}\right\}\quad(43.2)$$

[c.f. Eq. (41.7)] and the canonical commutation relations, we find for an arbitrary operator $F(x)$,

$$\left.\begin{aligned}[F(x),\,A_\mu^{(0)}(x')] = \int_{-\infty}^{x} D(x'-x'')\,[F(x),\,j_\mu(x'')]\,dx''-i\,D(x'-x)\,\frac{\partial F(x)}{\partial A_\mu(x)}+\\ +i\,\frac{\partial D(x'-x)}{\partial x_0'}\,\frac{\partial F(x)}{\partial\dfrac{\partial A_\mu(x)}{\partial x_0}}.\end{aligned}\right\}\quad(43.3)$$

In the last two equations $j_\mu(x)$ stands for the right side of the equation of motion for $A_\mu(x)$:

$$\Box A_\mu(x) = -j_\mu(x). \tag{43.4}$$

The detailed structure of the current operator is not important for (43.2) and (43.3). Just as in Sec. 41, the vacuum expectation value of (43.3) gives an expression for the matrix elements of $F(x)$ between the vacuum and a one-photon state. We have

$$\left.\begin{aligned} \langle 0|F(x)|k\rangle &= i\int_{-\infty}^{x} \langle 0|[F(x), j_\mu(x'')]|0\rangle \langle 0|A_\mu^{(0)}(x'')|k\rangle dx'' + \\ &+ \left[\langle 0|\frac{\partial F(x)}{\partial A_\mu(x)}|0\rangle - i\,k_0\langle 0|\frac{\partial F(x)}{\partial\frac{\partial A_\mu(x)}{\partial x_0}}|0\rangle\right]\langle 0|A_\mu^{(0)}(x)|k\rangle. \end{aligned}\right\} \tag{43.5}$$

If we take $F(x)$ as $A_\mu(x)$, then (43.1) follows from (43.5) and the use of invariance properties of the vacuum expectation values which appear.

The constant C in (43.1) corresponds closely to the constant N in (41.15) but the constant M has its origin in the particular vector character of the potential. If the state $|k\rangle$ contains a transverse photon, this term automatically drops out. The constant N of Sec. 42 was not eliminated from the theory by a counter term. [The only counter term in (41.2) contains the self-mass.] It was isolated by renormalization of the field operator $\psi(x)$ and it was then considered separately. The situation is somewhat different for the constant C of (43.1), since now the observable quantities are the electromagnetic field strengths and are therefore linear combinations of the potentials. The constant C has a "physical" significance in so far as it connects the units of the field strengths (i.e., also the charge) of the complete fields $A_\mu(x)$ with those of the in-fields $A_\mu^{(0)}(x)$. For simplicity, we shall use the same units for these two kinds of fields and therefore we have to take $C=1$. This can only be done by means of a counter term; accordingly, we shall add an extra term to the current operator:

$$j_\mu(x) = \frac{i\,e\,N^2}{2}[\overline{\psi}(x), \gamma_\mu\psi(x)] - L\Box A_\mu(x). \tag{43.6}$$

One can substitute this expression with the charge renormalization constant L into (43.4), transfer the term in L to the left side and divide by $1-L$. This factor can then be understood as a renormalization of the charge. The factor N^2 enters the first term in (43.6) because $\psi(x)$ is now the renormalized Dirac field. In what follows, it will be useful for us to define the charge renormalization so that the counter term is gauge invariant. This is most simply done by defining the renormalized current as

$$j_\mu(x) = \frac{i\,e\,N^2}{2}[\overline{\psi}(x), \gamma_\mu\psi(x)] - L\left(\Box A_\mu(x) - \frac{\partial^2 A_\nu(x)}{\partial x_\mu \partial x_\nu}\right). \tag{43.7}$$

Using

$$A_\mu(x) = A_\mu^{(0)}(x) + \int D_R(x-x')\, j_\mu(x')\, dx' , \qquad (43.8)$$

we obtain

$$\frac{\partial A_\mu(x)}{\partial x_\mu} = \frac{\partial A_\mu^{(0)}(x)}{\partial x_\mu} + \int D_R(x-x')\, \frac{\partial j_\mu(x')}{\partial x'_\mu}\, dx' = \frac{\partial A_\mu^{(0)}(x)}{\partial x_\mu} . \qquad (43.9)$$

It follows from this that the difference of (43.6) and (43.7) is of importance only for certain special matrix elements with scalar and longitudinal photons.

Since the new counter term in (43.7) contains the derivatives of the potentials, we must check to see if the canonical commutation relations are changed by these new terms in the Lagrangian. Our complete Lagrangian is now

$$\mathcal{L} = \mathcal{L}_\psi + \mathcal{L}_A + \mathcal{L}_W , \qquad (43.10)$$

$$\left.\begin{aligned} \mathcal{L}_\psi = &-\frac{N^2}{4}\left[\overline{\psi}(x), \left(\gamma \frac{\partial}{\partial x} + m\right)\psi(x)\right] - \\ &-\frac{N^2}{4}\left[-\frac{\partial \overline{\psi}(x)}{\partial x_\mu}\gamma_\mu + m\overline{\psi}(x), \psi(x)\right] + \frac{1}{2}\delta m\, N^2\, [\overline{\psi}(x), \psi(x)] , \end{aligned}\right\} \qquad (43.11)$$

$$\mathcal{L}_A = -\frac{1-L}{4}\left(\frac{\partial A_\mu(x)}{\partial x_\nu} - \frac{\partial A_\nu(x)}{\partial x_\mu}\right)\left(\frac{\partial A_\mu(x)}{\partial x_\nu} - \frac{\partial A_\nu(x)}{\partial x_\mu}\right) - \frac{1}{2}\frac{\partial A_\mu(x)}{\partial x_\mu}\frac{\partial A_\nu(x)}{\partial x_\nu} , \qquad (43.12)$$

$$\mathcal{L}_W = \frac{ie}{2} N^2 A_\mu(x)\, [\overline{\psi}(x), \gamma_\mu \psi(x)] . \qquad (43.13)$$

The usual rules for quantization give

$$\{\overline{\psi}(x), \psi(x')\}_{x_0 = x'_0} = \gamma_4 \frac{1}{N^2}\, \delta(\boldsymbol{x} - \boldsymbol{x}') , \qquad (43.14)$$

in complete agreement with (42.2). We also obtain

$$[A_\mu(x), A_\nu(x')]_{x_0 = x'_0} = 0 , \qquad (43.15)$$

$$\left[\frac{\partial A_\mu(x)}{\partial x_0}, A_\nu(x')\right]_{x_0 = x'_0} = \frac{-i}{1-L}\, \xi_{\mu\nu}\, \delta(\boldsymbol{x} - \boldsymbol{x}') , \qquad (43.16)$$

$$\xi_{\mu\nu} = \delta_{\mu\nu} - L\, \delta_{\mu 4}\, \delta_{\nu 4} , \qquad (43.16a)$$

$$\left[\frac{\partial A_\mu(x)}{\partial x_0}, \frac{\partial A_\nu(x')}{\partial x'_0}\right]_{x_0 = x'_0} = \frac{L}{1-L}\left(\delta_{\mu 4}\frac{\partial}{\partial x_\nu} + \delta_{\nu 4}\frac{\partial}{\partial x_\mu}\right)\delta(\boldsymbol{x} - \boldsymbol{x}') . \qquad (43.17)$$

For $L=0$, (43.15) through (43.17) go over into (17.8) and (17.9). The relations (17.10), (17.12) and (17.13) are still valid in the renormalized theory. Apart from the replacement of $\delta_{\mu\nu}$ by $\xi_{\mu\nu}$ in (43.16), we see that the constant $\sqrt{1-L}$ plays an analogous role for the electromagnetic field to the constant N for the electron field. Both appear in a similar way in the commutation relations for the renormalized fields. If we do not consider the last term in (43.12), we can take the renormalization of charge to mean the replacement of the field $A_\mu(x)$ by $\sqrt{1-L}\, A_\mu(x)$ and of the charge

e by $\frac{e}{\sqrt{1-L}}$. In this way we can understand why the interaction term (43.13) is formally not affected by L. The exceptional character of the last term in (43.12) has its origin in the gauge-invariant form used in the definition of the current (43.7). In order to obtain the expression (43.7) for the current, we need only multiply the gauge-invariant part of (17.3) by $1-L$. This form of charge renormalization was first given by Gupta.[1]

After these formal preparations, we are now ready to derive an explicit equation for L. To do this, we shall employ the same method used in Sec. 42 to derive (42.13). We therefore calculate the vacuum expectation value of the commutator of two potentials $A_\mu(x)$:

$$\left.\begin{aligned}\langle 0|[A_\mu(x), A_\nu(x')]|0\rangle &= \langle 0|[A_\mu^{(0)}(x), A_\nu^{(0)}(x')]|0\rangle + \\ &+ \langle 0|[A_\mu(x)-A_\mu^{(0)}(x), A_\nu^{(0)}(x')]|0\rangle + \\ &+ \langle 0|[A_\mu^{(0)}(x), A_\nu(x')-A_\nu^{(0)}(x')]|0\rangle + \\ + \iint dx''\,dx'''\, D_R(x-x'')\, &D_R(x'-x''')\langle 0|[j_\mu(x''), j_\nu(x''')]|0\rangle .\end{aligned}\right\} \quad (43.18)$$

As before, it is useful to introduce a new notation for the vacuum expectation value in the last term. We therefore write

$$\left.\begin{aligned}\langle 0|[j_\mu(x), j_\nu(x')]|0\rangle = \frac{-1}{(2\pi)^3}\int_{p_0>0} dp\, e^{ip(x'-x)}\, \Pi_{\mu\nu}^{(+)}(p) + \\ + \frac{1}{(2\pi)^3}\int_{p_0<0} dp\, e^{ip(x'-x)}\, \Pi_{\mu\nu}^{(-)}(p)\,,\end{aligned}\right\} \quad (43.19)$$

with

$$\Pi_{\mu\nu}^{(+)}(p) = V \sum_{p^{(z)}=p} \langle 0|j_\nu|z\rangle\langle z|j_\mu|0\rangle\,, \quad (43.20a)$$

$$\Pi_{\mu\nu}^{(-)}(p) = V \sum_{p^{(z)}=-p} \langle 0|j_\mu|z\rangle\langle z|j_\nu|0\rangle = \Pi_{\nu\mu}^{(+)}(-p)\,. \quad (43.20b)$$

From general principles of invariance, it follows that the two functions $\Pi_{\mu\nu}^{(\pm)}(p)$ must have the form

$$\Pi_{\mu\nu}^{(+)}(p) = A(p^2)\,\delta_{\mu\nu} + B(p^2)\,p_\mu p_\nu = \Pi_{\nu\mu}^{(-)}(-p) = \Pi_{\mu\nu}^{(-)}(p)\,. \quad (43.21)$$

From the continuity equation for the current, which is also satisfied by the renormalized current, we have

$$0 = p_\mu \Pi_{\mu\nu}^{(\pm)}(p) = [A(p^2) + p^2 B(p^2)]\,p_\nu\,, \quad (43.22)$$

and therefore

$$\Pi_{\mu\nu}^{(\pm)}(p) = (-p^2\delta_{\mu\nu} + p_\mu p_\nu)\,\Pi(p^2)\,, \quad (43.23)$$

$$\Pi(p^2) = \frac{V}{-3p^2}\sum_{p^{(z)}=p}\langle 0|j_\mu|z\rangle\langle z|j_\mu|0\rangle\,, \quad (43.24)$$

$$\langle 0|[j_\mu(x), j_\nu(x')]|0\rangle = \frac{-1}{(2\pi)^3}\int dp\, e^{ip(x'-x)}\varepsilon(p)\,[-p^2\delta_{\mu\nu} + p_\mu p_\nu]\,\Pi(p^2). \quad (43.25)$$

1. S. N. Gupta, Proc. Phys. Soc. Lond. A64, 426 (1951).

By using (43.1) and (43.25) we can now write (43.18) in the following form:

$$\left.\begin{aligned}\langle 0|[A_\mu(x), A_\nu(x')]|0\rangle = \frac{-1}{(2\pi)^3}\int dp\, e^{ip(x'-x)}\varepsilon(p) \times \\ \times \left[\delta_{\mu\nu}\left(\delta(p^2) - \frac{\Pi(p^2)}{p^2}\right) + p_\mu p_\nu\left(\frac{\Pi(p^2)}{(p^2)^2} - 2M\,\delta(p^2)\right)\right].\end{aligned}\right\} \quad (43.26)$$

If the two times x_0' and x_0 are equal, then the right side of (43.26) must vanish, according to (43.15). The first term in the square brackets vanishes identically on the basis of symmetry. The second term gives an equation for M. For $\mu=4$, $\nu \neq 4$, we have

$$\left.\begin{aligned}0 = \frac{-i}{(2\pi)^3}\int d^3p\, e^{i\boldsymbol{p}(\boldsymbol{x}'-\boldsymbol{x})}p_\nu \int dp_0 |p_0| \left(\frac{\Pi(p^2)}{(p^2)^2} - 2M\,\delta(p^2)\right) = \\ = -\frac{\partial}{\partial x'_\nu}\delta(\boldsymbol{x}'-\boldsymbol{x})\left[\int_0^\infty \frac{da\,\Pi(-a)}{a^2} - 2M\right],\end{aligned}\right\} \quad (43.27)$$

or

$$M = \frac{1}{2}\int_0^\infty \frac{da\,\Pi(-a)}{a^2}. \quad (43.28)$$

For other combinations of μ and ν the last term in (43.26) also vanishes identically.

By differentiating (43.26) with respect to the time x_0 and then setting the two times equal, we can also derive

$$\left.\begin{aligned}\langle 0|\left[\frac{\partial A_\mu(x)}{\partial x_0}, A_\nu(x')\right]|0\rangle_{x_0=x_0'} = -i\,\delta_{\mu\nu}[1+\overline{\Pi}(0)]\,\delta(\boldsymbol{x}'-\boldsymbol{x}) - \\ -\frac{i}{(2\pi)^3}\int d^3p\, e^{i\boldsymbol{p}(\boldsymbol{x}'-\boldsymbol{x})}\int_0^\infty da\int dp_0|p_0|\, p_\mu p_\nu\,\delta(p^2+a)\left[\frac{\Pi(-a)}{a^2} - 2M\delta(a)\right] \\ = -i\,\delta(\boldsymbol{x}-\boldsymbol{x}'))\,[\delta_{\mu\nu}(1+\overline{\Pi}(0)) - \delta_{\mu 4}\delta_{\nu 4}\overline{\Pi}(0)],\end{aligned}\right\} \quad (43.29)$$

$$\overline{\Pi}(p^2) = P\int_0^\infty \frac{da\,\Pi(-a)}{a+p^2}. \quad (43.30)$$

A comparison of (43.29) and (43.16) now gives

$$\frac{1}{1-L} = 1 + \overline{\Pi}(0). \quad (43.31)$$

This is the desired equation[1] for the constant L. As a by-product,

1. A similar equation for the charge renormalization was first given by H. Umezawa and S. Kamefuchi, Progr. Theor. Phys. 6, 543 (1951). For the complete system of the equations given here for the renormalization constants, see G. Källén, Helv. Phys. Acta 25, 417 (1952); H. Lehmann, Nuovo Cim. 11, 342 (1954); M. Gell-Mann and F. E. Low, Phys. Rev. 95, 1300 (1954).

we have a similar equation in (43.28) for the M of Eq. (43.1). After differentiating (43.26) with respect to x_0' as well as x_0 and using (43.28) and (43.31), it is not hard to show that (43.17) is also satisfied.

44. General Properties of the Functions $\Pi(p^2)$ and $\Sigma_i(p^2)$

In the earlier sections of this chapter we have given a formal system of equations for determining the renormalization constants of quantum electrodynamics. This system of equations is formally sufficient to determine the renormalization constants because there are the same number of equations and constants. We must still discuss the consistency of this system and if possible decide whether it has any physically useful solutions at all. We are not going to be able to answer these difficult questions completely; at least we shall try to give a basis for some future treatment.

In our formalism the renormalization constants are given as integrals over the weight functions $\Pi(p^2)$ and $\Sigma_i(p^2)$. Since these functions will play an extremely important role in what follows, we shall now discuss their properties. We start with the important observation, which will be of crucial significance, that the definitions (41.19) and (43.24) contain only sums with a finite number of terms. In order to show this, we first note that if we consider only states with one in-particle there are only a finite number $\frac{V}{(2\pi)^3} d^3p$ of them which have momentum between $\boldsymbol{p}$ and $\boldsymbol{p}+d\boldsymbol{p}$. At most, these terms contribute to the weight functions when $p^2=-m^2$ and, in fact, this contribution "vanishes" because[1] of the renormalization condition (41.16) and because of a similar condition for the current operator $j_\mu(x)$. For states with two in-particles we consider the weight functions in the particular coordinate system where the spatial $\boldsymbol{p}$ vanishes. Then the two particles have equal and oppositely directed spatial momenta $\boldsymbol{q}$ and if the two masses are m_1 and m_2, the total energy is $p_0=\sqrt{\boldsymbol{q}^2+m_1^2}+\sqrt{\boldsymbol{q}^2+m_2^2}$. This equation determines $|\boldsymbol{q}|$ as a function of p_0:

$$|\boldsymbol{q}|=\frac{p_0}{2}\sqrt{1-\frac{(m_1-m_2)^2}{p_0^2}}+\sqrt{1-\frac{(m_1+m_2)^2}{p_0^2}},$$

and the number of states of this type in the sums (41.19) and (43.24) is again finite and equals

$$\mathcal{N}=\frac{V}{(2\pi)^3}\frac{-p^2}{4}\sqrt{1+\frac{(m_1-m_2)^2}{p^2}}\sqrt{1+\frac{(m_1+m_2)^2}{p^2}}. \qquad (44.1)$$

In a similar fashion, we could proceed with three or more particles and show that there are a finite number of states with a finite number of in-particles which have a given total energy-momentum

1. The word "vanishes" is used in quotation marks here because these terms can still give certain contributions to the integrals. C.f. the discussion following Eq. (42.15).

vector p. On the other hand, in a state with a given finite p, there can be only a finite number of particles of non-zero mass. As elementary considerations show, this finite number is smaller than $\sqrt{-\frac{p^2}{m^2}}$, where m is the smallest mass of the particles which are present. In principle, an infinite number of photons (of zero mass) can occur in our states; the simplest way to avoid this difficulty is by introducing a small photon mass μ. At the same time, this automatically avoids all difficulties involving infrared divergences (c.f. Sec. 35). With these restrictions the definitions of our weight functions contain only finite sums, i.e., in the limit $V \to \infty$ they involve only integrals over finite regions in p-space. If the matrix elements of all the renormalized operators are finite, then the weight functions are finite quantities.[1] It has also been assumed here that there are at most a finite number of bound states of given binding energy. This restriction seems quite weak.

According to the assumptions made about the mass spectrum of the system following Eq. (41.12), the weight functions must vanish for space-like values of p^2. This fact has already been used several times. We can now go further and show that the function $\Pi(p^2)$ has no contribution from states of less than three photons (or one pair). This comes about for two reasons: According to the renormalization condition, the contribution of the one-photon state vanishes. Also, from the charge symmetry of the theory, it can be shown[2] that matrix elements of the current operator between the vacuum and states of two in-photons vanish. The conclusion is that for $-p^2 < 9\mu^2$, the function $\Pi(p^2)$ is zero. Because of the conservation of charge in our theory, we likewise conclude that

---

1. Here, by the word "finite" we mean that the integral of one of the weight functions over a finite region is finite; i.e., that the weight functions contain no singularities stronger than delta functions. Such delta functions, if they are present at all, occur only in connection with states which are not scattering states.

2. By a calculation similar to (43.3) one finds

$$\langle 0|[[j_\mu(x), A_\nu^{(0)}(x')], A_\lambda^{(0)}(x'')]|0\rangle =$$

$$= \int_{-\infty}^{x} dx''' \int_{-\infty}^{x'''} dx^{\mathrm{IV}} \Big[D(x'-x''')\,D(x''-x^{\mathrm{IV}}) \times$$

$$\times \langle 0|[j_\lambda(x^{\mathrm{IV}}), [j_\nu(x'''), j_\mu(x)]]|0\rangle +$$

$$+ D(x'-x^{\mathrm{IV}})\,D(x''-x''')\langle 0|[j_\nu(x^{\mathrm{IV}}), [j_\lambda(x'''), j_\mu(x)]]|0\rangle\Big].$$

Because of the charge symmetry [c.f. Eq. (14.21), which also holds in the theory with interacting fields] the right side vanishes, and hence also the matrix elements

$$\langle 0|\,j_\mu(x)|k, k'\rangle.$$

the states in (41.19) must have[1] charge e and hence that they must contain at least one electron. The states of only one particle give no contribution here also, and the first non-zero contribution comes from states of one electron and one photon, so that the functions $\Sigma_i(p^2)$ vanish for $-p^2<(m+\mu)^2$. The lower limits of integration in (42.20) and (42.30) can therefore be set equal to $(m+\mu)^2$ and $9\mu^2$, respectively.[2] Taking account of the small photon mass, $\overline{\Pi}(0)$ in (43.31) must be replaced by $\overline{\Pi}(-\mu^2)$.

Finally, we wish to show that the function $\Pi(p^2)$ is always positive. From the definitions (43.20) and (43.23) we have

$$V\sum_{p^{(z)}=p}\langle 0|\, j_x\,|z\rangle\langle z|\, j_x\,|0\rangle=(p_x^2-p^2)\,\Pi(p^2). \tag{44.2}$$

The x-component of the current operator has to have real expectation values, so that this operator must be self-adjoint, i.e.,

$$\langle z|\, j_x\,|0\rangle=(\langle 0|\, j_x\,|z\rangle)^*\,(-1)^{N_4^{(z)}}. \tag{44.3}$$

The symbol $N_4^{(z)}$ denotes the number of scalar in-photons in the state $|z\rangle$. From this we have

$$\Pi(p^2)=\frac{V}{p_x^2-p^2}\sum_{p^{(z)}=p}|\langle 0|\, j_x\,|z\rangle|^2\,(-1)^{N_4^{(z)}}. \tag{44.4}$$

In the sum on the right side of (44.4) all states without scalar photons (or, more generally, with an even number of scalar photons) give positive contributions. We shall now show that the negative contribution of a state with one scalar photon is just cancelled by the contribution of a corresponding state with a longitudinal photon. From Eqs. (43.3) through (43.5) we obtain the following expression for the matrix element of the current between the vacuum and a state $|a,k\rangle$:

$$\langle 0|\, j_\mu(x)\,|a,k\rangle=i\int\Theta(x-x')\,\langle 0|\,[j_\mu(x),j_\nu(x')]\,|a\rangle\,\langle 0|\,A_\nu^{(0)}(x')\,|k\rangle\,dx'. \tag{44.5}$$

---

1. It follows from the canonical commutation relations that

$$[j_4(x'),\psi(x)]_{x_0=x_0'}=-i\,e\,\psi(x)\,\delta(\boldsymbol{x}-\boldsymbol{x}')$$

and hence

$$[Q,\psi(x)]=-e\,\psi(x)\qquad\text{with}\qquad Q=-i\int d^3x\, j_4(x).$$

On computing the matrix element of this operator equation between the vacuum and a state $|z\rangle$ with charge $Q^{(z)}$, we find

$$-Q^{(z)}\langle 0|\psi(x)\,|z\rangle=-e\,\langle 0|\,\psi(x)\,|z\rangle.$$

Thus $Q^{(z)}=e$ if this matrix element of the Dirac field does not vanish.

2. In (42.13) the singular contribution in the last term must be taken into account, so that here one must integrate over the point $a=m^2$ also.

Here a stands for all the particles present except the photon. The contribution of the last three-dimensional integral in (43.2) is a c-number and therefore the corresponding matrix element vanishes if $|a\rangle$ is not the vacuum. After taking out the x-dependence, we can write (44.5) as

$$\langle 0|\, j_\mu \,|a, k\rangle = F_{\mu\nu}(a, k)\, \langle 0|\, A_\nu^{(0)}\, |k\rangle\,, \tag{44.6}$$

where

$$F_{\mu\nu}(a, k) = \frac{i}{(2\pi)^4} \int dx\, e^{ikx}\, \Theta(-x)\, \langle 0|\, [j_\mu(0), j_\nu(x)]\, |a\rangle. \tag{44.6a}$$

From the continuity equation for the current and from the vanishing of the appropriate matrix elements of the current commutator for space-like x, it follows that

$$F_{\mu\nu}(a, k)\, k_\nu = 0. \tag{44.7}$$

Equation (44.7) is clearly an expression of the gauge invariance of the theory even for virtual states. Together, the Eqs. (44.6) and (44.7) give the result that the absolute value of the matrix element (44.6) for scalar photons is equal to the corresponding quantity for longitudinal photons:

$$|\langle 0|\, j_\mu \,|a, k, \lambda = 3\rangle| = |\langle 0|\, j_\mu \,|a, k, \lambda = 4\rangle|\,. \tag{44.8}$$

The stated cancellation in (44.4) of the contributions of the scalar and longitudinal photons follows from (44.8). By similar arguments, this cancellation can readily be shown if more than one scalar photon is present. Thus we have shown that the function $\Pi(p^2)$ in (44.4) can be expressed as a finite sum of terms which are all positive.

A similar proof cannot be given for the functions $\Sigma_i(p^2)$, since the operator $f(x)$ in (41.19) is not gauge invariant, and therefore a cancellation of longitudinal and scalar photons cannot be expected.

Using the positive character of $\Pi(p^2)$ in (43.31) gives the important identity

$$0 < L \leq 1. \tag{44.9}$$

The charge renormalization constant L of quantum electrodynamics therefore lies between[1] zero and one. From this, $L=1$ is a singular point of the theory, for in this case the highest derivatives formally cancel out of the equations of motion. Also, as is evident from the remarks following Eq. (43.17), the case $L=1$ corresponds to an infinite charge renormalization.

---

1. The inequality (44.9) was first shown by J. Schwinger (unpublished). See also the additional references of footnote 1, p.211.

45. The Physical Significance of the Functions $\Pi(p^2)$ and $\overline{\Pi}(p^2)$. Connection with Previous Results

Although the treatment of renormalization given in Sec. 41 is quite clear, that of the charge renormalization of Sec. 43 had a fairly abstract form. There is no obvious connection with the lowest approximation of perturbation theory, given in Sec. 29, for charge renormalization in an external field. In this section we will study the vacuum polarization by introducing a weak external field into our system. At the same time we shall obtain a deeper understanding of the physical significance of the functions $\Pi(p^2)$ and $\overline{\Pi}(p^2)$.

In the renormalized theory the interaction Hamiltonian for an external field $A_\mu^{\text{äuss}}(x)$ and a system of quantized fields is

$$\delta E = -\int d^3x\, j_\mu(x)\, A_\mu^{\text{äuss}}(x)\,, \tag{45.1}$$

where $j_\mu(x)$ is the renormalized current. According to (43.7), this operator contains second derivatives of the field operators, so that (45.1) is not a suitable term to add to our Lagrangian. Rather than (45.1), we shall use the expression

$$\delta\mathcal{L} = \frac{i e N^2}{2}\left[\overline{\psi}(x), \gamma_\mu \psi(x)\right] A_\mu^{\text{äuss}}(x) + L A_\mu(x)\, j_\mu^{\text{äuss}}(x)\,, \tag{45.2}$$

which differs formally only by a four-divergence. In Eq. (45.2), $j_\mu^{\text{äuss}}(x)$ stands for the external current,

$$j_\mu^{\text{äuss}}(x) = -\left(\Box\, \delta_{\mu\nu} - \frac{\partial^2}{\partial x_\mu \partial x_\nu}\right) A_\nu^{\text{äuss}}(x)\,. \tag{45.3}$$

From (45.2), we obtain the following equations of motion:

$$\left(\gamma \frac{\partial}{\partial x} + m\right)\psi(x) = i e \gamma A(x)\,\psi(x) + \delta m\, \psi(x) + i e \gamma A^{\text{äuss}}(x)\,\psi(x), \tag{45.4}$$

$$\Box A_\mu(x) = -\frac{i e N^2}{2}\left[\overline{\psi}(x), \gamma_\mu \psi(x)\right] + L\left(\Box A_\mu(x) - \frac{\partial^2 A_\nu(x)}{\partial x_\mu \partial x_\nu} - j_\mu^{\text{äuss}}(x)\right). \tag{45.5}$$

We are going to consider only the case where the external field is very weak and regular enough so that we can expand our results in powers of it. One can readily show[1] that the first two terms in

---

1. By the use of the so-called functional derivative, these equations can be written as

$$\frac{\delta\psi(x)}{\delta A_\nu^{\text{äuss}}(x')} = -i\,\Theta(x-x')\left[j_\nu(x'), \psi(x)\right],$$

$$\frac{\delta A_\mu(x)}{\delta A_\nu^{\text{äuss}}(x')} = -i\,\Theta(x-x')\left[j_\nu(x'), A_\mu(x)\right] + \frac{L}{1-L}\left(\delta_{\mu\nu} - \delta_{\mu 4}\delta_{\nu 4}\right)\delta(x-x').$$

In a similar way, we can write Eq. (45.17) (to follow) as

$$\frac{\delta\langle 0|\, j_\mu(x)\,|0\rangle}{\delta j_\nu^{\text{äuss}}(x')} = -\frac{\delta_{\mu\nu}}{(2\pi)^4}\int dp\, e^{ip(x-x')}\left[\overline{\Pi}(p^2) - \overline{\Pi}(0) + i\pi\,\varepsilon(p)\,\Pi(p^2)\right].$$

This method of writing them has been suggested by several authors and has been used particularly by J. Schwinger, Proc. Nat. Acad. Sci. U.S.A. 37, 432, 435 (1951). See also V. Fock, Phys. Z. Sowjet. 6, 425 (1934).

such an expansion are

$$\psi(x) = \Psi(x) - i \int \Theta(x - x') \left[\mathrm{j}_\nu(x'), \Psi(x) \right] A_\nu^{\text{äuss}}(x')\, dx' + \cdots, \tag{45.6}$$

$$\left. \begin{aligned} A_\mu(x) &= \mathrm{A}_\mu(x) - i \int \Theta(x - x') \left[\mathrm{j}_\nu(x'), \mathrm{A}_\mu(x) \right] A_\nu^{\text{äuss}}(x')\, dx' + \\ &\quad + \frac{L}{1 - L} (\delta_{\mu\nu} - \delta_{\mu 4}\delta_{\nu 4}) A_\nu^{\text{äuss}}(x) + \cdots . \end{aligned} \right\} \tag{45.7}$$

Here $\Psi(x)$ and $\mathrm{A}_\mu(x)$ are the solutions of (45.4) and (45.5) for vanishing external field, i.e., they are just those operators studied in the previous sections. We prove (45.6) and (45.7) by substituting[1] them into the equations of motion (45.4) and (45.5). Substitution of (45.6) into (45.4) gives

$$\left. \begin{aligned} \left(\gamma \frac{\partial}{\partial x} + m\right) \psi(x) &= i e \gamma \mathrm{A}(x) \Psi(x) + \delta m \Psi(x) - \\ &\quad - i \int \Theta(x - x') \left[\mathrm{j}_\nu(x'), i e \gamma \mathrm{A}(x) \Psi(x) + \delta m \Psi(x) \right] \times \\ &\quad \times A_\nu^{\text{äuss}}(x')\, dx' - \gamma_4 \int_{x_0 = x_0'} d^3x' \left[\mathrm{j}_\nu(x'), \Psi(x) \right] A_\nu^{\text{äuss}}(x'). \end{aligned} \right\} \tag{45.8}$$

Using (45.6) and (45.7) we can combine the first two terms on the right side of (45.8) and the four-dimensional integral in the following way: (Terms which are quadratic in $A_\mu^{\text{äuss}}(x)$ may be dropped.)

$$\left. \begin{aligned} & i e \gamma \mathrm{A}(x) \Psi(x) + \delta m \Psi(x) - \\ & - i \int \Theta(x - x') \left[\mathrm{j}_\nu(x'), i e \gamma \mathrm{A}(x) \Psi(x) + \delta m \Psi(x) \right] A_\nu^{\text{äuss}}(x')\, dx' = \\ & = i e \gamma A(x) \psi(x) + \delta m \psi(x) - i e \frac{L}{1 - L} \gamma_k A_k^{\text{äuss}}(x) \psi(x). \end{aligned} \right\} \tag{45.9}$$

If we eliminate the time derivatives of second order from the current operator by the use of the equations of motion, the result can be written as

$$\mathrm{j}_\mu(x) = \frac{\xi_{\mu\lambda}}{1 - L} \left\{ \frac{i e N^2}{2} \left[\overline{\Psi}(x), \gamma_\lambda \Psi(x) \right] + L \frac{\partial^2 \mathrm{A}_\nu(x)}{\partial x_\lambda \partial x_\nu} \right\} - L \delta_{\mu 4} \Box \mathrm{A}_4(x), \tag{45.10}$$

$$\xi_{\mu\lambda} = \delta_{\mu\lambda} - L \delta_{\mu 4} \delta_{\lambda 4}, \tag{45.11}$$

Using this result and the canonical commutation relations, we can work out the commutator in the last term in (45.8). In this way we find

$$\gamma_4 \left[\mathrm{j}_\nu(x'), \Psi(x) \right]_{x_0 = x_0'} = - \frac{i e \xi_{\nu\lambda}}{1 - L} \gamma_\lambda \Psi(x) \delta(\boldsymbol{x} - \boldsymbol{x}'). \tag{45.12}$$

Substitution of (45.12) into (45.8) and application of (45.9) gives the right side of (45.4).

In a similar way (45.7) can be verified. Substitution into (45.5) and a transformation similar to (45.9) gives

1. For another proof see R. E. Peierls, Proc. Roy. Soc. Lond. A214, 143 (1952).

$$\left.\begin{aligned}\Box A_\mu(x) = &-\frac{i e N^2}{2(1-L)}[\overline{\psi}(x), \gamma_\mu \psi(x)] + i \int\limits_{x_0 = x_0'} d^3x' \Big([\mathrm{j}_\nu(x'), \mathrm{A}_\mu(x)] \frac{\partial A_\nu^{\text{äuss}}(x')}{\partial x_0'} + \\ &+ \Big[\mathrm{j}_\nu(x'), \frac{\partial \mathrm{A}_\mu(x)}{\partial x_0}\Big] A_\nu^{\text{äuss}}(x')\Big) + \frac{i L}{1-L} \int \Theta(x-x') \Big[\mathrm{j}_\nu(x'), \frac{\partial^2 \mathrm{A}_\lambda(x)}{\partial x_\mu \partial x_\lambda}\Big] \times \\ &\times A_\nu^{\text{äuss}}(x')\, dx' + \frac{L}{1-L}(\delta_{\mu\nu} - \delta_{\mu 4}\delta_{\nu 4}) \Box A_\nu^{\text{äuss}}(x) - \frac{L}{1-L} \frac{\partial^2 \mathrm{A}_\nu(x)}{\partial x_\mu \partial x_\nu}.\end{aligned}\right\} \quad (45.13)$$

From (45.10) it follows that

$$\left.\begin{aligned}i \int\limits_{x_0 = x_0'} d^3x' \Big([\mathrm{j}_\nu(x'), \mathrm{A}_\mu(x)] \frac{\partial A_\nu^{\text{äuss}}(x')}{\partial x_0'} + \Big[\mathrm{j}_\nu(x'), \frac{\partial \mathrm{A}_\mu(x)}{\partial x_0}\Big] A_\nu^{\text{äuss}}(x')\Big) = \\ = \frac{L}{1-L}\Big(-\frac{\partial^2 A_\nu^{\text{äuss}}(x)}{\partial x_\mu \partial x_\nu} + \delta_{\mu 4} \Box A_4^{\text{äuss}}(x)\Big),\end{aligned}\right\} \quad (45.14)$$

$$\left.\begin{aligned}&\int \Theta(x-x') \Big[\mathrm{j}_\nu(x'), \frac{\partial^2 \mathrm{A}_\lambda(x)}{\partial x_\mu \partial x_\lambda}\Big] A_\nu^{\text{äuss}}(x')\, dx' = \\ &= \frac{\partial^2}{\partial x_\mu \partial x_\lambda} \int \Theta(x-x') [\mathrm{j}_\nu(x'), \mathrm{A}_\lambda(x)] A_\nu^{\text{äuss}}(x') dx' + i \frac{L}{1-L} \frac{\partial}{\partial x_\mu} \frac{\partial A_k^{\text{äuss}}(x)}{\partial x_k}.\end{aligned}\right\} \quad (45.15)$$

Substitution of (45.14) and (45.15) into (45.13) and use of (45.7) gives the right side of (45.5). Thus the solution (45.6), (45.7) has been verified.

As in Sec. 29, we shall now consider the vacuum expectation value of the current operator. By differentiating (45.7) twice and using (45.14), we find

$$\left.\begin{aligned}\langle 0 | j_\mu(x) | 0 \rangle = -i \int \Theta(x-x') \langle 0 | [\mathrm{j}_\nu(x'), \mathrm{j}_\mu(x)] | 0 \rangle \times \\ \times A_\nu^{\text{äuss}}(x')\, dx' + \frac{L}{1-L} j_\mu^{\text{äuss}}(x).\end{aligned}\right\} \quad (45.16)$$

The first term in (45.16) contains the vacuum expectation value of the current commutator for the unperturbed problem, i.e., essentially the function $\Pi(p^2)$ of (43.25). Because of the Θ-function in (45.16), we do not obtain exactly the function $\Pi(p^2)$ on going over to p-space, but rather a linear combination of the functions $\Pi(p^2)$ and $\overline{\Pi}(p^2)$. [See the calculation in Sec. 41, especially Eq. (41.25).] Upon using (43.31), we find

$$\langle 0 | j_\mu(x) | 0 \rangle = \frac{1}{(2\pi)^4} \int dp\, e^{ipx} [-\overline{\Pi}(p^2) + \overline{\Pi}(0) - i\pi\, \varepsilon(p)\, \Pi(p^2)]\, j_\mu^{\text{äuss}}(p), \quad (45.17)$$

$$j_\mu^{\text{äuss}}(p) = \int dx\, e^{-ipx} j_\mu^{\text{äuss}}(x). \quad (45.17a)$$

The result (45.17) is a generalization of Eqs. (29.23), (29.33), and the functions $\Pi^{(0)}(p^2)$ and $\overline{\Pi}^{(0)}(p^2)$ introduced there are a first approximation in perturbation theory to the exact functions $\Pi(p^2)$ and $\overline{\Pi}(p^2)$. Just as in Sec. 29, (45.17) corresponds to a "dielectric constant" for the vacuum,

$$\varepsilon(p^2) = 1 - \overline{\Pi}(p^2) + \overline{\Pi}(0) - i\pi\, \varepsilon(p)\, \Pi(p^2). \quad (45.18)$$

This function $\varepsilon(p^2)$ is normalized to one for $p^2 = 0$ because of the last term in (45.16), i.e., because of the charge renormalization. This means that the vacuum has a dielectric constant of unity for a photon. In this way we have shown the connection of the charge renormalization of Sec. 43 to the results of Sec. 29. At the same time we have found the desired interpretation of the Π-functions by means of Eq. (45.18).

As was already noted in Sec. 29, the imaginary part of the dielectric constant corresponds to transfer of energy from the external field to the quantized fields, i.e., to the production of real particles. Likewise, the function $\Pi^{(0)}(p^2)$ was first introduced in Sec. 24 in connection with the production of particles by an external field. Here we can recall the calculation of Sec. 29, Eq. (29.38) and write the total energy given up by the external field as

$$\left.\begin{aligned} \delta E &= -\int \frac{\partial A_\mu^{\text{äuss}}(x)}{\partial x_0} \langle 0 | j_\mu(x) | 0 \rangle \, d x = \\ &= \frac{1}{(2\pi)^3} \int d p \, |p_0| \, \frac{j_\mu^{\text{äuss}}(p) \, j_\mu^{\text{äuss}}(-p)}{-2p^2} \, \Pi(p^2). \end{aligned}\right\} \quad (45.19)$$

A similar interpretation of the functions $\Sigma_i(p^2)$ could be given by considerations relating to the "polarization" of the vacuum by an "external spinor field". No doubt such a discussion is possible, although it would be of only limited interest physically, since "external spinor fields" are not available in nature.

46. Charge Renormalization for One-Electron States

The discussion of Sec. 45 has shown the relation between the charge renormalization and the problem of the vacuum polarization in an external field. In this connection there is another problem which is of major interest: the question of the charge of a state with one electron. It is not immediately evident whether the procedure given so far will yield the correct value for the charge of such a state or whether different renormalizations are necessary for an external field and for an electron. In Sec. 34 we have seen that in the lowest approximation of perturbation theory, the same renormalization solves both problems and now we are going to generalize (34.13) so that this result does not depend upon perturbation theory.

In order to do this, we first take the matrix element of the current operator between two one-electron states. We can do this by the same methods which we have already used several times before in working out matrix elements of various operators between the vacuum and one-particle states. We first compute the commutator of the current and the (Dirac) in-field. Using the complex conjugate of Eq. (41.7), we find

$$\left.\begin{aligned} [j_\mu(x), \psi^{(0)}(x')] = &- N \int \Theta(x - x''') \, S(x' - x''') \, [j_\mu(x), f(x''')] \, d x''' - \\ &- i N \int_{x_0''' = x_0} S(x' - x''') \, \gamma_4 [j_\mu(x), \psi(x''')] \, d^3 x'''. \end{aligned}\right\} \quad (46.1)$$

The last term can be simplified with the aid of (45.12):

$$[j_\mu(x), \psi^{(0)}(1)] = -N \int \Theta(x3) \, S(13) \, [j_\mu(x), f(3)] \, d x''' - \frac{e N}{1 - L} \xi_{\mu\lambda} \, S(1x) \, \gamma_\lambda \psi(x) \quad (46.2)$$

Here we have introduced a simplified notation for the different x-coordinates, which requires no further explanation. As the next step, we now compute the anticommutator of (46.2) with $\overline{\psi}^{(0)}(x'')$ and then take the vacuum expectation value which is of interest to us,

$$\left.\begin{aligned}&\langle 0|\{[j_\mu(x),\psi^{(0)}(1)],\overline{\psi}^{(0)}(2)\}|0\rangle=\\&=\frac{ieN}{1-L}\xi_{\mu\lambda}S(1x)\gamma_\lambda S(x2)-N\int\Theta(x3)\,S(13)\,dx'''\times\\&\quad\times\left(\langle 0|[j_\mu(x),\{f(3),\overline{\psi}^{(0)}(2)\}]|0\rangle-\langle 0|\{[j_\mu(x),\overline{\psi}^{(0)}(2)],f(3)\}|0\rangle\right).\end{aligned}\right\}\tag{46.3}$$

The first term in the last parenthesis can be rewritten as

$$\{f(3),\overline{\psi}^{(0)}(2)\}=N\int\Theta(34)\,\{f(3),\overline{f}(4)\}\,S(42)\,dx^{\mathrm{IV}}-\frac{i}{N}[ie\gamma A(3)+\delta m]\,S(32)\tag{46.4}$$

by the use of (41.8). It follows that

$$\left.\begin{aligned}\langle 0|[j_\mu(x),\{f(3),\overline{\psi}^{(0)}(2)\}]|0\rangle&=N\int\Theta(34)\,dx^{\mathrm{IV}}\langle 0|[j_\mu(x),\{f(3),\overline{f}(4)\}]|0\rangle\times\\&\times S(42)+\frac{e}{N}\gamma_\lambda S(32)\langle 0|[j_\mu(x),A_\lambda(3)]|0\rangle.\end{aligned}\right\}\tag{46.5}$$

The other term in (46.3) can be treated in a similar fashion:

$$\left.\begin{aligned}[j_\mu(x),\overline{\psi}^{(0)}(2)]&=N\int\Theta(x4)\,[j_\mu(x),\overline{f}(4)]\times\\&\times S(42)\,dx^{\mathrm{IV}}+\frac{eN}{1-L}\overline{\psi}(x)\gamma_\lambda S(x2)\,\xi_{\lambda\mu}\end{aligned}\right\}\tag{46.6}$$

and

$$\left.\begin{aligned}&N\int\Theta(x3)\,S(13)\,dx'''\langle 0|\{\overline{\psi}(x),f(3)\}|0\rangle=\\&=-\langle 0|\Big\{\overline{\psi}(x),\psi^{(0)}(1)+iN\int_{x_0'''=x_0}S(13)\gamma_4\psi(3)\,d^3x'''\Big\}|0\rangle=i\,S(1x)\frac{N-1}{N}.\end{aligned}\right\}\tag{46.7}$$

Collecting our results, we obtain

$$\left.\begin{aligned}&\langle 0|\{[j_\mu(x),\psi^{(0)}(1)],\overline{\psi}^{(0)}(2)\}|0\rangle=\frac{ie}{1-L}[1+2(N-1)]\,\xi_{\mu\lambda}S(1x)\gamma_\nu S(x2)-\\&\quad-e\int\Theta(x3)\,S(13)\gamma_\lambda S(32)\,dx'''\langle 0|[j_\mu(x),A_\lambda(3)]|0\rangle-\\&\quad-N^2\iint dx'''\,dx^{\mathrm{IV}}\,S(13)\left(\Theta(x3)\Theta(34)\langle 0|[j_\mu(x),\{f(3),\overline{f}(4)\}]|0\rangle-\right.\\&\quad\left.-\Theta(x3)\Theta(x4)\langle 0|\{f(3),[j_\mu(x),\overline{f}(4)]\}|0\rangle\right)S(42).\end{aligned}\right\}\tag{46.8}$$

For further calculation we express the commutator of the current and the potential in (46.8) by means of the functions $\Pi(p^2)$ and $\overline{\Pi}(p^2)$. Clearly we have

$$\left.\begin{aligned}\langle 0|[j_\mu(x),A_\lambda(3)]|0\rangle&=\int D_R(34)\langle 0|[j_\mu(x),j_\lambda(4)]|0\rangle\,dx^{\mathrm{IV}}=\\&=\frac{-1}{(2\pi)^3}\int dp\,e^{ip(3x)}\,\varepsilon(p)\,[p_\mu p_\lambda-p^2\delta_{\mu\lambda}]\,\frac{\Pi(p^2)}{p^2}\end{aligned}\right\}\tag{46.9}$$

and therefore

$$\left.\begin{aligned}&\Theta(x3)\langle 0|[j_\mu(x), A_\lambda(3)]|0\rangle = \\ &= \frac{i\,\delta_{\mu\lambda}}{(2\pi)^4}\int dp\, e^{ip(x3)}[\bar{\Pi}(p^2) + i\pi\,\varepsilon(p)\,\Pi(p^2)] + \Theta(x3)\frac{\partial^2\Phi(x3)}{\partial x_\mu\,\partial x_\lambda},\end{aligned}\right\} \quad (46.10)$$

$$\Phi(x) = \frac{1}{(2\pi)^3}\int dp\, e^{ipx}\,\varepsilon(p)\,\frac{\Pi(p^2)}{p^2}. \quad (46.10a)$$

The new function $\Phi(x)$ has the following properties:

$$\Phi(x)|_{x_0=0} = 0, \quad (46.11a)$$

$$\left.\frac{\partial\Phi(x)}{\partial x_0}\right|_{x_0=0} = -i\bar{\Pi}(0)\,\delta(\boldsymbol{x}). \quad (46.11b)$$

Therefore it follows that

$$\Theta(x3)\frac{\partial^2\Phi(x3)}{\partial x_\mu\,\partial x_\lambda} = \frac{\partial^2}{\partial x_\mu\,\partial x_\lambda}(\Theta(x3)\,\Phi(x3)) - i\bar{\Pi}(0)\,\delta_{\mu4}\delta_{\lambda4}\delta(x3), \quad (46.11c)$$

and we obtain

$$\left.\begin{aligned}&-e\int\Theta(x3)\,S(13)\,\gamma_\lambda\,S(32)\,dx'''\langle 0|[j_\mu(x), A_\lambda(3)]|0\rangle = \\ &= \frac{-ie}{(2\pi)^4}\int dp\int dx'''\,e^{ip(x3)}S(13)\,\gamma_\mu\,S(32)\,[\bar{\Pi}(p^2) + i\pi\,\varepsilon(p)\,\Pi(p^2)] + \\ &+ i\,\delta_{\mu4}\frac{L}{1-L}S(1x)\,\gamma_4\,S(x2).\end{aligned}\right\} \quad (46.12)$$

For the expression under the double integral in (46.8) we introduce a special notation:

$$\left.\begin{aligned}&N^2[\Theta(x3)\,\Theta(34)\langle 0|\{[f(3), j_\mu(x)], \bar{f}(4)\}|0\rangle + \\ &\quad + \Theta(x4)\,\Theta(43)\langle 0|\{f(3), [j_\mu(x), \bar{f}(4)]\}|0\rangle] - \\ &\quad - 2ie\frac{N-1}{1-L}L\,\delta_{\mu4}\gamma_4\,\delta(3x)\,\delta(x4) = \\ &= \frac{ie}{(2\pi)^8}\iint dp\,dp'\,e^{ip(3x)+ip'(x4)}\Lambda_\mu(p, p'),\end{aligned}\right\} \quad (46.13)$$

and using (46.12) and (46.13) we obtain from (46.8),

$$\left.\begin{aligned}&\langle q|j_\mu|q'\rangle = \langle q|j_\mu^{(0)}|q'\rangle\left[1 - \bar{\Pi}(Q^2) + \bar{\Pi}(0) - i\pi\,\varepsilon(Q)\,\Pi(Q^2) + 2\frac{N-1}{1-L}\right] + \\ &\quad + ie\langle q|\bar{\psi}^{(0)}|0\rangle\,\Lambda_\mu(q, q')\langle 0|\psi^{(0)}|q'\rangle,\end{aligned}\right\} \quad (46.14)$$

$$Q = q' - q. \quad (46.14a)$$

This is the desired expression for our matrix element. From it, we obtain for the charge of the state $|q\rangle$,

$$\langle q|Q|q\rangle = e\left[1 + 2\frac{N-1}{1-L} + \bar{u}(q)\,\Lambda_4(q, q)\,u(q)\right]. \quad (46.15)$$

In order to complete the proof, we compute $\Lambda_4(p, p')$ from (46.13):

$$\left.\begin{aligned}\Lambda_4(p, p') &= -2(N-1)\,\bar{\Pi}(0)\,\gamma_4 - \frac{iN^2}{e}\iint dx\,dx^{\mathrm{IV}}\,e^{ip(x3)+ip'(4x)}\times \\ &\times(\Theta(x3)\,\Theta(34)\langle 0|\{[f(3), j_4(x)], \bar{f}(4)\}|0\rangle + \\ &+ \Theta(x4)\,\Theta(43)\langle 0|\{f(3), [j_4(x), \bar{f}(4)]\}|0\rangle).\end{aligned}\right\} \quad (46.16)$$

If we put $p=p'$ here, we encounter very singular expressions. Because of this, we shall now set only the spatial components of the two vectors equal. It then follows that

$$\left.\begin{aligned} \Lambda_4(p,p')_{\mathbf{p}=\mathbf{p}'} = & -2(N-1)\,\overline{\Pi}(0)\,\gamma_4 + \\ & + \frac{N^2}{e}\int dx_0 \int dx^{\mathrm{IV}}\, e^{-i(p_0-p_0')(x_0-x_0''')+ip'(43)} \times \\ & \times \left(\Theta(x3)\,\Theta(34)\,\langle 0|\{[f(3),Q],\bar f(4)\}|0\rangle + \right. \\ & \left. + \Theta(x4)\,\Theta(43)\,\langle 0|\{f(3),[Q,\bar f(4)]\}|0\rangle\right). \end{aligned}\right\} \quad (46.17)$$

By the use of the canonical commutation relations and the fact that the charge Q is time independent,[1] we can easily show

$$[Q,f(3)] = -i \int_{x_0=x_0'''} d^3x\,[j_4(x),f(3)] = -e\,f(3) \qquad (46.18a)$$

and

$$[Q,\bar f(4)] = e\,\bar f(4)\,, \qquad (46.18b)$$

so that

$$\left.\begin{aligned} \Lambda_4(p,p')_{\mathbf{p}=\mathbf{p}'} = & -2(N-1)\,\overline{\Pi}(0)\,\gamma_4 + \\ & + N^2 \int dx_0 \int dx^{\mathrm{IV}}\, e^{-i(p_0-p_0')(x_0-x_0''')+ip'(43)} \times \\ & \times \left(\Theta(x3)\,\Theta(34) + \Theta(x4)\,\Theta(43)\right)\langle 0|\{f(3),\bar f(4)\}|0\rangle. \end{aligned}\right\} \quad (46.19)$$

The time integration can now be done easily and, after introducing the Σ-functions of (41.19), we obtain

$$\left.\begin{aligned} \Lambda_4(p,p')_{\mathbf{p}=\mathbf{p}'} = & -2(N-1)\,\overline{\Pi}(0)\,\gamma_4 + \\ & + N^2\left[P\frac{1}{p_0-p_0'} + i\pi\,\delta(p_0-p_0')\right][\Sigma^R(p') - \Sigma^A(p)]\,, \end{aligned}\right\} \quad (46.20)$$

with

$$\left.\begin{aligned} \Sigma^{R,A}(p) = & \,\overline{\Sigma}_1(p^2) \pm i\pi\,\varepsilon(p)\,\Sigma_1(p^2) + \\ & + (i\gamma p + m)\,[\overline{\Sigma}_2(p^2) \pm i\pi\,\varepsilon(p)\,\Sigma_2(p^2)]. \end{aligned}\right\} \quad (46.20a)$$

Equation (46.20) can be written in a formally invariant form as

$$(p-p')\,\Lambda(p,p') = -2(N-1)\,\overline{\Pi}(0)\,\gamma(p-p') - i\,N^2[\Sigma^A(p) - \Sigma^R(p')]. \quad (46.21)$$

The derivation of (46.21) is valid only if the vector $p-p'$ is time-like.

In (46.20) we can now let p_0' approach p_0 and we obtain

---

1. From the time independence of Q one might conclude that the charge of the physical electron has to be the same as that of the in-electron, and hence that the calculation of this section is unnecessary. It must be remembered that the charge need not be a constant of the motion during the adiabatic switching. In the calculations of this section all that is used is the time independence of the charge for finite time intervals.

$$\bar{u}(q)\,\Lambda_4(q,q)\,u(q) = -2(N-1)\,\bar{\Pi}(0) + N^2\bar{u}(q)\,F(q)\,u(q)\,, \quad (46.22)$$

$$\left.\begin{aligned} F(q) &= \lim_{\varepsilon\to 0}\frac{1}{\varepsilon}\left[\bar{\Sigma}_1(-m^2+2q_0\varepsilon) - \bar{\Sigma}_1(-m^2) + \gamma_4\,\varepsilon\,\bar{\Sigma}_2(-m^2)\right] = \\ &= \gamma_4\,\bar{\Sigma}_2(-m^2) + 2q_0\,\bar{\Sigma}_1'(-m^2)\,. \end{aligned}\right\} \quad (46.23)$$

Because

$$\bar{u}(q)\,q_0\,u(q) = m\bar{u}(q)\,\gamma_4\,u(q) = m\,, \quad (46.24)$$

Eq. (46.22) gives

$$\bar{u}(q)\,\Lambda_4(q,q)\,u(q) = -2(N-1)\left(\bar{\Pi}(0)+1\right) = -2\,\frac{N-1}{1-L}\,, \quad (46.25)$$

where (42.13) has been used. From this and (46.15) the charge of the electron is

$$\langle q|\,Q\,|q\rangle = e\,. \quad (46.26)$$

This shows that our definition of the charge renormalization not only normalizes the dielectric constant of the vacuum for a photon to be one, but that it also makes the charge of the one-electron state equal to e.[1]

47. Proof that the Theory Contains At Least One Infinite Quantity

Up to now our discussion has not been concerned with possible infinite values for the renormalization constants. In Chap. VI we saw that the renormalization constants were infinite in several examples using lowest order perturbation theory. Despite this, all the (renormalized) observable quantities were finite and in good agreement with the experimental measurements. In principle we can go to higher orders of perturbation theory and hence develop in powers of e the renormalization formalism given here. One can show that we would obtain[2] finite results for all observable quantities in every order of perturbation theory. If one could also show that by using this procedure the series we would obtain is convergent, then we would have a satisfactory theory--at least satisfactory for the observable quantities. Unfortunately, up to now, it has not been possible to give an adequate discussion of the convergence of this series. Investigations using certain simplified models of a quantum field theory[3] have shown that the per-

1. One can also show that a state with n electrons has a charge $n\cdot e$. See E. Karlson, Proc. Roy. Soc. Lond. A 230, 382 (1955).

2. Dyson and coworkers have proved this using methods somewhat different from those developed here. See F. J. Dyson, Phys. Rev. 75, 1736 (1949); 82, 428 (1951); 83, 608 (1951); A. Salam, Phys. Rev. 82, 217 (1951); J. C. Ward, Proc. Phys. Soc. Lond. A 64, 54 (1951).

3. C. A. Hurst, Proc. Cambridge Phil. Soc. 48, 625 (1952); W. Thirring, Helv. Phys. Acta 26, 33 (1953); A. Petermann, Phys. Rev. 89, 1160 (1953); A. Petermann, Arch. Sci. Phys. Nat. 6, 5 (1953); R. Utiyama and T. Imamura, Progr. Theor. Phys. 9, 431 (1953).

turbation series diverges (even after renormalization) for these models. One cannot rule out the possibility that the perturbation series of quantum electrodynamics also diverges. We shall not discuss these calculations further, except to note that even if we were able to prove the divergence of the perturbation series, not very much would be settled. Indeed, it is quite possible that there is a solution with finite renormalization constants, but one which has such a form that the renormalization constants would appear as formal power series with infinite coefficients. It is therefore of some interest to discuss these questions without using perturbation theory.

We shall show in this section that the last mentioned possibility can be disregarded. That is, we shall show that if there is any solution at all of the equations given here, this solution contains at least one infinite renormalization constant. In fact, it will turn out that among the constants which are infinite is either N^{-1} or $(1-L)^{-1}$ or both. In order to show this, we start with the assumption that there is a completely finite solution. This implies certain assumptions about the weight functions $\Pi(p^2)$ and $\Sigma_i(p^2)$ for large values of $-p^2$. With these assumptions, we shall then show that the formalism contains a contradiction, and so our assertion is then proved.

The principal tool in our method is the function $\Pi(p^2)$ which was defined in (43.24). In Sec. 44 it was shown that this function could be written as a sum of a finite number of positive terms, even though the definition (43.24) contains certain terms with negative signs. We therefore obtain a lower bound for the function $\Pi(p^2)$ if we neglect a few terms in (43.24). In particular, we can write

$$\Pi(p^2) > \frac{V}{-3p^2} \sum_{q+q'=p} \langle 0|j_\mu|q,q'\rangle\langle q',q|j_\mu|0\rangle. \tag{47.1}$$

The state $|q,q'\rangle$ is a state with an in-pair and the matrix elements of the current operator which enter (47.1) can be taken from the calculation of Sec. 46. From Eqs. (46.8), (46.12) and (46.13) it follows that

$$\left.\begin{aligned}\langle 0|j_\mu|q,q'\rangle = \langle 0|j_\mu^{(0)}|q,q'\rangle\Big[1-\overline{\Pi}(Q^2)+\overline{\Pi}(0)-i\pi\Pi(Q^2)-2\frac{N-1}{1-L}\Big]+\\ +ie\langle 0|\overline{\psi}^{(0)}|q'\rangle\Lambda_\mu(-q',q)\langle 0|\psi^{(0)}|q\rangle,\end{aligned}\right\} \tag{47.2}$$

$$Q = q+q'. \tag{47.2a}$$

By using considerations of invariance, the function $\Lambda_\mu(q',q)$ can be written in the following form:

$$\left.\begin{aligned}\Lambda_\mu(q',q) = \sum_{\varrho,\varrho'=0,1} (i\gamma q'+m)^{\varrho'}\times\\ \times[\gamma_\mu F^{\varrho'\varrho}(q',q)+q'_\mu G^{\varrho'\varrho}(q',q)+q_\mu H^{\varrho'\varrho}(q',q)](i\gamma q+m)^{\varrho}.\end{aligned}\right\} \tag{47.3}$$

Here the functions F, G, and H are invariant and are independent of the γ-matrices. From the charge symmetry of the theory or from the explicit definition (46.13) it follows that the function $\Lambda_\mu(q', q)$ must have a certain symmetry in q and q'. The following relation can easily be shown by using the method of Sec. 41, Eq. (41.21): [See also Eq. (14.6) for the γ-matrices.]

$$\Lambda_\mu^T(q', q) = - C^{-1} \Lambda_\mu(-q, -q')\, C. \tag{47.4}$$

From (47.4) it follows that the functions F, G, and H of (47.3) satisfy

$$F^{\varrho'\varrho}(q', q) = F^{\varrho\varrho'}(-q, -q') , \tag{47.5}$$

$$G^{\varrho'\varrho}(q', q) = H^{\varrho\varrho'}(-q, -q') . \tag{47.6}$$

If (47.3) is substituted into (47.2) or (46.14), only terms with $\varrho = \varrho' = 0$ contribute, since the others vanish by the equation of motion for the in-field. Since $q^2 = q'^2 = -m^2$, we have

$$\left.\begin{aligned} &\langle q | j_\mu | q' \rangle = \langle q | j_\mu^{(0)} | q' \rangle \times \\ &\times \left[1 - \bar{\Pi}(Q^2) + \bar{\Pi}(0) - i\pi\varepsilon(Q)\, \Pi(Q^2) + 2\frac{N-1}{1-L} + \bar{R}(Q^2) + i\pi\varepsilon(Q)\, R(Q^2) \right] - \\ &- \frac{e}{2m}(q_\mu + q'_\mu) \langle q | \bar{\psi}^{(0)} | 0 \rangle \langle 0 | \psi^{(0)} | q' \rangle [\bar{S}(Q^2) + i\pi\varepsilon(Q)\, S(Q^2)] , \end{aligned}\right\} \tag{47.7}$$

with

$$Q = q' - q , \tag{47.7a}$$

and

$$\left.\begin{aligned} &\langle 0 | j_\mu | q, q' \rangle = \langle 0 | j_\mu^{(0)} | q, q' \rangle \times \\ &\times \left[1 - \bar{\Pi}(Q^2) + \bar{\Pi}(0) - i\pi\Pi(Q^2) + 2\frac{N-1}{1-L} + \bar{R}(Q^2) + i\pi R(Q^2) \right] - \\ &- \frac{e}{2m}(q_\mu - q'_\mu) \langle 0 | \bar{\psi}^{(0)} | q' \rangle \langle 0 | \psi^{(0)} | q \rangle [\bar{S}(Q^2) + i\pi\, S(Q^2)] , \end{aligned}\right\} \tag{47.8}$$

with

$$Q = q + q' . \tag{47.8a}$$

There are the following relations between the new functions $R(Q^2)$, $\bar{R}(Q^2)$, $S(Q^2)$, $\bar{S}(Q^2)$, and the old functions F, G, H:

$$\left.\begin{aligned} &\bar{R}(Q^2) + i\pi\varepsilon(Q)\, R(Q^2) = F^{00}(q, q'), \\ &-\frac{e}{2m}[\bar{S}(Q^2) + i\pi\varepsilon(Q) S(Q^2)] = H^{00}(q, q') = G^{00}(-q', -q), \\ &\text{for } Q = q' - q,\ \ q^2 = q'^2 = -m^2 \ \text{ and } \ \varepsilon(q) = \varepsilon(q') = 1 . \end{aligned}\right\} \tag{47.9}$$

Equations (47.9) give the most general form which the right side can have on the basis of invariance. From the reality properties of the current operator, it then follows that the four new functions must be real.

We now return to (46.13). First we note that this equation has a structure similar to Eq. (45.16), i.e., it contains a vacuum expectation value multiplied by Θ-functions. If we define a function $\mathscr{F}_\mu(p, p')$ by

$$\langle 0|\{[f(3), j_\mu(x)], \bar{f}(4)\}|0\rangle = \frac{-1}{(2\pi)^6}\iint dp\, dp'\, e^{ip(3x)+ip'(x4)}\mathscr{F}_\mu(p, p'), \tag{47.10}$$

then we can show as in Sec. 44 that this definition (47.10) contains a sum over only a finite number of intermediate states. If the renormalized operators exist, then the function $\mathscr{F}_\mu(p, p')$ is finite. With the aid of the integral representation

$$\Theta(x3) = \frac{1}{2\pi i}\int_{-\infty}^{+\infty}\frac{d\tau}{\tau - i\varepsilon}e^{i\tau(x3)_0}, \tag{47.11}$$

we then obtain

$$\left.\begin{aligned}&\Theta(x3)\langle 0|\{[f(3), j_\mu(x)], \bar{f}(4)\}|0\rangle = \\ &= \frac{i}{(2\pi)^7}\iint dp\, dp'\, e^{ip(3x)+ip'(x4)}\bar{\mathscr{F}}_\mu(p, p'),\end{aligned}\right\} \tag{47.12}$$

with

$$\bar{\mathscr{F}}_\mu(p, p') = \int_{-\infty}^{+\infty}\frac{d\tau}{\tau - i\varepsilon}\mathscr{F}_\mu(p - \eta\tau, p') = -\int_{-\infty}^{+\infty}\frac{dx}{x - (p_0 - i\varepsilon)}\mathscr{F}_\mu(\mathbf{p}, x; p'). \tag{47.12a}$$

Here η is an arbitrary time-like vector with a positive time component. In a similar way we obtain from the second Θ-function,

$$\left.\begin{aligned}&\Theta(x3)\,\Theta(34)\langle 0|\{[f(3), j_\mu(x)], \bar{f}(4)\}|0\rangle = \\ &= \frac{1}{(2\pi)^8}\iint dp\, dp'\, e^{ip(3x)+ip'(x4)}\bar{\bar{\mathscr{F}}}(p, p'),\end{aligned}\right\} \tag{47.12b}$$

$$\bar{\bar{\mathscr{F}}}_\mu(p, p') = \int_{-\infty}^{+\infty}\frac{d\tau}{\tau - i\varepsilon'}\bar{\mathscr{F}}_\mu(p - \eta'\tau, p' - \eta'\tau). \tag{47.12c}$$

In this way we can regard the functions F, G, H in (47.3) as a type of generalized Hilbert transform of the finite function $\mathscr{F}_\mu(p, p')$ in (47.10) plus a similar contribution from the second term in (46.13). The real part of these expressions gives the functions $\bar{R}$ and $\bar{S}$ in (47.9), while the imaginary part gives the functions R and S. [See the similar calculation in Sec. 29, especially Eqs. (29.27) through (29.32).]

From (46.25) we also have

$$\left.\begin{aligned}&\langle q|j_\mu^{(0)}|q\rangle\left[2\frac{N-1}{1-L} + \bar{R}(0)\right] - \frac{e}{m}q_\mu\langle q|\bar{\psi}^{(0)}|0\rangle\langle 0|\psi^{(0)}|q\rangle\bar{S}(0) = \\ &= \langle q|j_\mu^{(0)}|q\rangle\left[2\frac{N-1}{1-L} + \bar{R}(0) - \bar{S}(0)\right] = 0,\end{aligned}\right\} \tag{47.13}$$

or

$$-2\frac{N-1}{1-L} = \bar{R}(0) - \bar{S}(0). \tag{47.14}$$

At this point we must recall that the functions $\bar{R}(p^2)$ and $\bar{S}(p^2)$ are defined by means of (47.9), (47.3) and (46.13) in terms of the operators $f(x)$ of (41.2). These operators contain singular expressions of the type (42.18). We must therefore expect corresponding singular pieces[1] of R and S. The calculations of Sec. 46 allow us to isolate these singular pieces in a simple way. The argument that leads from (46.16) to (46.25) was originally carried through for $f(x)$, but it can obviously be carried through for just the regular part of $f(x)$. If $R^{\text{reg}}(p^2)$ and $S^{\text{reg}}(p^2)$ are the regular parts of the functions $R(p^2)$ and $S(p^2)$, then we conclude that in addition to (47.14) we must also have

$$\bar{R}^{\text{reg}}(0)-\bar{S}^{\text{reg}}(0)=\frac{N^2}{1-L}\left(\bar{\Sigma}_2^{\text{reg}}(-m^2)+2m\bar{\Sigma}_1^{\prime\,\text{reg}}(-m^2)\right)=\frac{1-N^2}{1-L}. \qquad (47.15)$$

On the other hand, the result (46.26) is independent of the special treatment of the singular terms, and by a transformation similar to that of (33.2) through (33.6) we can rewrite (47.8) as

$$\left.\begin{aligned}\langle 0|\,j_\mu\,|q,q'\rangle=\langle 0|\,j_\mu^{(0)}\,|q,q'\rangle\,[1+\bar{\Pi}(Q^2)-\bar{R}^{\text{reg}}(Q^2)+\bar{S}^{\text{reg}}(Q^2)-\\ -i\pi(\Pi(Q^2)-R^{\text{reg}}(Q^2)+S^{\text{reg}}(Q^2))]+\\ +i\,Q_\nu\langle 0|\,m_{\mu\nu}^{(0)}\,|q,q'\rangle\,[\bar{S}^{\text{reg}}(Q^2)-i\pi\,S^{\text{reg}}(Q^2)]\,.\end{aligned}\right\} \qquad (47.16)$$

So far we have not made use of our assumption that the renormalization constants are to be finite and Eq. (47.16) has been derived using only general assumptions. Now we shall take explicit account of the fact that $\frac{1-N^2}{1-L}$ and $\frac{1}{1-L}$ have been assumed to be finite. This means convergence for the integral (43.30) for the function $\bar{\Pi}(0)$ as well as for the integrals which enter the definition (47.9) according to (47.12). With this assumption we find

$$\lim_{-Q^2\to\infty}\bar{\Pi}(Q^2)=\bar{\Pi}(0)-\lim_{-Q^2\to\infty}\int_0^\infty\frac{\Pi(-a)\,da}{a\left(1+\frac{a}{Q^2}\right)}=\bar{\Pi}(0)-\bar{\Pi}(0)=0 \qquad (47.17)$$

and

$$\left.\begin{aligned}&\lim_{|p_0-p_0'|\to\infty}\bar{\bar{\mathscr{F}}}_\mu(p,p')=\\ &=\lim_{|p_0-p_0'|\to\infty}\iint\frac{d\tau\,d\tau'}{(\tau-i\varepsilon)(\tau'-i\varepsilon')}\mathscr{F}_\mu(\mathbf{p},p_0-\tau-\tau';\mathbf{p}',p_0'-\tau')=\\ &=\lim_{|p_0-p_0'|\to\infty}\iint\frac{dx\,dy\,\mathscr{F}_\mu(\mathbf{p},x;\mathbf{p}',y)}{[x-y-(p_0-p_0'-i\varepsilon)]\,[y-(p_0'-i\varepsilon')]}=0.\end{aligned}\right\} \qquad (47.18)$$

---

1. In an earlier work these terms were overlooked. [G. Källén, Dan. Mat. Fys. Medd. 27, No. 12 (1953).] This would give a factor of $2N-1$ in (47.19) rather than a factor of N^2.

If $-Q^2$ is very large, we therefore obtain[1] from (47.16),

$$\langle 0| j_\mu |q, q'\rangle \to \frac{N^2}{1-L} \langle 0| j_\mu^{(0)} |q, q'\rangle. \qquad (47.19)$$

Equation (47.19) can be interpreted physically in the following way: If $-Q^2$ is very large, that is, if the kinetic energy of one particle is very large in the rest frame of the other, then the interaction of the particles does not play an essential role, but rather the current is practically the current of two free particles. We shall, momentarily, regard a "physical" particle as a mixture of "free particles at time zero". Then the factor $N^2(1-L)^{-1}$ can be understood as an expression for the amplitude that at very large energies the two particles can penetrate the "clouds of virtual photons and pairs" and interact with each other as "bare" particles. If we introduce the unrenormalized electromagnetic field $A_\mu^{nr}(x) = \sqrt{1-L}\, A_\mu(x)$, the unrenormalized Dirac field $\psi^{nr}(x) = N\psi(x)$, and the "bare" charge $e_0 = \frac{e}{\sqrt{1-L}}$, then we can write (47.19) as follows:

$$\lim_{-Q^2\to\infty} \Box \langle 0| A_\mu^{nr}(x) |q, q'\rangle = i e_0 \langle 0| \bar{\psi}^{nr(0)}(x) \gamma_\mu \psi^{nr(0)}(x) |q, q'\rangle. \qquad (47.19a)$$

Equation (47.19a) is identical to the first approximation to the equations of motion of the unrenormalized fields.

We substitute this result into the inequality (47.1) and find

$$\left.\begin{aligned} \lim_{-p^2\to\infty} \Pi(p^2) &\geq \left(\frac{N^2}{1-L}\right)^2 \lim_{-p^2\to\infty} \frac{V}{-3p^2} \sum_{q+q'=p} \langle 0| j_\mu^{(0)} |q, q'\rangle \langle q', q| j_\mu^{(0)} |0\rangle \\ &= \left(\frac{N^2}{1-L}\right)^2 \lim_{-p^2\to\infty} \Pi^{(0)}(p^2) = \frac{e^2}{12\pi^2}\left(\frac{N^2}{1-L}\right)^2. \end{aligned}\right\} \qquad (47.20)$$

---

1. In (47.19) it has also been assumed that the weight functions [like $\Pi(p^2)$] vanish for large values of $-p^2$. The necessary assumption is only that integrals like (43.30) converge and this is a weaker condition. As an example, consider the integral $\int_0^\infty \frac{\sin a}{a}\, da$. For the moment, if we denote $F = R^{\text{reg}} - S^{\text{reg}} - \Pi$, then from (47.20) we see that the essential point is the divergence of the integral

$$J = \int_0^\infty \frac{da}{a} \left| \frac{N^2}{1-L} + \bar{F} - i\pi F \right|^2.$$

Clearly

$$J \geq \left(\frac{N^2}{1-L}\right)^2 \int_0^\infty \frac{da}{a} + 2\,\frac{N^2}{1-L} \int_0^\infty \frac{da}{a}\, \bar{F}.$$

For large values of a, $\bar{F}$ vanishes according to (47.17) and (47.18) and the integral J is divergent because of the first term. This is sufficient for the proof and the result (47.19) is to be understood in this sense.

We have therefore proved that if N^{-1} and $(1-L)^{-1}$ are both finite, then for large values of $-p^2$, the function $\Pi(p^2)$ is larger than the positive number $\frac{e^2}{12\pi^2}\left(\frac{N^2}{1-L}\right)^2$. Consequently the integral (43.30) cannot converge--in contradiction to our assumption about the constant L. We must therefore conclude that either N^{-1} or $(1-L)^{-1}$ or both are infinite, i.e., that at least one of the renormalization constants is infinite. It should be emphasized that this result has been obtained <u>without</u> the use of perturbation theory.

48. Concluding Remarks

We have now come to the end of our discussion of the general theory of renormalization. In retrospect, we have succeeded in isolating the renormalization constants present in the theory and in writing them as integrals over certain weight functions. In addition, we have shown that even in a theory with infinite renormalization constants these weight functions must be finite if only the matrix elements of the renormalized operators are finite quantities. In Sec. 45 it has also been shown that integrals over the same weight functions, but with an additional factor involving a in the denominator, arise as observable quantities. [See, for example, Eq. (45.17), where the function

$$\bar{\Pi}(p^2) - \bar{\Pi}(0) = -p^2 P \int_0^\infty \frac{da\,\Pi(-a)}{a(a+p^2)}$$

enters as an observable quantity.] If our theory actually has physically usable solutions, we must therefore require that the observable integrals such as

$$\int_0^\infty \frac{\Pi(-a)\,da}{a^2} \tag{48.1}$$

converge, even if integrals like

$$\int_0^\infty \frac{\Pi(-a)\,da}{a} \tag{48.2}$$

are divergent. <u>No one has yet succeeded in giving a proof for the assertion (48.1)</u>. The fact that the integrals converge in every approximation of perturbation theory is of no great significance in this connection--at least as long as the convergence or divergence of the perturbation series has not been considered. One may ask whether it is not possible to generalize the argument of Sec. 47 so that one would have to assume only the convergence of (48.1), rather than that of (48.2), in order to be able to obtain asymptotic conditions like (47.19) for certain matrix elements. Consequently, by considering several states in (43.24) one might finally obtain a sharper bound for the weight functions than (47.20). With this new bound, if a contradiction to (48.1) resulted, then this would show that the theory had no physically useful solution at all. The details of such a program are quite complicated, and up to the present it has not been possible to give[1] substantial results.

In this connection, it should be noted that T. D. Lee has recently constructed[2] an interesting model of a renormalizable field theory. Although the model is non-relativistic, it is significant because it contains[3] a renormalization of the coupling constants as well as a renormalization of the mass and yet is exactly solvable, in part. A detailed study of this model has shown[4] that the solution obtained after renormalization does not have a sensible energy spectrum, but that a "ghost state" ("pathological state") of negative probability appears in the theory. It is quite possible that something similar is present in other theories with infinite charge renormalization, i.e., possibly in quantum electrodynamics. We must recall that in our discussion we made certain assumptions about the mass spectrum of the theory. [See the remarks following Eq. (41.12).] Therefore our arguments are valid only for solutions of this kind. In particular, it has been shown in Sec. 47 that there is no solution with a physically sensible mass spectrum which has all renormalization constants finite. In no way can we exclude the possibility that there exist other unphysical solutions with more arbitrary properties.

One can certainly say that no matter how interesting they are in a fundamental sense, the unresolved questions of quantum electrodynamics discussed here are of less significance practically. This is so because in quantum electrodynamics we already have a theory which enables us to compute observable quantities with great precision and to compare them with experimental results. Indeed, it is just this excellent agreement of the perturbation theory of quantum electrodynamics and the observations which hints that the present formalism might be the limit of some future more complete theory and that study of the present theory will be rewarding One can hardly say the same thing about meson theories[5] in their present form!

---

1. In this connection, see G. Källén, Proc. CERN Symposium, Geneva 2, 187 (1956).

2. T. D. Lee, Phys. Rev. 95, 1329 (1954).

3. The other solvable models which are known (e.g., the example worked out in Sec. 11) sometimes contain a mass renormalization but no charge renormalization.

4. G. Källén and W. Pauli, Dan. Mat. Fys. Medd. 30, No. 7 (1955).

5. In this connection see Vol. XLIII of this handbook. (Handbuch der Physik, edited by S. Flügge, Springer-Verlag, Heidelberg.)

INDEX